ERRATUM

The Subject Index on pp. 315–333 is incorrectly paginated. We apologize
for this error in binding. In case of any difficulty locating references because
of it, please refer to pp. 1233–1251 in Volume 3B which contain the
identical information, paginated in the correct sequence.

FRACTURE 1977

ADVANCES
in
RESEARCH
on the
STRENGTH
and
FRACTURE
of
MATERIALS

TITLES IN THIS SERIES

FRACTURE 1977

ADVANCES in RESEARCH on the STRENGTH and FRACTURE of MATERIALS

D.M.R. TAPLIN
Editor

Vol. 4—Fracture and Society

Fourth International Conference on Fracture
June 1977
University of Waterloo, Canada

PERGAMON PRESS
New York / Oxford / Toronto / Sydney / Frankfurt / Paris

Pergamon Press Offices:

U.S.A.	Pergamon Press Inc., Maxwell House, Fairview Park, Elmsford, New York 10523, U.S.A.
U.K.	Pergamon Press Ltd., Headington Hill Hall, Oxford OX3, OBW, England
CANADA	Pergamon of Canada, Ltd., 75 The East Mall, Toronto, Ontario M8Z 5W3, Canada
AUSTRALIA	Pergamon Press (Aust) Pty. Ltd., 19a Boundary Street, Rushcutters Bay, N.S.W. 2011, Australia
FRANCE	Pergamon Press SARL, 24 rue des Ecoles, 75240 Paris, Cedex 05, France
WEST GERMANY	Pergamon Press GmbH, 6242 Kronberg/Taunus, Pferdstrasse 1, West Germany

Copyright © 1978 Pergamon Press Inc.

Library of Congress Cataloging in Publication Data

International Conference on Fracture, 4th, University
of Waterloo, 1977.
Fracture 1977.

Includes indexes.
CONTENTS: v. 1. An overview.--v. 2A. The physical
metallurgy of fracture.--v. 2B. Fatigue.--v. 3A. Analysis and mechanics. [etc.]
1. Fracture mechanics--Congresses. 2. Strength of
materials--Congresses. I. Taplin, David M. R.
II. Title
TA409.I44   1977   620.1'126   77-15623
ISBN 0-08-022136-X Vol. 1
       0-08-022138-6 Vol. 2A
       0-08-022140-8 Vol. 2B
       0-08-022142-4 Vol. 3A
       0-08-022144-0 Vol. 3B
       0-08-022146-7 Vol. 4
       0-08-022130-0 6 -vol-set

Printed in the United States of America

To Diana and Justin

Contents

Foreword

The Fourth International Conference on Fracture, or ICF4 as it came to be known, was planned over a period of about four years. The Conference was intended as a state-of-the-art summary of our understanding of fracture in a wide variety of materials. In this respect ICF4 was very successful, and the presence of approximately 750 participants from 38 countries attested to the drawing power of the subject.

If we compare the present Conference with those preceding, several long-range trends may be deduced. There is now less concern with micromechanisms of the cleavage of iron, but there is more emphasis on effects of the environment. There is a growing realization that fracture of real materials may be dominated by the presence of inclusions and chemical segregates. We now have more work on polymers and ceramics and the beginnings of some efforts on biological materials. There is a vast range of subject matter in these papers, which will become a primary reference for workers in the field.

ICF4 had features which were absent in the earlier conferences. First was the emphasis on fracture in large structures. This has become exceedingly important with the proliferation of big ships, big aircraft, big nuclear reactors, big pipelines, big bridges and big buildings, where fractures can become major catastrophes. It was interesting to observe the different concerns of those who deal with large structures and those who are accustomed to working on a laboratory scale. The interaction was useful, and we must find ways of making it better.

Other innovations were the sessions on *Fracture, Education and Society* and on *Fracture, Politics and Society*. Public attention is now being directed towards questions of safety and of the environmental consequences of failures in large structures. These public concerns are being translated into legislation, regulation and lawsuits. We are thus being propelled, willy nilly, into one side or the other of questions of public policy, and few of us are adequately prepared to cope with this situation. If we shirk this responsibility, other more legal minds will assume this role, and we will lose the opportunity to make an important contribution to society. Our discussions in this area probe only the outer bounds of the problem and we must now learn how to become more effective in matters where the public is directly involved.

All ICF meetings are sponsored by the parent organization, The International Congress on Fracture, and the growth of these Conferences is a tribute to the vision of the Founder-President, Professor Takeo Yokobori of Tohoku University. ICF4 was organized by Professor D.M.R. Taplin of the University of Waterloo with the assistance of a Canadian Organizing Committee. Professor Taplin was also the Editor-in-Chief of the Editorial Board. The success of the Conference and the excellence of the Proceedings are the consequence of their hard work, and I would like to express my personal appreciation as well as the thanks of the International Congress on Fracture.

B. L. Averbach
President (1973-77)
International Congress on Fracture

June 24, 1977

Foreword

My first duty as the new President of the International Congress on Fracture turns out to be one of the most pleasant — to write this short message for the permanent record of the 1977 Waterloo Conference, ICF4. A pleasant task because ICF4 was such a pleasant and successful conference, due in the main to the dedication and hard work shown by Professor David Taplin and all his co-workers. The pattern of the meeting, with its effective plenary lectures each morning and the several workshop sessions running in parallel later in the day, allowed one at the same conference to obtain both the detailed discussion of a particular interest and the general overview of many fracture disciplines which is so much a feature of the concept of ICF.

The siting together of almost all of the delegates on the beautiful and comfortable campus, together with the alternating afternoon and evening free period, actively encouraged free and informal technical discussions. Anyone who looked into the Village Bar or the Faculty Club any evening will have seen the strange paradox of the bringing together of many people whose main technical interest is that of separation.

It would be invidious, and indeed virtually impossible, for me to single out particular technical contributions for praise, but important technical contributions there were, and I am sure these volumes will be the reference works on Fracture for some years to come. In addition to the more usual form of technical papers, I commend the two panel discussions on education and on the relationship of fracture to politics and society, both of which contain much hard sense and emphasise the important role that this Congress can play in the improvement of the standards of life. In this respect it is worth adding that Waterloo also provided the opportunity for a number of meetings of the ICF Council and Executive, in which discussions took place which will encourage various activities in these areas. If any readers have ideas, please be sure that your Council and I would be pleased to hear from you.

But to return to the pleasures of ICF4, one of the features that contributed so greatly to this will not be found explicitly in these volumes. The friendliness of our Canadian hosts, the effectiveness of their social arrangements, the high standard of the catering and domestic arrangements and the major contributions made by the wives of the organising team, were all factors that made this meeting one that will be long remembered as a happy occasion.

So I will close by thanking all those concerned with making ICF4 such a success that we all look forward with pleasant anticipation to meeting in France at ICF5 in 1981.

Roy W. Nichols
President (1977-1981)
International Congress on Fracture

June 28, 1977

Preface

The International Congress on Fracture was founded by Professor Takeo Yokobori at the First International Conference on Fracture held at Sendai, Japan in 1965. This was followed by the Second Conference in Brighton, England, 1969, the Third Conference in Munich, West Germany, 1973 and the Fourth Conference in Waterloo, Canada, 1977. The purpose of the Conference is to foster research in the mechanics & mechanisms of fracture, fatigue and strength of materials; to promote co-operation amongst scientists and engineers covering the many disciplines of fracture research and to assist in making available the results of research and development.

This General Edition of the Proceedings of the Fourth International Conference on Fracture differs in content and format from the Conference Edition, *Fracture 1977*, which was published prior to the Conference for the registered delegates. The expanded title, *Fracture 1977 — Advances in Research on the Strength and Fracture of Materials*, was used to ensure a clear distinction between this edition and its antecedent. The General Edition incorporates a full Subject and Citation Index, in addition to the Author Index, plus corrections of textual and typographical errors. *Overviews* of the individual Parts of the Workshop Programme have been incorporated and appear in Volume 4. Messages from the incoming and outgoing Presidents of ICF are also included in a *Foreword*, plus certain crucial papers and documents received after publication of the Conference Edition.

In order to produce books of more manageable size, the General Edition of the Proceedings appears in six volumes, the original page numbering being retained; thus the content of these six volumes is as follows:

Volume 1, *An Overview*, comprises all the invited plenary papers received when the Conference Edition went to press, and is thus similar in content in the two editions. The same page numbering and Citation Index for the plenary papers is retained and the full Author Index has been added.

Volume 2A, *The Physical Metallurgy of Fracture*, consists of the papers presented in Parts I and II of the Workshop Programme, which appeared in the first half of *Fracture 1977*, Volume 2; hence it contains pages 1 through 678 of this volume. The full Author Index is also included.

Volume 2B, *Fatigue*, consists of the papers presented in Parts III and IV of the Workshop Programme, which appeared in the second half of *Fracture 1977*, Volume 2; hence it contains pages 679 to 1392 of this Volume, which includes the full Author Index.

Volume 3A, *Analysis and Mechanics*, consists of the papers presented in Part V of the Workshop Programme, which appeared in the first half of *Fracture 1977*, Volume 3 .hence it contains pages 1 through 522 of this Volume. This volume also includes the full Author Index.

Volume 3B, *Applications and Non-Metals*, consists of the papers presented in Parts VI and VII of the Workshop Programme, which appeared in the second half of *Fracture 1977*; hence it contains pages 523 through 1232 of this Volume. Volume 3B contains a full Subject and Citation Index to the Proceedings in addition to the Author Index.

Volume 4, *Fracture and Society*, contains the papers issued in a softbound supplementary volume, published a few hours before the Conference began, plus the edited transcript of the two Plenary Panel Discussions *Fracture, Education and Society* and *Fracture, Politics and Society* held under this general title. Included are the ICF4 Interview with Sir Alan Cottrell FRS, the paper *Political and Social Decision Making in Relation to Fracture, Failure, Risk Analysis and Safe Design*, by Max Saltsman MP and the full text of the general survey paper *Fracture,* presented at the conclusion of the Plenary Programme by Professor Bruce Bilby FRS. Also included are the plenary and workshop papers received too late for publication in the earlier volumes. This volume also incorporates a full Subject and Citation Index to all the papers presented at the Conference and published in these Proceedings.

A further book, *Conference Theory and Practice,* by D.M.R. Taplin and R.F. Smith will also be published. This will be a full report of the Waterloo Conference, also providing some general guidelines for the planning of large-scale Technical Conferences.

It is recommended that references to papers in the volumes be cited in the following way:

Reference to the General Edition –

King, J.E., Smith, R.F. and Knott, J.F., "Fracture 1977 – Advances in Research on the Strength and Fracture of Materials", ed. D.M.R. Taplin, Vol. 2A, Pergamon Press, New York, 1977, page 279 (Conference Edition, University of Waterloo Press).

Reference to the Conference Edition –

Rabotnov, Yu. N. and Polilov, A.N., "Fracture 1977", ed. D.M.R. Taplin, Vol. 3, University of Waterloo Press, 1977, page 1059 (General Edition, Pergamon Press, New York).

These Proceedings will serve as a very substantial physical reminder of the large and significant technical content of ICF4. It is hoped that the memory of other aspects of the Conference, the friends and acquaintances made and renewed, the formal and informal technical discussions, the planned and impromptu social activities, will prove equally enduring and valuable to the 700 participants from some 40 countries who assembled in Waterloo.

A final innovation at ICF4 was the distribution of a detailed questionnaire inviting criticism and comment on the organization of the Conference, to aid the planning of ICF5 and other similar conferences. Responses were generally complimentary about the technical pro-gramme and many kind comments were received on the quality of the hospitality and accommodation offered and on the beauty and compactness of the facilities available on the Waterloo Campus. The structure of the Workshop Sessions came under some criticism, perhaps not surprisingly in view of its innovative nature. Some authors seemed unable or unwilling to describe the main points of their work brought up to date (June 1977) in the eight minutes allotted, preferring to attempt a full formal presentation delivered at a gallop. Problems seemed to arise only where speakers did not study the very full instructions provided. This is a common failing of us all. However, this aspect of the Conference also gained many very positive comments and most speakers came extremely well-prepared. Certainly this approach merits repetition in a similar form at ICF5.

The Plenary Sessions were positively received – indeed plenary speakers came extremely well prepared and chairmen were strict in control of the sessions. The *essence* of each paper was presented as required under the instructions, with full up-dating of the work to June 1977, such that it was possible to cover virtually the whole field of fracture in an up-to-date way at the highest possible level. To have had fewer Plenary papers with more time for each presentation, as suggested by some respondents to the questionnaire, would have left significant gaps and failed in this purpose. Plenary speakers are surely to be highly complimented on the unusually commanding quality of the presentations. The fact that little time was available for immediate discussion in Plenary Sessions, a criticism of others, is, frankly, hardly avoidable. With an audience of about 750, controlled and effective discussion is impossible. Discussion of plenary papers, in fact, occurred in the appropriate Workshops. It should be recorded that these Workshop Discussions were often extremely wide and effective and many very positive comments to this effect were received. Two other points seem worth mentioning. It would perhaps have been beneficial to have scheduled the Plenary Panel Discussions earlier in the Conference – perhaps even on the first two days. Also, earlier and stronger measures could perhaps have been taken to involve the national and international media, and thereby the public at large, in the problems of fracture and failure in our advanced technological society. This suggestion in fact came from Mr. Robert Maxwell MC, the Publisher of this General Edition and it was also emphasized in the comments of Max Saltsman MP. These items are worthy of consideration for ICF5.

The opening and closing ceremonies of the Conference were designed to be formal but very brief; this was welcomed by delegates. The closing ceremony is outlined at the conclusion of the Panels on Fracture and Society. As Chairman of ICF4, I formally declared the Fourth International Conference on Fracture in Session at 8:50 am, June 20th, 1977 and called upon Professor B.L. Averbach, President of the International Congress on Fracture and Dr. B.C. Matthews, President of the University of Waterloo, to make speeches of welcome to the 700 assembled delegates and guests from some 40 countries. The founding father of ICF, Professor Takeo Yokobori, was also introduced on the stage of the Humanities Theatre. The Conference was described as the *Olympics of Fracture* where the whole topic of Fracture and all its ramifications are under discussion in a World Assembly. The whole opening ceremony was concluded in a matter of ten minutes including general announcements from the Conference Secretary Dr. Richard Smith. The first Plenary Session commenced at 9:00 am, under the Chairmanship of Professor A.J. McEvily and Professor A.N. Sherbourne, with the opening lecture by Professor M.F. Ashby.

The various social events both formal and informal proved to be very successful. A complete Family Programme was organized jointly by Ms Diana Theodores Taplin and Mrs. Sherry Pick and the Social Programme was organized by Dr. Kon Piekarski and Dr. Roy Pick. The visits to the Shakespeare Festival at Stratford to see Midsummer Night's Dream (on midsummer night itself), to Niagara Falls, Douglas Point, Elora Gorge, and the Mennonites and the Conference Banquet and Cabaret were particularly memorable. The most important feature of the Social Programme was, however, its informality — this included the many spontaneous parties which developed in the Faculty Club and elsewhere on the campus — during a week blessed with no less than perfect midsummer's weather. Surely the *Force* smiled upon us at ICF4!

A point mentioned in Dr. Nichols' Foreword is worth re-emphasis here. The names of all the new Executive Officers of ICF (1977-1981 term) are listed in each Volume of these Proceedings. To be sure, any of these individuals would welcome suggestions on the organization of further Conferences and any other activities which ICF might usefully initiate or co-ordinate, particularly in regard to Publications. The success of ICF4 derived from the whole-hearted participation of many people. ICF wishes to serve all those around the world working on Fracture Problems. This purpose can be achieved effectively only by the further active involvement of us all and the continuing recognition of ICF as the appropriate world organization and "umbrella" for coordinating work on fracture.

Waterloo, Canada
September 30, 1977

Acknowledgements

On behalf of the Canadian Fracture Committee I would like to thank all those who contributed to the preparations of the Fourth International Conference on Fracture. We are particularly grateful to the University of Waterloo and the International Congress on Fracture under whose joint auspices the Conference was organized. Financial assistance from the following organizations is gratefully acknowledged:

National Research Council of Canada
Ontario Hydro
Babcock and Wilcox
University of Waterloo
Atomic Energy of Canada
ALCAN
Westinghouse Canada Limited
International Nickel Company Limited
Canadian Welding Development Institute
Consumers' Gas Company
Trans Canada Pipelines
Noranda Mines Limited
Alberta Gas Trunk Line Company
Canadian Vickers Limited
Atlas Steels Company
Dominion Foundries and Steel Company
STELCO
International Congress on Fracture
Dominion Bridge Company Limited
General Electric
Algoma Steel
American Society for Metals
Union Carbide of Canada
Pratt and Whitney of Canada
MTS Corporation
Instron of Canada

On behalf of the Editorial Board I wish to record our gratitude to the authors of the workshop papers published in these proceedings (Volumes 2 and 3). The standard was high and yet authors were forbearing with critical editorial comments. All papers were extensively reviewed by the Board through the assistance and cooperation of a large body of expert referees. The technical quality of the publication is also directly related to these efforts, which are greatly appreciated but, according to custom, remain anonymous.

I am pleased to record a special thanks in these proceedings to the Publications Group of the Solid Mechanics Division of the University of Waterloo — Professor D.E. Grierson, Technical Editor; Mrs Pam McCuaig and Miss Linda Heit, Editorial Staff; and Mrs. Cynthia Jones — to the secretarial staff, Miss Elizabeth Krakana, Mrs. Daniela Michiels and Mrs. Jana Karger — and to Mr. David Bartholomew, Graphic Designer. I am also pleased to acknowledge my appreciation to Mr. Robert N. Miranda, Senior Vice President and Mrs. Sylvia M. Halpern, Chief Manuscript Editor of Pergamon Press, Inc., Elmsford, New York, and to Mrs. Patty Patrick of LithoCrafters, Inc.

International Congress on Fracture

EXECUTIVE COMMITTEE 1973 - 1977

Founder-President	Takeo Yokobori	Japan
President	B. L. Averbach	U.S.A.
Vice-Presidents	A. Kochendörfer	West Germany
	R. V. Salkin	Belgium
	S. N. Zhurkov	U.S.S.R.
Directors	R. W. Nichols	United Kingdom
	C. J. Osborn	Australia
	Yu. N. Rabotnov	U.S.S.R.
	D. M. R. Taplin	Canada
	H. C. van Elst	Netherlands
Treasurer	J. Nemec	Czechoslovakia
Secretary-General	T. Kawasaki	Japan
Members	P. Haasen	West Germany
	A. K. Head	Australia
	N. J. Petch	United Kingdom
	J. L. Swedlow	U.S.A.
	M. L. Williams	U.S.A.

Executive Officers Elected at ICF4 for 1977-1981

Founder-President	T. Yokobori	Japan
President	R. W. Nichols	United Kingdom
Vice-Presidents	Yu. N. Rabotnov	U.S.S.R.
	D.M.R. Taplin	Canada
	H.C. van Elst	Netherlands
Directors	A. J. Carlsson	Sweden
	W. Dahl	West Germany
	D. Francois	France
	C. J. McMahon, Jr.	U.S.A.
	J. Pelczynski	Poland
Treasurer	H. H. Kausch	Switzerland
Secretary-General	T. Kawasaki	Japan

MEMBERS OF COUNCIL 1977 - 1981

Australia

A. K. Head*
M. Murray
C. J. Osborn
C. M. Perrott

Belgium

R. V. Salkin
W. Soete*
A. Vinckier

Canada

M. R. Piggott
L. A. Simpson
D. M. R. Taplin*

Czechoslovakia

M. Klesnil
V. Linhart
J. Nemec*

Denmark

F. I. Niordson*
N. Olhoff

France

M. Brunetaud
D. Francois
J. Philibert
J. Plateau
J. Friedel*

Hungary

L. F. Gillemot*

India

S. N. Bandyopadhyay
P. Rama Rao*
A. K. Seal

Israel

A. Buch*
A. Libai
J. Tirosh

Italy

F. Gatto*
L. Lazzarino
F. Manna

Japan

T. Kawasaki
T. Yokobori*

Luxembourg

E. A. Hampe*

The Netherlands

D. Broek
H. C. van Elst*
C. A. Verbraak

Norway

H. Wintermark*

*People's Republic
of China*

Chun Tu Liu*

Poland

Z. Pawlowski*
T. Pelczynski

South Africa

Z. T. Bieniawski*
S. B. Luyckx
L. O. Nicolaysen

Spain

Sistiaga Aguirre*

Sweden

B. Broberg
J. Carlsson*
J. Hult

Switzerland

E. Amstutz
M. J. Briner*

United Kingdom

M. J. May
K. J. Miller
P.L. Pratt
R.W. Nichols*
N. J. Petch

U.S.A.

B. L. Averbach*
J. L. Swedlow
M. L. Williams

U.S.S.R.

Yu. N. Rabotnov
S. N. Zhurkov*

West Germany

W. Dahl
H. H. Kausch
F. Kerkhof
P. Hassen
A. Kochendörfer*

*voting member

Honorary Fellows of ICF (Elected at ICF4)

T. Yokobori

R. W. Nichols

A. Kochendörfer

D.M.R. Taplin

B. L. Averbach

ICF Nominating Committee 1977-1981

Chairman	R. V. Salkin	Belgium
	Z. T. Bieniawski	South Africa
	K. B. Broberg	Sweden
	A. G. Evans	U.S.A.
	A. K. Head	Australia
	T. Kunio	Japan
	Liu Chun Tu	People's Republic of China
	J. Nemec	Czechoslovakia
	P. Rama Rao	India
	E. Smith	United Kingdom
	V. I. Vladimirov	U.S.S.R.

ICF Publications and Finance Committee (1977-1981)

Chairman	D.M.R. Taplin
	B. L. Averbach
	D. Francois
	H. H. Kausch
	J. F. Knott
	J. L. Swedlow
	T. Yokobori

ICF Committees on Regional Liaison and a Standing Committee on Policy (Chairman R.W. Nichols) are also being commissioned by the President. Through the work of these various committees we look forward to a new phase in the development of the International Congress on Fracture as an influential world body.

Fourth International Conference on Fracture

Waterloo, June 19 - 24, 1977

CANADIAN FRACTURE COMMITTEE

Chairman	D. M. R. Taplin	(Waterloo)
	C. M. Bishop	(de Havilland)
	D. J. Burns	(Waterloo)
	J. Dunsby	(NRC)
	W. H. Erickson	(Defence Dept)
	J. D. Embury	(McMaster)
	J. Hood	(Stelco)
	R. R. Hosbons	(AECL)
	H. H. E. Leipholz	(Waterloo)
	I. Le May	(Saskatchewan)
	J. T. McGrath	(CWDI)
	D. Mills	(Ontario Hydro)
	K. R. Piekarski	(Waterloo)
	M. R. Piggott	(Toronto)
	L. A. Simpson	(AECL)
	T. A. C. Stock	(Alcan)
	T. H. Topper	(Waterloo)
	G. C. Weatherly	(Toronto)
Conference Secretary	R. F. Smith	(Waterloo)

LOCAL ARRANGEMENTS COMMITTEE

Chairman:	D. M. R. Taplin
	J. R. Cook
	K. D. Fearnall
	H. W. Kerr
	H. H. E. Leipholz
	D. Mills
	R. J. Pick
	Sherry Pick
	K. R. Piekarski
	M. R. Piggott
	A. Plumtree
	Betty Statham
	Diana Theodores Taplin
	T. H. Topper
Conference Manager:	R. F. Smith

Plate 1. *Professors Nemec, Yokobori and Averbach at the ICF Executive Meeting, June 19, 1977*

Plate 2. *Professors Taplin, Osborn and Rabotnov and Mr. Cook at the ICF Executive Meeting*

Plate 3.　*First Plenary Session - Professor McEvily introducing Professor Ashby - from left, Professors Embury, Knott, Sherbourne, McEvily, Rice, Ashby*

Plate 4.　*The Delegation from The People's Republic of China*

Plate 5.   *The Conference Office Staff - Dr. Smith, Linda Heit, Jana Karger, Tina Sully and Nancy Nelson*

Plate 6.   *Dr. Richard Smith, the Conference Secretary, at work*

Plate 7.   *Coffee break during Workshop Session*

Plate 8. *Coffee Break - Professors Rama Rao and Armstrong, and Dr. Woodford*

Plate 9. *The Humanities Court*

Plates 10 - 17 : Conference Lecturing Styles

Plate 10. Professor Michael Ashby

Plate 11. Professor Dominique Francois

Plate 12.   Professor Max Williams

Plate 13.   Professor Ted Smith

Plate 14. Professor Bruce Bilby

Plate 15. Professor David Taplin

Plate 16. Dr. Roy Nichols

Plate 17. Professor Dinah Christie

Plate 18.   The Delegation from the U.S.S.R.

Plate 19.   Professors Liebowitz, Rabotnov, Kochendörfer and Radon

Plate 20. Professor Francois being encouraged in his work ahead for ICF5

Plate 21. At the Banquet "... if you believe that you'll believe anything ..." Mr. Fearnall, Professors Argon, Poturaev, Burns, Kitagawa, Osborn

Plate 22. *The Welcome Reception - Dr. Knott, Professor Topper, Mrs. Diana Theodores Taplin and Professor Burns*

Plate 23. *Professor Kerr, the Transport Officer*

Plate 24. *Coffee Break during Plenary Session - Professor Pick, Registration Chairman, with Drs. Rau and Tetelman*

Plate 25.   *The Final Plenary Session - Dr. Nichols, Max Saltsman,
Dr. Mills, Dr. Hahn*

Plate 26.   *Aerial View of a corner of the Campus*

Standard Nomenclature List

In order to minimize unnecessary confusion, a standard nomenclature for commonly used quantities has been adopted for ICF4. This coincides closely with other developing nomenclatures in the field of fracture and it is hoped that this notation will become widely used. SI units have been used throughout the Proceedings with fracture toughness reported as $MPam^{1/2}$. It was originally thought that this quantity might be designated the "griffith". Whilst we surely wished to honour the father of the science of fracture in this way, we thought better of taking any unilateral action at this time. Thus only informal use of the griffith is recommended at ICF4.

A	Area of Cross-Section of a Specimen
A_o	Area of Cross-Section of a Specimen at the Start of Testing
A_f	Area of Cross-Section of a Specimen at Fracture
a	Crack Length - One-Half the Total Length of an Internal Crack or Depth of a Surface Crack
a_o	Original Crack Length - One-Half of Total Length of an Internal Crack at the Start of a Fracture Toughness Test, or Depth of a Surface Crack at the Start of a Fracture Toughness Test
a_p	Measured Crack Length - One-Half the Total Length of an Internal Crack or Depth of a Surface Crack as Measured by Physical Methods
a_e	Effective Crack Length - One-Half the Effective Total Length of an Internal Crack **or** Effective Depth of a Surface Crack (Adjusted for the Influences of a Crack-Tip Plastic Zone)
Δa, Δa_p, ...	Crack Growth Increment
da/dN	Rate of Fatigue Crack Propagation
B	Test Piece Thickness
b	Atomic Interval (Burgers Vectors Magnitude)
d	Average Grain Diameter
D_L	Lattice Diffusion Rate
D_B	Grain Boundary Diffusion Rate
D_S	Surface Diffusion Rate
E	Young's Modulus of Elasticity
exp	Exponential Base of Natural Logarithms

G	Strain Energy Release Rate with Crack Extension per unit length of Crack Border of Crack Extension Force
G_I G_{II} G_{III}	Crack Extension Forces for Various Modes of Crack Opening
h	Planck's Constant
I	Moment of Inertia
J	Path-Independent Integral Characterizing Elastic/Plastic Deformation Field Intensity at Crack Tip; also, Energy Release Rate for Non-Linear Elastic Material
K	Stress Intensity Factor - A Measure of the Stress-Field Intensity near the Tip of a Perfect Crack in a Linear-Elastic Solid
K_c	Fracture Toughness - The Largest Value of the Stress-Intensity Factor that exists prior to the Onset of Rapid Fracture
K_{max}	Maximum Stress-Intensity Factor
K_{min}	Minimum Stress-Intensity Factor
K_{th}	Threshold Stress Intensity Factor Below which Fatigue Crack Growth Will Not Occur
K_I	Opening Mode Stress Intensity Factor
K_{IC}	Plane-Strain Fracture Toughness as Defined by ASTM Standard Designation E 399-74
K_{Ii}	Elastic Stress-Intensity Factor at the Start of a Sustained-Load Flaw-Growth Test
K_{ISCC}	Plane-Strain K_I Threshold Above Which Sustained-Load Flaw-Growth Occurs
K_{II}	Edge-Sliding Mode Stress Intensity Factor
K_{III}	Tearing Mode Stress Intensity Factor
$\dot{K}$	Rate of Change of Stress-Intensity Factor with Time
ΔK	Stress Intensity Range
k	Boltzmann Constant
k_y	Parameter that Determines Grain-Size Dependence of Yield Strength
l_o	Gauge Length
ln	Natural Logarithm
log	Common Logarithm

Nomenclature

m	Strain-Rate Sensitivity Exponent
N_f	Number of Cycles to Failure
n	Strain Hardening Exponent
P	Force
P_{max}	Maximum Force
P	Pressure
Q	Activation Energy
Q_a	Activation Energy for Crack Growth
Q_c	Activation Energy for Creep
Q_d	Activation Energy for Self Diffusion
T	Temperature
T_M	Absolute Melting Temperature
T_D	Brittleness Transition Temperature
t	Time
t_o	Time at the Onset of a Test
t_f	Fracture Time
U	Potential Energy
δz	Thickness of Grain Boundary Layer
γ_s	True Surface Energy
γ_B	Grain Boundary Surface Energy
γ_p	Effective Surface Energy of Plastic Layer
δ	Value of Crack Opening Displacement
δ_c	Critical Crack Opening Displacement, Being One of the Following:
	(1) Crack Opening Displacement at Fracture
	(2) Crack Opening Displacement at First Instability or Discontinuity
	(3) Crack Opening Displacement at Which an Amount of Crack Growth Commences
δ_m	Crack Opening Displacement at First Attainment of Maximum Force
ε	Normal Strain

ε_e	Normal Strain, Elastic
ε_p	Normal Strain, Plastic
ε_T	Normal Strain, Total
ε_{max}	Normal Strain, at Maximum Tensile Load
ε_E	Engineering Normal Strain
ε_f	Normal Strain, Critical Value at Fracture
ε_i	Principal Strains (i = 1, 2, 3)
ε_{pi}	Principal Strains, Plastic
$\varepsilon_x\ \varepsilon_y\ \varepsilon_z$	Cartesian Strain Components
ε_{ij}	Strain Tensor
$\dot{\varepsilon}$	Strain Rate
$\dot{\varepsilon}_e$	Strain Rate, Elastic
$\dot{\varepsilon}_p$	Strain Rate, Plastic
$\dot{\varepsilon}_o$	Strain Rate, Initial Value
$\Delta\varepsilon$	Strain Range
$\Delta\varepsilon_p$	Plastic Strain Range
ν	Poisson's Ratio
σ	Normal Stress
σ_y	Yield Stress Under Uniaxial Tension
$\sigma_1\ \sigma_2\ \sigma_3$	Principal Normal Stresses
σ_e	Fatigue Strength, Endurance Limit
σ_f	Fracture Stress
σ_{max}	Maximum Stress
$\sigma_x\ \sigma_y\ \sigma_z$	Cartesian Components of Normal Stress
$\dot{\sigma}$	Stress Rate
τ	Shear Stress
τ_o	Critical Shear Stress
$\tau_1\ \tau_2\ \tau_3$	Principal Shear Stresses
τ_{max}	Shear Stresses, Maximum Value
Ω	Atomic Volume

Conversion Units

To Convert From	To	Multiply By
inch	meter (m)	2.54×10^{-2}
pound force	newton (N)	4.448
kilogram force	newton (N)	9.807
kilogram force/meter^2	pascal (Pa)	9.807
pound mass	kilogram mass(kg)	4.536×10^{-1}
ksi	pascal (Pa)	6.895×10^6
ksi $\sqrt{in}$	MPam$^{1/2}$ (Gr)	1.099
ton	pascal (Pa)	1.333×10^2
torr	pascal (Pa)	1×10^5
angstrom	meter (m)	1×10^{-10}
calorie	joule (J)	4.184
foot-pound	joule (J)	1.356
degree Celsius	kelvin (K)	$T_K = T_C + 273.15$

Important Multiples

Multiplication Factor	Prefix	Symbol
10^{-12}	pico	p
10^{-9}	nano	n
10^{-6}	micro	μ
10^{-3}	milli	m
10^3	kilo	k
10^6	mega	M
10^9	giga	G

ABBREVIATIONS - *a poem by Roy Nichols*
It started out with Griffith who took to using L,
George Irwin used his squiggly G, but there's much more to tell -
There soon came K and COD, and R both large and small
And J for Joe, and H and B - I'm sure you know them all.
They come in ones or multiples like K_{ISCC},
But this meeting's made it easy, just like an ABC.
So if you want to know what's what, or how or why or when
To use your Q's and W's, or t, D, m, and n
Try Taplin's slimmer volume under "Nomenclature" -
Just one of many blessings that's from ICF4.

Fracture 1977, Volume 4, ICF4, Waterloo, Canada, June 19 - 24, 1977

FRACTURE

B. A. Bilby FRS*

INTRODUCTION

There are three important processes whereby a condensed phase can be separated into two parts. *Cracking*, in which rows of atoms or molecules are
pulled apart normal to their centres of mass; *sliding off*, in which finite
rows of them slide over one another until they ultimately part company;
and *the removal of individual atoms*, as in vacancy migration or electro-
chemical attack. These processes, and, in crystals, those of deformation
twinning and martensitic transformation also, interact on a microscale
during the manufacture, assembly and use of materials to produce each
other. Inhomogeneities of material and structure may lead to cracking and
voids; cracking is relaxed or blunted by local sliding while sliding and
twinning themselves can cause cracking. So our engineering structures
generally contain many small cracks and voids, as well as inhomogeneities
in material and structure which readily generate them under loads. In
fracture we are mostly interested in the conditions under which these small
discontinuities can grow and propagate as macroscopic cracks. For this
propagation to proceed, two conditions must be satisfied. It is necessary
that the decrease of total energy (the elastic energy of the body plus the
potential energy of the loading system) be at least equal to the energy
required to drive this separation process; and it is necessary that some
physical mechanism can take place permitting this separation to occur.
It may be convenient to consider the separation process on many different
scales. We may look at a catastrophic failure occurring in a massive
structure; at a specimen undergoing a fracture toughness test; at a small
region near the tip of a larger crack where there is subcritical stable
growth or slow extension during fatigue or creep; or at slip occurring on
a microscale during the formation of craze nuclei. Whatever the scale
however, these two principles govern the extension process.

The engineer must design and build structures and keep them safely in
service. So there arises a continual need for practical tests to characterize the properties of materials. As understanding of these properties
increases, these tests become more discerning and reliable, but their
development must go hand in hand with more fundamental studies. The
history of fracture and fracture mechanics is yet another example of how
theory and practice interact to their mutual advantage. At the present
time, when science is a little unfashionable, it is well to remember that
we cannot go against Nature and that our progress will be faster if we
learn a little to understand her ways.

---

*Department of the Theory of Materials, University of Sheffield,
Mappin Street, Sheffield, S1 3JD, U.K.

FRACTURE CRITERIA

In a linear elastic material the singular field p_{ij} near the tip of a sharp crack is characterized by the stress intensity factors K_1, K_2, K_3 and has the form

$$p_{ij} = (2\pi r)^{-1/2} K_s f_{sij} \tag{1}$$

where the f_{sij} depend on θ. If the advance of the crack is governed by the stress field in this region, then we can determine by a test, for example in mode I, the critical value K_{1c} at which the crack will advance catastrophically. Then, if the service environment and other conditions are similar to those of the test, a cracked structure is secured against catastrophic failure until the K_1 for some crack in it reaches K_{1c}. This is the basis of linear elastic fracture mechanics. In the test, the departures from linear elasticity at the crack tip are carefully controlled. Corrections which make the slightly relaxed crack appear a little longer can be applied to extend the approach to small scale yielding. However, difficulties begin to multiply when we recognise that fractures in practice are generally accompanied by considerable departures from linearity, and when we try to make small scale tests on tough materials. The problems are compounded by the facts that fractures in structures occur under combined stresses and by the necessity of making proper allowance for chemical reactions, temperature, and varying stress.

For a sharp crack in an ideally brittle elastic material the critical K criterion embraces both of the fundamental conditions for fracture. The physical process condition is automatically satisfied, in a continuum model by the 'infinite' stress, or in one which is more realistic [1] by the existence of some bond at the crack tip which is always on the point of breaking. Although the situation at the tip of a macroscopic crack is much more complicated than this we have to remember that separation processes of this kind are occurring on a microscale in this region. It is thus important to model them, and particularly to study how the rate at which they occur is affected by temperature and the chemical environment. This is because whether such micro-cracking occurs or not may greatly influence the nature of the whole macroscopic fracture itself.

The energy condition we now formulate in terms of the *energy release rate* or *crack extension force* G, introduced in 1948 by Irwin [2] and shown by him in 1957 [3, 4] to be determined by K. Griffith's energy condition [5, 6] may then be written $G = 2\gamma$, where γ is an effective surface energy for fracture; alternatively we have (in mode I) $G = G_{1c} = K_{1c}^2/2M$, where G_{1c} is a critical value of G and M is an elastic modulus [7]. From this point of view K_{1c} is an indirect way of describing the effective surface energy for fracture. In 1960, G was expressed as a path-independent integral [8], and in 1968 a number of authors [1, 9 - 11] independently gave related expressions for G, one of which is now widely known as the J integral. The generalisation to the dynamic case was also considered at this time [12], and it was shown [1, 12] that the theory of the crack extension force followed naturally from the general theory of forces on elastic singularities developed in 1951 using the *energy-momentum tensor* [13].

Many computations and experiments have been devoted recently to the examination of J and quantities related to it as candidates for fracture criteria in post yield fracture mechanics. This work is not always easy to follow because of variations in the terminology and interpretation of different authors. Moreover, the confusion is worse confounded by the use of similar symbols for the integrals themselves (which are mathematical entitites in their own right, and which can be calculated as numbers without any interpretation if desired), and other quantities. These latter quantities are derived from the experimentally determined load-deflection curves of specimens containing cracks of various lengths, or sometimes, with the help of approximate theories, by other experimental methods. They, and the integrals themselves, are also calculated theoretically entirely from model experiments, by numerical methods using large computers. If the specimens were non-linear elastic these quantities (as well as the integrals themselves) would be crack extension forces, but in the usual practical and model situations they are not (and neither are the integrals). It might be helpful [14] to use some symbols other than J for these pseudo-crack-extension forces, retaining J for the integral defined by Rice [9]. We shall now try to discuss some of the problems arising in this work.

FUNDAMENTAL INTEGRALS

For the linear or non-linear elastic body the quantity

$$F_\ell = \int_S P_{\ell j} \, dS_j \tag{2}$$

is such that $-F_\ell \delta \xi_\ell$ is the free energy change when all singularities inside a closed surface S drawn in the body are displaced by $\delta \xi_\ell$ [13] (we limit our discussion here to the static case; for the dynamic see [12, 15, 16]). Here

$$P_{\ell j} = W \, \delta_{\ell j} - p_{ij} \, u_{i,\ell} \tag{3}$$

is the energy-momentum tensor of the elastic field for which the stresses p_{ij} are given by $\partial W/\partial u_{i,j}$ and $+W(u_i, u_{i,j}, X_i)$ is the strain energy density, assumed, for generality, to depend not only on the field quantities, but also explicitly on X_i, the initial coordinates. (We use a notation which makes (2) valid for the finite deformation of a non-linear elastic material; p_{ij} is the (unsymmetrical) nominal or Boussinesq or second Piola-Kirchhoff stress-tensor, the commas denote differentiation with respect to the X_i and S is a surface in the undeformed body. -W is, in the static case, the Lagrangian density function from which the field equations are derived from a variational principle [17 - 21]. The treatment can readily be extended if need be to a material of grade n [19, 20]. Using the field equations, we can show that

$$\frac{\partial P_{\ell j}}{\partial X_j} = \left(\frac{\partial W}{\partial X_\ell}\right)_{exp} \tag{4}$$

where "exp" denotes the explicit derivative, with $u_i, u_{i,j}$ and $X_{,j}, j \neq \ell$, held constant. Putting $\ell = 1$, regarding the crack either as a distribution of dislocations [1] or as a singularity in its own right [15], and letting $dS_j = n_j ds$ for $j = 1,2$, we get for the crack extension force,

$$F_1 = \int_\Gamma (W \, \delta_{ij} - p_{ij} \, u_{i,1}) \, n_j ds \tag{5}$$

3

The divergence of the integrand vanishes if $(\partial W/\partial X_1)_{exp} = 0$; that is, if the material is homogeneous in the direction X_1 of the crack extension. However, it may be inhomogeneous in the X_2 direction; for example, the crack might lie between two different media. The integral J [9] is of the same form as (5), but W may be replaced by W', the density of stress working

$$W'(X_m,t) = \int_0^t P_{ij}(X_m,t') \left\{\partial u_{i,j}(X_m,t')/\partial t'\right\} dt' \tag{6}$$

where t is a parameter denoting the progress of the deformation. The derivation of F_1 and the proof of its path-dependence involves the assumption of the existence of the function W. J and F_1 are thus identical and independent of Γ in linear and non-linear elasticity. Deformation plasticity, provided there is no unloading, can be regarded as a kind of non-linear elasticity. Thus, if the same strains and displacements are used in both J and F_1, they are again identical and path-independent. In a region of plasticity modelled by the incremental theory, J may be evaluated with du interpreted as the total shape displacement giving the shape change of the solid; that is, $du_i = du_i^E + du_i^P$, where E and P denote the elastic and plastic contributions. No general proof that it is then path-independent has been given, although, as discussed at ICF3 [22], it may be approximately so [23]. It is clear that the arguments leading to the path-independence of F_1 depend on the existence of the function W. Now, if in the actual loading the density of stress working is independent of the stress-strain path, W' is a function only of the current state and not of the strain history. Thus W' can be used for W in F_1; then if u is the total shape displacement, $J\delta$ and F_1 are the same and are independent of Γ, for we cannot tell that the field quantities were not derived from a density function [24]. A steadily moving plastic-elastic crack is an example of this kind [25]. The matter has been discussed recently in terms of the DBCS model [26, 27]. It is emphasized that, in general, J is path-independent in any situation where W' is independent of the stress-strain path by which the current state is reached [27]. It will be evident that by using various combinations of the elastic, plastic (or total) strains and displacements appearing in the two terms of the integrands a considerable number of integrals resembling J and F_1 can be obtained. It would be helpful, when these are evaluated numerically, if the quantities being evaluated were very clearly defined. Studies of the path-dependence of J in incremental plasticity are continuing [14, 28, 29].

If plastic flow has occurred at the crack tip, the integral F_ℓ gives the resultant force on the crack tip and on all the dislocations inside S [13, 26], but it is defined only for paths in the elastic region. However, an integral Q_ℓ may be derived [22, 30] which reduces to F_ℓ in the elastic region and which can be taken through a continuous distribution of dislocations representing the crack tip plasticity (and any micro-cracking there). This integral Q_ℓ was given at ICF3 and is [22]

$$Q_\ell = \int_S \left(W\delta_{\ell j} - P_{ij}\ \beta_{\ell i}^E\right) dS_j \tag{7}$$

where W is the elastic energy density and $\beta_{\ell i}^E$ the elastic distortion tensor giving the spatial increments of elastic displacement $du_i^E = dx_\ell \beta_{\ell i}^E$ in a continuous distribution of dislocations (the elastic displacement u_i^E does not exist [31, 32, 33]). Q_ℓ reduces to zero when shrunk on to the crack tip [22] for the small scale yielding from an edge slit in anti-plane strain [34]. As discussed at ICF3 [22], it would not be surprising if a realistic model of crack tip plasticity showed that the crack and its plastic zone were in neutral equilibrium, in the sense that any energy released by crack advance is absorbed by plastic work. This may be shown to be so for the quasi-static DBCS model [1, 35, 36], for the dynamic DBCS model [15, 37], and, more generally, [38], for elastic-plastic materials having a flow stress tending to a constant value at large strains; see also [39] for further discussion. These questions raise problems about the use of quantities like F_1 and J for the characterization of crack extension [22]. The interpretation of F_1 shows that $F_1 \delta \xi$ is the energy released when the crack tip and all dislocations representing the plasticity are displaced in the X_1 direction for $\delta \xi$. This is not (necessarily) an equilibrium displacement of the crack and its plasticity, but the significance of this energy release, and how much of it is mopped up by plastic work in an actual movement of the crack, are not clear. Attention has again been focussed on the matter by recent numerical work [40, 41] confirming the result [25, 38] that there is no energy release rate for a growing crack in plastic-elastic material. If $-\Delta W$ is the work of unloading the initially stressed segments of crack face it is suggested that a crack tip energy release rate $G^\Delta = \Delta W/\Delta a$ calculated over a finite crack growth step Δa should be considered [42, 43] (the quantity $G^\Delta \to 0$ as $\Delta a \to 0$).

Disregarding heat fluxes, we can write for an imposed small extension Δa of the crack tip,

$$-\Delta E_{POT} = \Delta E_{EL} + \Delta w + G\Delta a + 0(\Delta a^2)$$

Here $-\Delta E_{POT}$ is the work done by the loading system, ΔE_{EL} the increase in stored elastic energy, Δw the work dissipated in plastic flow and $G\Delta a$ the energy released at the crack tip. In linear and non-linear elasticity, when $\Delta w = 0$, it is the essential property of the integrals F_1 and J that they give G directly. For the fracture condition we then put $G\Delta a = 2\gamma'\Delta a$, where $2\gamma'$ is the effective surface energy for fracture. The result that G = 0 when plastic flow is allowed really shows that the plastic elastic continuum models considered are too simple. We have to make a more realistic representation of the fracture process allowing for rate effects, micro-cracking and mechanical instabilities in the fracture zone. At the simplest level, we simply lump some of the plastic work into the fracture energy, recognising it as part of the failure process; this is the extension of the Griffith theory originally proposed by Irwin [2] and Orowan [44]. In a semi quantitative way, as we discuss below, the DBCS model can be used to develop this idea. As has been noted [22], its developments to include rate effects [11, 45 - 47] show some of the qualitative features required to describe slow stable growth and the transition to fast fracture.

From any numerical solution for the plastic-elastic crack, we can calculate not only the integrals F_1 and J and the increment $\Delta E_{POT} + \Delta E_{EL} + \Delta w$ (tending to zero, perhaps, as $\Delta a \to 0$), but also by suitable integration over the developing field, the quantities $+E_{POT}$, E_{EL} and w as we load up a crack of fixed length a. We can then find the derivative $-\partial(E_{POT} + E_{EL} + w)/\partial a$ and compare it with F_1 and J. There is evidence [14] that these quantities are not the same. This we should indeed expect, for, as in the corresponding experimental procedure when we load up specimens with cracks of increasing length [48 - 50], we are not dealing with perfect differentials. The states obtained after loading a specimen of

crack length a and allowing it to extend to a + Δa or alternatively loading a specimen of crack length a + Δa are different [21, 51]. It is not
established even that ΔA, the area between the load extension curves for
the cracks of lengths a and a + Δa, is -JΔa, nor is the connection with
crack extension at all straightforward [24]. It is thus a matter for
experimental study whether these or related methods [52 - 54] will yield a
satisfactory characterisation of the onset of fracture.

THE CRITICAL DISPLACEMENT CRITERION

An interesting development using a critical displacement criterion for
fracture began with the appearance of the BCS fracture theory [55, 56],
which uses a highly simplified model of the crack tip plasticity consisting
of a linear array of dislocations. A similar model to remove the elastic
crack tip singularity and represent the plasticity was used by Dugdale
[57]; a closely related (though not quite equivalent) procedure for eliminating the crack tip singularity is central to Barenblatt's work [58].
The procedure has also been used by Vitvitskii and Leonov (see [59]); a
similar idea was employed by Prandtl [60]. The DBCS model has been
elaborated in various ways and very widely applied to discuss many aspects
of fracture [61 - 89]; recent reviews have discussed some of these developments [1, 22, 51]. An aspect which is currently receiving increasing
attention is the BS (Bilby-Swinden [70]) model in which two (or more)
dislocation arrays inclined to the crack are used in an attempt to make a
slightly more realistic representation of the plasticity [42, 88 - 91].

The DBCS model has also been used in discussions of the COD concept in
post-yield fracture mechanics [92 - 95]. There has been a considerable
development of this criterion on the engineering side, but, like its
rivals, its status as a single characterizing parameter is still a matter
for further elucidation and debate.

Nevertheless the use of the criterion in the BCS theory has been very
useful in providing a two-parameter model fo the energy expended in the
fracture process, and an interpolation between the Griffith theory (or
linear elastic fracture mechanics) and failure after considerable yielding
or plastic collapse. If ϕ_c is the critical displacement at the crack tip
and σ_1 the stress in the relaxed zone, the fracture stress σ_f is related
to the crack length c by the equation [65, 66].

$$\sigma_f/\sigma_1 = (2/\pi) \cos^{-1}\{\exp(-c*/\pi c)\} \tag{8}$$

where $c* = M \phi_c/4\sigma_1$, M being an elastic modulus. The condition $c = c*$
defines the crack length at which the material becomes *notch-sensitive*
[65]. The equation (8) reduces to the Griffith condition when $c \gg c*$;
we then have a "low-stress" failure with $\sigma_f \ll \sigma_1$. When $c < c*$, the
fracture stress approaches σ_1, the strength of the layer ahead of the
crack. This equation is successful in describing the stresses at which
failures occur below general yield in large structures and may be used
to estimate dangerous notch sizes in them [64 - 66, 70, 83 - 85]. Also,
with σ_1 identified with the ultimate tensile strength or the collapse
stress, and with an appropriate stress intensity factor for the geometry
considered it is remarkably effective in correlating post yield fractures
with defect size in a wide class of materials [68, 69, 96] and can be
used to estimate K_{1c} values from "invalid" ASTM tests. It has also been
suggested as an interpolation between failure by plastic collapse and

linear elastic fracture mechanics, of potential use in assessing critical defect sizes in large structures and in design [97]. The engineer cannot afford to make mistakes and, if he must, he will test his actual structures to destruction. His inclination is frequently for the simplest approach, based on large-scale tests [98]. Although the great simplifications in the BCS theory are obvious, it is not wholly empirical, and so may be of some value in the correlations which he has nevertheless to make.

The theory gives for the fracture energy $2\gamma'$ the expression $\sigma_1 \phi_c$. We distinguish two *modes* of fracture [22, 61]; a stable, *non-cumulative* or *non-localised* mode which occurs when $\sigma_1 \sim \sigma_f$; the material is not notch-sensitive. The non-linearity represented by the dislocations spreads through the specimen much faster than the crack. The second mode is *cumulative* or *localised* and is unstable; in this type of failure the material is notch-sensitive for $c \gg c^*$. A similar localised set of dislocations representing the non-linearity moves with the crack as it grows, so that it can advance without the non-linearity spreading through the whole net section ahead of it; the fracture is a low-stress one with $\sigma_f \ll \sigma_1$ and $c \gg r$, the extent of the non-linearity ($r \sim \pi c^* \sim E\phi_c/\sigma_1$). We can see with this classification the mechanical similarity of fractures with very different values of $2\gamma' = \sigma_1 \phi_c$. Thus ideal brittle fracture, discontinuous ductile-cleavage, mode I plane stress necking and mode III ductile tearing, and the 45° shear mode in steel plates are all cumulative. Except for the first all involve a mechanical instability because the capacity to harden has been exhausted, non-linear flow has concentrated, and large strains have occurred. These large strains are possible whenever free surfaces allow large geometry changes, on a microscale at blunting crack tips or in the internal necks between cracks and voids, and on a macroscale when the specimen is (relatively) small in one dimension.

OTHER CRITERIA

A number of other proposals for the characterisation of post-yield fractures have been made, some of which are reviewed at this meeting [52 - 54, 99 - 103]. Their use and applicability are still a matter of active current research. However, the concept of the R curve [100, 104 - 106] perhaps deserves special mention. It touches upon the fundamental conditions for fracture referred to at the beginning of this paper, and also discussed in one of the plenary sessions here [107]. The crack will not run until the total free energy of the whole system begins to decrease as it advances. The R-curve gives explicit recognition to the idea that the physical processes for the advance of the crack (cracking and sliding off in combination on a microscale) can occur, but recognises that these processes are, temporarily, self equilibrating. Just as a material work hardens, so the resistance R to crack propagation rises. Many workers have considered this phenomenon [10]. Here we wish only to draw attention to the fact that it again forces us to think in detail about the processes of sliding, blunting and microcracking which are going on at the crack tip in all the materials we consider [107 - 115]. It may well be that we shall not achieve a complete understanding of these processes without considering their sensitivity to the strain-rate and the environment.

OTHER PATH-INDEPENDENT INTEGRALS

There are other path-independent integrals of use in the theory of fracture besides the J and the F_1 we have already discussed. Before doing so we

make a few comments on the rather profligate introduction of "new"
integrals currently in fashion. This is not the place for the detailed
critique of individual proposals, but we believe that those planning to
launch a vessel of this kind should first study carefully the background
theory and bear the following points in mind.

Firstly, a complicated integral expression may be path-independent because,
in any example of interest, it is identically zero. Secondly, we have to
distinguish two types of path-independence. If the two-dimensional
divergence of the integrand vanishes, the integral will have the same
value for two paths each beginning at a point A on the lower crack surface
and ending at a point B on the upper crack surface. However the value
may be different for a path beginning at another point A_1 on the lower
surface and ending at another point B_1 on the upper. Indeed it will be
unless the sum of the contributions from the paths A_1A and BB_1 is zero.
If this sum is zero for all A_1 and B_1, then the integral has the same
value for *all paths beginning at any point on the lower surface and ending
at any point on the upper*; we can slide the points A and B along the crack
faces in any manner without changing its value. It is this kind of path-
independence which is of real value, since we can deduce values for paths
close to the tip from those placed far away at our convenience, where the
field quantities are easier to find. Of course, we can always make an
expression "path-independent" by subtracting from it the contributions
from the paths BB_1 and A_1A. But then, if we wish to use the expression,
we still have to evaluate these contributions, and this requires a knowledge of the field close to the crack tip; we have made no real progress.

The general theory of path-independent integrals stems from the work of
Noether [116]. They arise for any field when the Lagrangian density
function from which the field equations are derived is invariant under
the operations of a continuous group. The general consequences for
elastic singularities and cracks have been discussed in a number of papers
by Eshelby [13, 15, 19, 20]. Gunther [117] was the first to apply
Noether's theorem systematically to elastostatics. In addition to F_ℓ he
found the integrals

$$L_{k\ell} = \int_S (X_k P_{\ell j} - X_\ell P_{kj} + u_k P_{\ell j} - u_\ell P_{kj})\, dS_j \tag{9}$$

and

$$M = \int_S \left(X_\ell P_{\ell j} - \frac{1}{2} u_\ell P_{\ell j}\right) dS_j \tag{10}$$

also given by Budiansky and Rice [118]. F_ℓ, $L_{k\ell}$ and M are path independent because a picture of a general elastic field remains one after it
has been respectively translated, rotated and enlarged. Consequently
[19, 20] F_ℓ is valid for finite deformation and a non-linear material,
provided only that it is homogeneous, while for $L_{k\ell}$ the material must in
addition be isotropic. For M we must have linearity in the displacement
gradients, but we may have anisotropy. There are some special cases in
which these requirements may be relaxed [9]. Arguments have been given
[119] that F_ℓ, $L_{k\ell}$ and M are the only path-independent integrals of
Noether's type and that in plane situations the only new feature is that
(10) reduces to

$$M = \int_S X_\ell P_{\ell j}\, dS_j \tag{11}$$

a transformation which results from Gauss's theorem [19]. However, in two dimensions, several infinite classes of path-independent integrals have been found [19].

It is interesting to calculate the force F_2 given by (2). We may think loosely of F_2 as the force normal to the crack tip, but its interpretation requires some care. If we evaluate F_2 using the singular stresses (1), that is, for a small circuit about the crack tip, we find that $F_2 = -2K_1K_2$ [120, 121]. This is an example where the integral (2) has a different value for a large circuit round the crack tip; that is, the integral is not path-independent in the really useful sense because there are, outside the singular field, non-vanishing contributions along the crack faces. We cannot make a useful path-independent integral simply by subtracting the crack face terms [121], because we still have to know the field along the crack if we wish to use such an integral. It is indeed [24] easy to show that F_2 is the limit of $(\pi/2)p_{11}u_2$ as the tip is approached along either the top or bottom surface of the crack.

Loosely, we expect F_2 to push the crack sideways, and this raises the interesting question of what determines the path of a crack. This problem also arises in considering fracture under combined stresses and in crack forking. It is a subtle one because the crack constantly alters the field as it proceeds. A possible criterion is that the crack moves so as to keep $F_2 = 0$ [24]. This has been used by Kalthoff [122] in the form $K_2 = 0$ to discuss the angle at which a crack forks. See also [123] for an equivalent proposal. There has been considerable interest for some time both in the "angled-crack" problem and in the more general problem of crack initiation under combined stress and a number of theories have been proposed [124 - 129]; for a selection of earlier references, see [51]. A discussion of these problems based on an analysis [130] of a crack under general loading with a small kink at its tip making an angle α with the main crack was given at a recent meeting [51]; several authors have published analyses of this kind [130 - 135]. However, to discuss the *onset* of deviation, or for *initiation* of the kink under combined stress, the most suitable results are those for the limit when the kink is vanishingly small compared with the main crack. Several criteria for the path and the initiation have been examined using results of this kind [51, 130, 135].

The integral $L_{k\ell}$ enables an alternative interpretation of the force F_2 to be given when the crack with a kinked tip is considered. If f_1 and f_2 are the crack extension forces determined by evaluating (2) round the tip of the kink, then it may be shown [20] that

$$f_2 = \left[\frac{df_1}{d\alpha}\right]_{\alpha=0} \tag{12}$$

That is, if the tip of the main crack deviates through a small angle $d\alpha$, the change in the crack extension force f_1 is $f_2 d\alpha$. Several authors have also recently examined the problem of the forked crack [134 - 135]; the results show some discrepancies [135]. Again, for the initiation of forking, the case when the forks are vanishingly small is of most interest [135]. Using a $k_2 = 0$ criterion, the predicted branching angle does not differ very much from that observed and calculated by Kalthoff [122].

Reference [19] contains the expression for $P_{\ell j}$ for a material of grade 2; there seems to be some uncertainty in its application to crack problems

[20, 136]. Other examples of the application of path-independent integrals to problems of fracture are also given. They include a discussion of Obreimoff's experiments on mica; of the "trouser test" for rubber using (2) with finite deformation; and of the two-dimensional analogue of the "conical crack", and of the edge crack wedged open by concentrated forces, using (11). The static version of equation (59) of [15] furnishes an integral which is path-independent in the presence of certain types of body forces.

Before leaving the topic of path independent integrals we wish to repeat the brief comment we have made [137] about the use of the quantities J or C* in creep crack growth; for a selection of references see [138]. If the material is linear, viscous and incompressible and the flow is slow, then we can make the usual analogy with linear elasticity by replacing the displacement by the velocity, the shear modulus by the viscosity and by setting Poisson's ratio equal to one half. Then an integral of the form of F_1 is path-independent. However, what it represents is the following [20]. A body instantaneously contains a crack of length a under some load and a state of viscous flow is established. All the work done by the external forces is being dissipated by the viscosity and there is a certain dissipation rate. Now the crack is lengthened by Δa; then the elastic-viscous analogy shows that $2F_1\Delta a$ is the *increase* in the rate of dissipation when the boundary loading is held fixed, but the decrease in it if the boundary velocities are kept constant [24]. In other contexts, this kind of integral can perhaps be used [20] to select from a class of slow viscous flows depending on parameters a flow which is actually observed, by requiring that the dissipation be stationary (although the principle involved is not easy to justify). It is not, however, clear how relevant the integral is to creep crack growth. Of course, as is often the case, it is not the value of an integral which is usually compared with experiments, but some quantity which would be the G derived from the compliance of a specimen if we were dealing with elasticity. Moreover, the viscosity is non-linear. Nevertheless, if creep crack growth can be satisfactorily characterised in this way, there will clearly be a need for some re-interpretations.

In fact, one must expect a crack in a linear viscous material to elongate in the direction of the stress [139]. A hole is a special case of an inhomogeneity, and there has been some recent progress in the theory of the deformation of ellipsoidal viscous inhomogeneities [140, 141], a process of interest in glass manufacture, geology and in the interpretation of phenomena in inhomogeneous fluids. The growth of voids at crack tips is, of course, one of the phenomena we must understand if we are to improve our model of the processes going on there [142 - 146].

MOVING CRACKS

We refer only briefly to moving cracks; for more detailed accounts see [11, 12, 15, 16, 51, 147 - 151]. In a general dynamic elastic field there is no path-independent integral for the force on a moving crack. The best we can do [12] is to write the elastic field in the form

$$u_i = u_i^0(X_1 - vt, X_2) + u_i'(X_1, X_2, t) \tag{13}$$

and try to arrange that near the tip $u_i' \ll u_i^0$. Here the crack tip is moving with instantaneous velocity v in the X_1 direction. Then we can write

$$G = \lim_{S \to 0} \int_S H_{\ell j} \, dS_j \tag{14}$$

where S is a surface moving with the crack tip and

$$H_{\ell j} = (W + T) \, \delta_{ij} - p_{ij} u_{i,\ell} \tag{15}$$

Here T is the kinetic energy density. It should be noted that $H_{\ell j}$ is not the dynamic 4 x 4 energy-momentum tensor $P_{\ell j}$. The integral of the dynamic $P_{\ell j}$ gives the force on the crack tip, plus the rate of change of "quasi-momentum" inside S [15].

The integral (14) is, in general, path-independent only when $S \to 0$. If the dynamic elastic field is a special one which moves rigidly with the crack tip, then G is independent of S, but special simple fields can be used to show that this independence cannot be true for arbitrary finite S in a general dynamic field [12]. It may be shown [12] that G vanishes at the Rayleigh velocity for the uniformly expanding crack in plane strain [152, 153], and at the shear velocity for a similar crack in anti-plane strain [154].

The equation of motion may be found by allowing the crack tip to move arbitrarily so that at time t its tip is at $x = \xi(t)$, say, and then calculating the field, as was first done by Kostrov [155] and Eshelby [156, 157]; Freund [158] has extended the work to plane strain. We can then calculate G, which turns out to be a function of ξ and $\dot{\xi}$, but not of $\ddot{\xi}$; the crack tip behaves as if it had no inertia [156]. If $2\gamma(\xi,\dot{\xi})$ is the fracture energy as a function of ξ and $\dot{\xi}$, the equation of motion is

$$G(\xi,\dot{\xi}) = 2\gamma(\xi,\dot{\xi}) \tag{16}$$

We find that the velocity dependences of G and K are different, and that G contains a factor which increases as the velocity falls. We can thus understand how the conservation of energy can be maintained during crack branching. For instance, a lower limit to the velocity of crack branching might be set by requiring that the crack momentarily stops [15, 158]. With the continuing interest in the solution of dynamic crack problems [159 - 165], we can look forward to further progress in our understanding of discontinuous crack propagation, crack paths and crack branching.

DISCUSSION

We have been able to refer to a few only of the many interesting papers at this meeting. About specific materials and their behaviour in composites [107] we have said very little. We have not mentioned the effect of the environment [166 - 167], the phenomenon of fatigue [168, 169], or the growth of cracks in creep [170], all topics in the front line of practical interest. Nor have we touched on the efforts now being made to take a more three-dimensional look at the fracture problem, which stretches our analytical and computing powers and also raises some interesting topological questions [114]. Among the papers also, there are several references to probability and statistics, applied both to flaws and to microstructure [108, 171] and to failure probabilities [172, 173]. These methods will be with us increasingly, and the analysis of reliability will help to identify more quantitatively some of the critical factors to which we should devote our current attention.

In cavitation during high temperature creep [174] we have to consider thermally assisted motion of individual atoms. This provides us with a gentle introduction to the phenomena where mechanics alone will not do and we have to consider the combined effect of stress, strain, temperature and electro-chemical processes. Cracking, sliding and individual atom movements all play their part in contributing to creep damage. Johnson's work [170] reminds us that the damage may be general or local and that we may relieve stresses both by sliding, and by void formation and microcracking, a process occurring also in brittle materials [115]. Let us remember too that although the macroscopic creep rates of interest to the engineer are very slow, these rates may be much faster in local regions where stress is concentrated. Thus we must consider a range of mechanisms, and the deformation and failure map [175] helps us to view things as a whole, and warns of the pitfalls associated with the long extrapolations that have sometimes to be made. These maps remind us too that as the strain rate rises, the problem of creep fracture passes into that of workability [176, 177]. The emphasis on mechanism also recalls that although the macroscopic behaviour may be viscous, in small regions of crystalline materials at least, we have blocks of *elastic* material, in which atoms and dislocations are moving and material is separating.

Particularly challenging are those fractures where we have to think of the transport of impurities across internal and external surfaces, as these are exposed and films on them reform [166]; of the migration of these impurities and the effects produced when they are segregated, adsorbed on surfaces, or associated with defects; and of their trapping and precipitation as condensed phases or bubbles of gas within the solid [174, 178, 179]. In thinking of these phenomena, as of fatigue, which is also influenced by them [168, 169], we are led yet again to focus on the detail of the microstructure and the processes occurring at the crack tip [108, 114, 115, 180 - 183]. In trying to characterize the intrinsic ductility of a material, the delicate balance between microcrack propagation and slip must be studied [184, 185]. In processes on this scale, the true surface energy (or a modified surface energy which is still quite small) is important, and it has been widely argued (for example, [38, 39, 186]) that when this is changed, the whole macroscopic toughness may be affected. The large observed fracture energy thus depends critically on a much smaller surface energy, important for the microprocesses. Adsorption can make a radical change in the true surface energy (see [187, 188], for example), so that in this way we can explain the effect of trace impurities. Our model will not be complete however unless it takes account of the rates at which the various competing processes occur. Despite the preoccupation with macroscopic toughness, the critical experiments and the interpretation necessary to formulate such a rate-sensitive model of these crack tip processes must be continued.

The subject of fracture embraces the full range of the study of condensed matter; it is necessary both to test large engineering structures, and to use the most refined techniques for detecting the presence of individual atoms. We have to consider the deformation and flow of many types of microstructure, and the effect of the environment upon these processes.

The work touches on earthquakes and the failure of rocks and masses of ice [189]; on the integrity of pressure vessels, pipelines, aircraft and electrical generators; and on the structure of our very selves [190]. There are many interesting phenomena to be investigated and some formidable problems to be solved. We look forward to learning more about them at an interesting meeting.

ACKNOWLEDGEMENTS

I am greatly indebted to Professor J. D. Eshelby, Dr. I. C. Howard and Mr. G. E. Cardew for many valuable discussions.

REFERENCES

1. BILBY, B. A. and ESHELBY, J. D., In: "Fracture", (ed. H. Liebowitz), Academic Press, New York, I, 1968, 99.
2. IRWIN, G. R., In: "Fracturing of Metals", ASM, Cleveland, 1948, 147.
3. IRWIN, G. R., J. Appl. Mech., 24, 1957, 361.
4. IRWIN, G. R., Handbuch der Physik, VI, 1958, 551.
5. GRIFFITH, A. A., Phil. Trans. Roy. Soc., A221, 1920, 163.
6. GRIFFITH, A. A., Proc. First International Congress on Applied Mechanics, (ed. C. B. Biezeno and J. M. Bergers), Delft, Waltman, 1924, 55.
7. ESHELBY, J. D., In: "Fracture Toughness", ISI Publication 121, 1968, 30.
8. SANDERS, J. L., J. Appl. Mech., 27, 1960, 352.
9. RICE, J. R., J. Appl. Mech., 35, 1968, 379.
10. RICE, J. R., In: "Fracture", (ed. H. Liebowitz), Academic Press, New York, II, 1968, 191.
11. CHEREPANOV, G. P., Int. J. Solids Structures, 4, 1968, 811.
12. ATKINSON, C. and ESHELBY, J. D., Int. J. Fract. Mech., 4, 1968, 3.
13. ESHELBY, J. D., Phil. Trans., A244, 1951, 87.
14. CARLSSON, A. J. and MARKSTROM, K. M., "Fracture 1977", (ed. D. M. R. Taplin), University of Waterloo Press, Canada, I, 1977.
15. ESHELBY, J. D., In: "Inelastic Behaviour of Solids", (ed. M. F. Kanninen et al.), McGraw Hill, New York, 1970, 77.
16. BILBY, B. A., In: "Amorphous Materials", Proceedings of the Third International Conference on the Physics of Non-Crystalline Solids, Sheffield (ed. R. W. Douglas and B. Ellis), Wiley, New York, 1972, 489.
17. ESHELBY, J. D., In: "Internal Stresses and Fatigue in Metals", (ed. R. M. Rassweiler and W. L. Grube), Elsevier, Amsterdam, 1959, 41.
18. ESHELBY, J. D., Solid State Physics, 3, 1956, 79.
19. ESHELBY. J. D., In: "Prospects of Fracture Mechanics", (ed. G. C. Sih et al.), Noordhoff International, Leyden, 1975, 69.
20. ESHELBY, J. D., Journal of Elasticity, 5, 1975, 321.
21. BILBY, B. A., Advanced Seminar on Fracture Mechanics, Commission of the European Communities, Ispra 1975, Paper ASFM/75, No. 6.
22. BILBY, B. A., Papers Presented to the Third International Congress on Fracture, Munich, Part XI, 1973, 1.
23. HOWARD, I. C., Private communication, 1973: see [22].
24. ESHELBY, J. D., Private communication, 1974.
25. HUTCHINSON, J. W., Report DEAP S-8, Division of Engineering and Applied Physics, Harvard University, 1974.
26. CHELL, G. C. and HEALD, P. T., Int. Journal Fract., 11, 1975, 349.
27. RICE, J. R., Int. Journal Fract., 11, 1975, 352.

28. NEALE, B. K., CEGB Report No. RD/B3253, 1975.
29. ROCHE, R. L., "Fracture 1977", (ed. D. M. R. Taplin), University of Waterloo Press, Canada, II, 1977.
30. BILBY, B. A. and ESHELBY, J. D., Unpublished work.
31. BILBY, B. A., Progr. Solid Mech., 1, 1960, 331.
32. BILBY, B. A., IUTAM Symposium Freudenstadt-Stuttgart, Mechanics of Generalized Continua, (ed. E. Kröner), Springer, 1968, 180.
33. KRÖNER, E., "Kontinuumstheorie der Versetzungen und Eigenspannungen", Springer, 1958.
34. HULT, J. A. H. and McCLINTOCK, F. A., Proceedings of the Ninth International Congress on Applied Mechanics, 8, 1957, 51.
35. SWINDEN, K. H., Ph.D. Thesis, University of Sheffield, 1964.
36. YOKOBORI, T. and ICHIKAWA, M., Reports of the Research Institute for Strength and Fracture of Materials, Tohoku University, Sendai, 2, 1966, 21.
37. ATKINSON, C., Ark. Fys., 35, 1967, 469.
38. RICE, J. R., Proceedings First International Congress on Fracture, Sendai, (ed. T. Yokobori et al.), Jap. Soc. Strength and Fracture of Materials, Tokyo, 1, 1966, Paper A-18, 309.
39. RICE, J. R. and DRUCKER, D. C., Int. Journal Fract., 3, 1967, 19.
40. KFOURI, A. P. and MILLER, K. J., Int. J. Pres. Ves. and Piping, 2, 1974, 179.
41. KFOURI, A. P. and MILLER, K. J., "Crack Separation Energy Rates in Elastic-Plastic Fracture Mechanics", to be published in Proc. Inst. Mech. Eng. (London).
42. KFOURI, A. P. and RICE, J. R., "Fracture 1977", (ed. D. M. R. Taplin), University of Waterloo Press, Canada, I, 1977.
43. KFOURI, A. P. and MILLER, K. J., "Fracture 1977", (ed. D. M. R. Taplin), University of Waterloo Press, Canada, II, 1977.
44. OROWAN, E., Reports on Progr. in Physics, 12, 1949, 214.
45. CHEREPANOV, G. P., Prikl. Mat. Mech., 33, 1968, No. 3.
46. WNUK, M. P., Int. Journal Fract. Mech., 7, 1971, 217.
47. KNAUSS, W. G., Appl. Mech. Rev., 26, 1973, 1.
48. BEGLEY, J. A. and LANDES, J. D., ASTM STP 514, 1971, 1.
49. LANDES, J. D. and BEGLEY, J. A., ASTM STP 514, 1971, 24.
50. BUCCI, R. J., PARIS, P. C., LANDES, J. D. and RICE, J. R., ASTM STP 514, 1971, 40.
51. BILBY, B. A., Conference on Mechanics and Physics of Fracture, Cambridge, January 1975, 1/1-1/11.
52. RICE, J. R., PARIS, P. C. and MERKLE, J. G., ASTM STP 536, 1973, 231.
53. ADAMS, N. J. I. and MUNRO, H. G., Engng Fract. Mech., 6, 1974, 119.
54. BEGLEY, J. A. and LANDES, J. D., ASTM STP 536, 1973, 246.
55. COTTRELL, A. H., Symposium on Steels for Reactor Pressure Circuits, 1960, Special Report No. 69, I.S.I., London, 1961, 281.
56. BILBY, B. A., COTTRELL, A. H. and SWINDEN, K. H., Proc. Roy. Soc., A272, 1963, 304.
57. DUGDALE, D. S., Journal Mech. Phys. Solids, 8, 1960, 100.
58. BARENBLATT, G. I., Prikl. Mat. Mech., 23, 1959, 434,706,893; Advan. Appl. Mech., 7, 1962, 55.
59. VITVITSKII, P. M., PANASYUK, V. V. and YAREMA, S. Ya., Engng. Fract. Mech., 7, 1975, 305.
60. PRANDTL, L., Z. f. angew, Math. und Mech., 13, 1933, 129.
61. COTTRELL, A. H., In: "Properties of Reactor Materials and the Effects of Radiation Damage", (ed. D. J. Littler), Butterworths, London, 1962, 5.
62. COTTRELL, A. H., Proc. Roy. Soc., A276, 1963, 1.
63. COTTRELL, A. H., Proc. Roy. Soc., A282, 1964, 2.
64. COTTRELL, A. H., Proc. Roy. Soc., A285, 1965, 10.

65. COTTRELL, A. H., In: "Fracture", Proceedings of the First Tewkesbury Symposium, 1963 (ed. C. J. Osborn), Butterworths, London, 1965, 1.
66. BILBY, B. A., COTTRELL, A. H., SMITH, E. and SWINDEN, K. H., Proc. Roy. Soc., A272, 1963, 304.
67. SMITH, E., In: Proceedings of the First International Congress on Fracture, Sendai (ed. T. Yokobori et al.), Japanese Society for Strength and Fracture of Materials, Tokyo, I, 1965, 133.
68. HEALD, P. T., SPINK, G. M. and WORTHINGTON, P. J., Mat. Sci. Engng., 10, 1972, 129.
69. WORTHINGTON, P. J., SPINK, G. M. and HEALD, P. T., In: Proceedings of the Third International Congress on Fracture, Munich, IX, 1973, Paper 515.
70. BILBY, B. A. and SWINDEN, K. H., Proc. Roy. Soc., A285, 1965, 22.
71. ATKINSON, C. and KAY, T. R., Acta Met., 19, 1971, 679.
72. ATKINSON, C. and CLEMENTS, D. L., Acta Met., 21, 1973, 55.
73. SMITH, E., Proc. Roy. Soc., A299, 1967, 455.
74. SMITH, E., Int. J. Engng Sci., 5, 1967, 791.
75. BILBY, B. A. and HEALD, P. T., Proc. Roy. Soc., A305, 1968, 429.
76. HEALD, P. T. and BILBY, B. A., In: "Fracture Toughness of High Strength Materials", ISI Publication No. 120, 1968, 63.
77. WEERTMAN, J., Int. Journal Fract. Mech., 2, 1966, 460.
78. WEERTMAN, J., Int. Journal Fract. Mech., 5, 1969, 13.
79. FLEWITT, P. E. J. and HEALD, P. T., In: "Fracture Toughness of High Strength Materials", ISI Publication No. 120, 1968, 66.
80. KOSTROV, B. V. and NIKITIN, L .V., Prikl. Mat. Mekh., 31, 1967, 334.
81. ARTHUR, P. F. and BLACKBURN, W. S., In: "Fracture Toughness of High Strength Materials", ISI Publication No. 120.
82. HOWARD, I. C. and OTTER, N. R., J. Mech. Phys. Solids, 23, 1975, 139.
83. SMITH, E., Proc. Roy. Soc., A285, 1965, 46.
84. HAYES, D. J. and WILLIAMS, J. G., Int. Journal Fract. Mech., 8, 1972, 239.
85. FENNER, D. N., Int. Journal Fract., 10, 1974, 71.
86. CHELL, G. C., Int. Journal Fract., 10, 1974, 128.
87. CHELL, G. C., Int. Journal Fract., 12, 1976, 135.
88. RICE, J. R., J. Mech. Phys. Solids, 22, 1974, 17.
89. KOCHENDORFER, A., "Fracture 1977", (ed. D. M. R. Taplin), University of Waterloo Press, Canada, I, 1977.
90. RIEDEL, H., Journal Mech. Phys. Solids, 24, 1976, 277.
91. VITEK, V., Journal Mech. Phys. Solids, 24, 1976, 263.
92. WELLS, A. A., Proceedings of the Crack Propagation Symposium, Cranfield, 1, 1961, 201.
93. WELLS, A. A., Brit. Welding Journal, 12, 1965, 1.
94. BURDEKIN, F. M., STONE, D. E. W. and WELLS, A. A., In: "Fracture Toughness Testing, ASTM STP 381, 1965, 400.
95. EGAN, G. R., Eng. Fract. Mech., 5, 1973, 167.
96. HEALD, P. T. and EDMONSON, B., In: "Periodic Inspection of Pressurized Components", Institution of Mechanical Engin-ers, London, 1974, 95.
97. DOWLING, A. R. and TOWNLEY, C. H. A., Int. J. Pres. Ves. and Piping, 3, 1975, 77.
98. SOETE, W., "Fracture 1977", (ed. D. M. R. Taplin), University of Waterloo Press, Canada, I, 1977.
99. LIEBOWITZ, H., EFTIS, J. and JONES, D. L., "Fracture 1977", (ed. D. M. R. Taplin), University of Waterloo Press, Canada, I, 1977.
100. IRWIN, G. R. and PARIS, P. C., "Fracture 1977", (ed. D. M. R. Taplin), University of Waterloo Press, Canada, I, 1977.
101. WITT, F. J. and MAGER, J. R., Nucl. Eng. Design, 17, 1971, 91.
102. ÖSTENSSON, B., Eng. Fract. Mech., 6, 1974, 473.

103. MERKLE, J. G., ASTM STP 536, 1973, 264.
104. IRWIN, G. R., Report 5486, U.S. Naval Research Lab., 1960.
105. KRAFFT, J. M., SULLIVAN, A. M. and BOYLE, R. W., Proceedings of the Crack Propagation Symposium, Cranfield, I, 1961, 8.
106. SRAWLEY, J. E. and BROWN, W. F., ASTM STP 381, 1965, 133.
107. COOPER, G. A. and PIGGOTT, M. R., "Fracture 1977", (ed. D. M. R. Taplin), University of Waterloo Press, Canada, I, 1977.
108. KNOTT, J. F., "Fracture 1977", (ed. D. M. R. Taplin), University of Waterloo Press, Canada, I, 1977.
109. RICE, J. R. and JOHNSON, M. A., In: "Inelastic Behaviour of Solids", (ed. M. F. Kanninen et al.), McGraw Hill, New York, 1970, 511.
110. SMITH, E., Proc. Conf. on Physical Basis of Yield and Fracture, Inst. of Phys. and Phys. Soc., 1966, 36.
111. RITCHIE, R. O., KNOTT, J. F. and RICE, J. R., J. Mech. Phys. Solids, 21, 1973, 395.
112. COWLING, M. J. and HANCOCK, J. W., "Fracture 1977", (ed. D. M. R. Taplin), University of Waterloo Press, Canada, II, 1977.
113. HANCOCK, J. W. and COWLING, M. J., "Fracture 1977", (ed. D. M. R. Taplin), University of Waterloo Press, Canada, II, 1977.
114. ARGON, A. S., HANNOOSH, J. G. and SALAMA, M. M., "Fracture 1977", (ed. D. M. R. Taplin), University of Waterloo Press, Canada, II, 1977.
115. EVANS, A. G., HEUER, A. H. and PORTER, D. L., "Fracture 1977", (ed. D. M. R. Taplin), University of Waterloo Press, Canada, I, 1977.
116. NOETHER, E., Göttinger Nachrichten (Math.-Phys. Klasse), 1918, 235, (English translation by M. A. Tavel), Transport Theory and Statistical Physics, 1, 1971, 183.
117. GÜNTHER, W., Abh. braunsch. wisch. Ges., 14, 1962, 54.
118. BUDIANSKY, B. and RICE, J. R., J. Appl. Mech., 40, 1973, 201.
119. KNOWLES, J. K. and STERNBERG, E., Arch. rat. Mech. Anal., 44, 1972, 187.
120. CARLSSON, J., In: "Prospects of Fracture Mechanics", (ed. G. C. Sih et al.), Noordhoff International, Leyden.
121. BERGEZ, D., Revue de Phys. Appliquée, 276, 1973, 1425.
122. KALTHOF, J., Proc. Third International Congress on Fracture, Munich, X, 1973, Paper 325.
123. GOL'DSTEIN, R. V. and SALGANIK, R. L., Int. Journal Fract. 10, 1974, 507.
124. SHAW, M. C. and KOMANDURI, R., "Fracture 1977", (ed. D. M. R. Taplin), University of Waterloo Press, Canada, II, 1977.
125. PISARENKO, G. S. and LEBEDYEV, A. A., "Fracture 1977", (ed. D. M. R. Taplin), University of Waterloo Press, Canada, II, 1977.
126. MAREK, P., "Fracture 1977", (ed. D. M. R. Taplin), University of Waterloo Press, Canada, II, 1977.
127. JAYATILAKA, A de S., JENKINS, I. J. and PRASAD, S. V., "Fracture 1977" (ed. D. M. R. Taplin), University of Waterloo Press, Canada, II, 1977.
128. HAHN, G. T., HOAGLAND, R. G. and ROSENFIELD, A. R., "Fracture 1977", (ed. D. M. R. Taplin), University of Waterloo Press, Canada, II, 1977.
129. KIRCHNER, H. P. and GRUVER, R. M., "Fracture 1977", (ed. D. M. R. Taplin), University of Waterloo Press, Canada II, 1977.
130. BILBY, B. A., and CARDEW, G. E., Int. Journal Fract., 11, 1975, 708.
131. PALANISWAMY, K. and KNAUSS, W. G., Int. Journal Fract., 8, 1972, 114.
132. COUGHLAN, J. and BARR, B. I. G., Int. Journal Fract., 10, 1974, 590.
133. CHATTERJEE, S. N., Int. Journal Solids Struct., 11, 1975, 521.
134. KITAGAWA, H., YUUKI, R. and OHIRA, T., Engng Fract. Mech., 7, 1975, 521.
134a. KITAGAWA, H. and YUUKI, R., "Fracture 1977", (ed. D. M. R. Taplin), University of Waterloo Press, Canada, II, 1977.

134b. CHEREPANOV, G. P. and KULIEV, V. D., Int. Journal Fract., 11, 1975, 29.

134c. THEOCARIS, P. S. and IOAKIMIDIS, N., J. Appl. Math. and Phys., 27, 1976, 801.

134d. MONTULLI, L. T., Ph.D. Thesis, University of California, 1975.

134e. HUSSAIN, M. A., PU, S. L. and UNDERWOOD, J., ASTM STP 560, 1973, 2; also, Graduate Aeronautical Laboratories, California Institute of Technology, Report No. 74-8, to appear in Mechanics Today (ed. S. Nemat-Nasser), Pergamon.

134f. VITEK, V., "Plane Strain Stress Intensity Factors for Branched Cracks", CEGB Report No. RD/L/N210/76.

135. BILBY, B. A., CARDEW, G. E. and HOWARD, I. C., "Fracture 1977", (ed. D. M. R. Taplin), University of Waterloo Press, Canada, II, 1977.

136. ATKINSON, C. and LEPPINGTON, F. G., Int. Journal Fract., 10, 1974, 599.

137. BILBY, B. A., Conference on Creep Crack Growth, Sheffield, The Metals Society, January 1976, unpublished.

138. NIKBIN, K. M., WEBSTER, G. A. and TURNER, C. E., "Fracture 1977", (ed. D. M. R. Taplin), University of Waterloo Press, Canada, II, 1977.

139. BERG, C. A., Proc. Fourth U.S. National Congress of Appl. Mechanics, Berkeley, (ed. R. M. Rosenberg), ASME, 2, 1962, 885.

140. BILBY, B. A., ESHELBY, J. D. and KUNDU, A. K., Tectonophysics, 28, 1975, 265.

141. BILBY, B. A. and KOLBUSZEWSKI, M. L., Proc. Roy. Soc., 1977, to be published.

142. McCLINTOCK, F. A., J. Appl. Mech., 35, 1968, 363.

143. HANCOCK, J. W. and MACKENZIE, A. C., J. Mech. Phys. Solids, 24, 1976, 147.

144. RICE, J. R. and TRACEY, D. M., J. Mech. Phys. Solids, 17, 1969, 201.

145. NAGPAL. V., McCLINTOCK, F. A., BERG, C. A. and SUBDHI, M., "Foundations of Plasticity", International Symposium, Warsaw, 1972, (ed. A. Sawczuk), Noordhoff, Leyden, 365.

146. PERRA, M. and FINNIE, I., "Fracture 1977", (ed. D. M. R. Taplin), University of Waterloo Press, Canada, II, 1977.

147. Advanced Seminar on Fracture Mechanics, Commission of the European Communities, Ispra, 1975. Papers by IRWIN, G. R. (ASFM/10 and 13), GROSS, D. (ASFM/11), TURNER, C. E. (ASFM/12) and KERKHOF, F. (ASFM/18).

148. BARENBLATT, G. I., ENTOV, V. M. and SALGANIK, R. L., In: "Inelastic Behaviour of Solids", (ed. M. F. Kanninen et al.), McGraw-Hill, 1970, 635.

149. ATKINSON, C. and WILLIAMS, M. L., Int. Journal Solids Struct., 9, 1973, 237.

150. ROSE, L. R. F., Int. Journal Fract., 12, 1976, 799.

151. WILLIAMS, J. G. and BIRCH, J. W., "Fracture 1977", (ed. D. M. R. Taplin), University of Waterloo Press, Canada, I, 1977.

152. BROBERG, K. B., Ark. Fys., 18, 1960, 159.

153. BROBERG, K. B., J. Appl. Mech., 31, 1964, 546.

154. AUSTWICK, A., M.Sc.(Tech.) Dissertation, Sheffield, 1968.

155. KOSTROV, B. V., Prikl. Mat. Mekh., 30, 1966, 1042.

156. ESHELBY, J. D., J. Mech. Phys. Solids, 17, 1969, 177.

157. ESHELBY, J. D., In: "Physics of Strength and Plasticity", (ed. A. S. Argon), M.I.T. Press, 1969, 263.

158. FREUND, L. B., J. Mech. Phys. Solids, 20, 1972, 129,141; J. Mech. Phys. Solids, 21, 1973, 47.

159. ROSE, L. R. F., Proc. Roy. Soc., A349, 1976, 497.

160. ACHENBACH, J. D., Int. Journal Solids Structures, 11, 1975, 1301.

161. ACHENBACH, J. D. and BAZANT, Z. P., Journal Appl. Mech., 42, 1975, 183.

162. ACHENBACH, J. D., "Fracture 1977", (ed. D. M. R. Taplin), University of Waterloo Press, Canada, II, 1977.

163. ACHENBACH, J. D. and VARATHARAJULU, V. K., Quart. Appl. Math., 32, 1974, 123.

164. FREUND, L. B., Int. Journal Engng. Science, 12, 1974, 179.

165. ROSE, L. R. F., Int. Journal Fract., 12, 1976, 829.

166. SCULLY, J. C., "Fracture 1977", (ed. D. M. R. Taplin), University of Waterloo Press, Canada, II, 1977.

166a. McMAHON, C. J. Jr., BRIANT, C. L. and BANERJI, S. K., "Fracture 1977", (ed. D. M. R. Taplin), University of Waterloo Press, Canada, II, 1977.

167. KAMDAR, H. H., "Fracture 1977", (ed. D. M. R. Taplin), University of Waterloo Press, Canada, I, 1977.

168. COFFIN, L. F., "Fracture 1977", (ed. D. M. R. Taplin), University of Waterloo Press, Canada, I, 1977.

169. BEEVERS, C. J., "Fracture 1977", (ed. D. M. R. Taplin), University of Waterloo Press, Canada, I, 1977.

170. McLEAN, D., DYSON, B. F. and TAPLIN, D. M. R., "Fracture 1977", (ed. D. M. R. Taplin), University of Waterloo Press, Canada, I, 1977.

171. RICKERBY, D. G., "Fracture 1977", (ed. D. M. R. Taplin), University of Waterloo Press, Canada, II, 1977.

172. TETELMAN, A. S. and BESUNER, P., "Fracture 1977", (ed. D. M. R. Taplin), University of Waterloo Press, Canada, I, 1977.

173. NEMEC, J., DREXLER, J. and KLESNIL, M., "Fracture 1977", (ed. D. M. R. Taplin), University of Waterloo Press, Canada, I, 1977.

174. GREENWOOD, G. W., "Fracture 1977", (ed. D. M. R. Taplin), University of Waterloo Press, Canada, I, 1977.

175. ASHBY, M. F., "Fracture 1977", (ed. D. M. R. Taplin), University of Waterloo Press, Canada, I, 1977.

176. KUHN, H. A. and DIETER, G. A., "Fracture 1977", (ed. D. M. R. Taplin), University of Waterloo Press, Canada, I, 1977.

177. EMBURY, J. D. and LeROY, G. H., "Fracture 1977", (ed. D. M. R. Taplin) University of Waterloo Press, Canada, I, 1977.

178. GOODS, S. H. and NIX, W. D., "Fracture 1977", (ed. D. M. R. Taplin), University of Waterloo Press, Canada, II, 1977.

179. ALLEN-BOOTH, D. M., ATKINSON, C. and BILBY, B. A., Acta Met., 23, 1975, 371.

180. YOKOBORI, T., KONOSU, S. and YOKOBORI, A. T., "Fracture 1977", (ed. D. M. R. Taplin), University of Waterloo Press, Canada, I, 1977.

181. SMITH, E., COOK, T. S. and RAU, C. A., "Fracture 1977", (ed. D. M. R. Taplin), University of Waterloo Press, Canada, I, 1977.

182. KAUSCH, H. H., "Fracture 1977", (ed. D. M. R. Taplin), University of Waterloo Press, Canada, I, 1977.

183. PETERLIN, A., "Fracture 1977", (ed. D. M. R. Taplin), University of Waterloo Press, Canada, I, 1977.

184. KELLY, A., TYSON, W. and COTTRELL, A. H., Phil. Mag., 15, 1967, 567.

185. RICE, J. R. and THOMSON, R., Phil. Mag., 29, 1974, 73.

186. WILLIAMS, M. L., Int. Journal Fract. Mech., 1, 1965, 292.

187. PETCH, N. J. and STABLES, P., Nature, 169, 1952, 842.

188. PETCH, N. J., Phil. Mag., 1, 1956, 331.

189. SMITH, R. A., "Fracture 1977", (ed. D. M. R. Taplin), University of Waterloo Press, Canada, II, 1977.

190. PIEKARSKI, K. R., "Fracture 1977", (ed. D. M. R. Taplin), University of Waterloo Press, Canada, I, 1977.

FRACTURE PROBLEMS IN NUCLEAR REACTOR TECHNOLOGY[+]

R. W. Nichols*

1. INTRODUCTION

The above title was to have been that for a plenary paper by Professor Corten, who unfortunately has had to withdraw at the last minute. The conference organisers felt that the topic was of such importance that a plenary contribution on fracture problems in nuclear technology was most desirable. I was therefore asked to give this off-the-cuff presentation, for the roughness of which I apologise. Time limitations will make it necessary to limit my treatment of the title to that of large-scale structural components which is, indeed, the focus of this plenary session : nuclear *fuel* behaviour is covered in Workshop Session VI.3 of the Conference. The following discussion thus refers to coolant boundary aspects - the *pressure vessels* and *piping* of thermal reactors and the *primary tanks* of Liquid Metal Fast Breeder Reactors (LMFBR). Since the given title refers to *Problems*, I will confine my paper to outlining areas where future work in the fracture field is likely to be most rewarding.

2. THERMAL REACTOR PRESSURE VESSELS

Since the early discussion of this topic at ICF1 in Sendai, there has been major development of the application of Linear Elastic Fracture Mechanics (LEFM) to the large, thick-walled steel pressure vessels of Light Water Reactor (LWR) systems, to the extent that assessment techniques based on LEFM are codified in the non-mandatory appendices to the ASME Boiler & Pressure Vessel Code, Section III Appendix G and Section XI Appendix A. LEFM also played a prominent role in the acceptance of the novel concept of allowing flaws to remain in pressure vessels if found during service, and if they were smaller than sizes calculated as permissible on LEFM principles. These are major steps forward : such codification of procedures in what is in effect a legal document, having required much first rate technical work. In order to indicate what are the most important problems still remaining I will sketch in the procedure.

The first step is to define what conditions the vessel and circuit is expected to withstand. This is perhaps one of the most difficult steps of all. Automatic control during normal operation leads to large numbers of small variations in coolant conditions on which we need more information to calculate

[+]The text of this paper represents an edited transcript of the lecture Dr. Nichols, President of ICF 1977-1981, gave on short notice. We are particularly grateful to Dr. Nichols for consenting to step in and fill what otherwise would have been an unfortunate gap in this session and indeed the Conference as a whole.

*Risley Nuclear Power Development Laboratories, U.K. Atomic Energy Authority, Warrington, U.K.

fatigue behaviour. But more important is to establish what reactor fault and engineering conditions must be examined, and what are the resulting conditions of temperature and pressure. This work can involve extensive calculations of hydrodynamics and heat flow dynamics for such situations as a major coolant pipe failure or breaks in steam lines, pump bodies and heat exchanger tubes. Having made reasonable assessments of each of these, the next step is that of converting them into stress intensity values at a defect, which may be assumed to occur at any point in the system. The problem here is twofold - the *first* being that of *design geometry*, and indeed there have been major strides in the development of computer calculation to cover the various difficult geometries around nozzles, flanges and supports. The *second* point is that the actual geometry can be different from that of the design, and there is need for more work on the effect on local stresses of errors in alignment, of local weld profiles and of non-regular geometries.

The next step is to feed in the appropriate fracture toughness value for the local material and here we strike major problems. As George Irwin said in his plenary paper, almost all the interest, except perhaps for some accident calculations for breaks in steam-line calculations, is at temperatures when the material used displays upper shelf toughness values, whereas most of the K_I curves relate to sub-transitional temperatures. Moreover, the upper shelf toughness may decrease with increasing temperature, so that it is *not* sufficient to assume that measurements at lower temperatures give lower values. Here we need development of techniques, more measurements, and in particular, decision on the instability criterion. There is a large difference in values between the results for the point of *first growth of a crack*, possibly in a stable manner, and the point of maximum load, and use of the former is believe to provide a considerable margin of safety. The difference in instability conditions for different geometries makes it likely that the computation problems alone will make it necessary for the safety assessment of reactors to be based on *first initiation* for a long time to come. Having developed a method, we still need to know more about variations due to fabrication, welding, strain aging and environmental effects, and how to measure and control production.

Recently in the U.K. I have been involved in carrying out such an assessment. With all these limitations, the best estimates we can make show that the critical defect size, even in stress concentration areas, is > 50 mm under operating, test and upset conditions. Only under certain reactor fault conditions is the critical size ~ 25 mm, and in such cases there is a strong thermal gradient through the wall thickness, resulting in a gradient in fracture toughness. Thus, one can well argue that even if an unstable crack initiates it should be arrested. To prove that such arrest will, in fact, occur needs a greater understanding of the conditions controlling crack arrest than we have now. The recent Oak Ridge work on this aspect should be of great interest. Another area where further work is required is that of the effect of environment and mode of stressing on the rates of crack growth.

At present we can conclude that it is desirable to show that there are no defects in the component bigger than 25 mm deep. In principle this is well above the sensitivity of ultrasonic testing techniques, but there are occasion in practice when such defects may be rejected with only rather poor reliability. In part this is due to *physical* difficulties of ultrasonic examination (UE) - dependence on coupling, surface effect, defect orientation, attenuation or obstruction effects and local geometry, and in part due to *human* aspects inherent in using manual operators. Indeed, a recent review of various trials with UE suggests that manual techniques, unless very closely specified and practised by experienced personnel, may have only about 50% chance of finding such a defect. A major problem is thus both to improve the reliability of UE (e.g. by the use of mechanical recording multi-probe techniques) and to know more accurately this reliability. The reason for this last aspect is that there is considerable interest in assessing the risk of failure quantitatively using the technique of probabilistic analysis of the various fracture mechanics parameters. It would appear that the two parameters in which there is most uncertainty, relate to the number of defects above a given size likely to remain in the structure, and to the fact that the currently low failure rate depends more on getting a fabrication route which shows intrinsically a low rate of defect production, rather than relying on detecting and repairing all such defects.

Another approach to the same problem is to reduce the number of welds and to make them easier to fabricate and inspect. In this respect recent developments of pressure vessel design in Europe and the U.S.A. involve heavy forgings with integral flanges, nozzles and supports, which, together with fewer welds in the pressure vessel rings and heads, results in the vessel having only 20% of the welds of earlier designs. Moreover, in these designs, such welds can be positioned where they can be more readily inspected.

A similar advantage is associated with the thermal reactor design which uses pressure tubes rather than pressure vessels. In these a simple inspectable geometry is combined with the absence of welding. Mention of this reactor reminds me of another aspect of fracture research, that of determining the likely results of fracture - e.g. whether the component will fragment. One of the papers in Part VI [1] describes work to demonstrate that such pressure tubes will not fragment. On the other hand, one could argue that a better understanding of the events after fracture initiation is needed if one is to design protection against the possibility of pressure vessel failure as is currently suggested in Germany.

Time does not permit me to say much about fracture problems in the pipework, except that there is need for considerably better understanding of the detailed parameters controlling stress corrosion cracking in austenitic stainless steel welds, and how to inspect for such cracks, particularly in transition welds between austenitic and low alloy steels. In this respect Acoustic Emission methods are proving very useful.

3. FAST BREEDER REACTORS

In conclusion I will mention some of the fracture problems associated with the nuclear reactor which many regard as providing the only real hope of maintaining our existing standard of living into the next century, that is the LMFBR. The problem is very different from that in the steel pressure vessel of the light water reactors, as the pressure loads are small. The problems should thus be less and of a different nature, which will depend on design details. I will give two examples based on current U.K. work in this area. First there is the need to ensure that the primary vessel wall will not completely fracture, so that part of the vessel could drop from its supports. The second aspect is the need to demonstrate that the core support cannot crack, as this may allow the fuel positioning to change adversely. Both of these problems relate to fracture mechanics assessment of austenitic stainless steel, for which LEFM is not really applicable. Various alternatives are being studied, including COD, J and the 2-criterion approach described at this conference by Milne [2]. The lower mechanical stresses in these components are in some situations supplemented by high thermal stress and high residual welding stresses, such that even with these relatively ductile stainless steels, care is needed in design, or small defect sizes could lead to some degree of crack extension from a defect. However, such crack extension would in the main be due to the strain-limited thermal stresses and residual stress, which together make up almost the whole load-ing. An important area needing further analysis is whether such loading can cause significant crack extension. A further series of fracture problems arises from the sodium cooling itself. Firstly there is the effect of the sodium in any crack on all of the aspects discussed in Section 2 - crack detectability, fracture toughness, creep ductility and fatigue life - considerable work is in hand on these aspects. Then the ability of high heat transfer of the sodium brings fluctuating surface temperatures caused by flowing pockets of coolant which have different temperatures. This "thermal striping" problem puts great importance on the design assessment of thermal fatigue and the need to avoid undue conservatism in our estimates of permissible stress/cycles under these conditions. Finally, these condi-tions lead to creep/fatigue interactions where the sort of approaches out-lined by Ashby [3] need more application.

However, to conclude, it is perhaps appropriate to emphasise that in most of these problems the best protection is to design for a situation where one can get warning of potential trouble by a small leak in sufficient time to take preventive action before a major failure occurs - the so-called *leak-before-break* [4] or *drip-before-flood* concept. Perhaps the most important area for future fracture work is in the proving of the criteria which define when leak-before-break occurs - how does the fatigue crack grow, what is the effect of mixed mode loading, does the crack change shape as it enlarges, what is the rate of bulging in giving leak before break? Perhaps at the next Conference in this series we may find some of the answers to these important questions.

4.   REFERENCES

1.   PICKLES, B.W., JOHNSON, E.R., and COCKS, H., "Fracture 1977 - Advances in Research on Fracture and Strength of Materials", ed. D.M.R. Taplin, Vol. 3B, Pergamon Press, New York, 1977, page 713 (Conference Edition, University of Waterloo Press).

2.   MILNE, I., "Fracture 1977 - Advances in Research on Fracture and Strength of Materials", ed. D.M.R. Taplin, Vol. 3A, Pergamon Press, New York, 1977, page 419 (Conference Edition, University of Waterloo Press).

3.   ASHBY, M.F., "Fracture 1977 - Advances in Research on Fracture and Strength of Materials", ed. D.M.R. Taplin, Vol. 1, Pergamon Press, New York, 1977, page 1 (Conference Edition, University of Waterloo Press).

4.   COTTRELL, A.H., "Fracture 1977 - Advances in Research on Fracture and Strength of Materials", ed. D.M.R. Taplin, Vol. 4, Pergamon Press, New York, 1977, page 85 (Conference Edition, University of Waterloo Press).

PLASTIC FLOW AROUND A CRACK UNDER FRICTION AND COMBINED STRESS

Frank A. McClintock*

INTRODUCTION

A boundary integral relaxation method was used to calculate the plasticity
at the tip of a small horizontal crack buried in a rail head. The Hertz
equations were used as boundary conditions, along with axial residual stress.
A wheel passage gives initial sliding, followed by locking, squeezing the
plastic zone, reversed sliding, locking, and finally unloading.

Computing four cycles of this rail stress history cost $51. At steady
state (only three cycles), the plastic zone extended in the shear direction
almost exactly as far as predicted from linear elastic fracture mechanics,
in spite of the compression being 8 times the net shear tending to produce
Mode II. The crack always remained closed. Reversed shearing on cross-
slip planes was no more than 10% of that on the segment directly in front
of the crack.

Some speculations are given about a fracture criterion in terms of the
displacements and the compressive stresses.

THE PROBLEM

About 800 trains per year are derailed in the United States due to broken
rails. The damage is over $60,000,000.00 [1]. This occurs in spite of con-
tinuous inspection by a fleet of 15 rail cars, which leads to the replacement of 200,000 defective rails per year [2]. Thus, one might say that
the inspection system is 99.6% perfect. In turn, these 200,000 defective
rails are only 0.2% of the rails in service, suggesting that the rails
themselves are of generally good quality. Their replacement rate is very
low, so that existing rails will be used for perhaps 50 years. The fundamental question giving rise to the detailed study reported here is whether
reasonable improvements in inspection procedures, taking account of service
loads and roadbed compliance, could reduce the loss due to derailments
without too great an additional cost.

This paper is concerned with the specific problem of estimating whether or
when the relatively benign shell fractures that run parallel to the rail
surface will turn and become transverse breaks. The tendency of the crack
to run straight or to suddenly turn a corner is presumably controlled by
the macroscopic stresses due to loading, by the residual stresses in the
head due to contact loads, and by the anisotropy of the fracture strength
in the material itself. Other workers are making three-dimensional finite
element studies of the macroscopic stress distribution. Here we shall assume
a simple stress field, such as might be found from such a study, and consider some aspects of the plasticity of crack growth within that field.

* Department of Mechanical Engineering, Massachusetts Institute of Technology, Cambridge, Massachusetts, USA.

The results of this study should also be of interest in contact fatigue and also in wear, which in many cases appears to be due to progressive deformation and fracture under repeated contacts of the micro-asperities. Furthermore the numerical methods may be of interest in other cases where history or high nominal stress distorts the usual linear Irwin-Williams or non-linear Hutchinson, and Rice and Rosengren, (HRR) stress and strain fields (see e.g. [3]-[9]).

The contact area between a rail and a wheel is typically of the order of 10 mm diameter, with the point of maximum equivalent stress below the surface by 1/4 the contact diameter. Overloads cause larger contact diameter and plastic flow, which will shake down at loads up to 70% above that for initial yielding (Johnson [10]). The compressive residual stress leaves tensile stress at greater depths. There are also stresses due to bending and shear of the rail. For definiteness, neglect these and consider only the residual stress at the shakedown limit, superimposed on a repeated rolling load small enough for small scale yielding around a horizontal crack buried at the point of maximum shakedown stress. Choose the current cracking load so that the extent of the plastic zone due to the shear stress R_{II}, taking friction into account, will not exceed some given fraction of the crack half-length a. Furthermore, consider the contact to be plane strain and the crack to be short enough so that the stress is uniform over the crack. This assumption appears reasonable for the relatively long time that any crack is short.

The elastic stress field is obtained from the known solution of Hertz. In the coordinates of Figure 1 with load per unit thickness P, half-width of contact b, and z = (x+iy)/b, from Radzimovsky [11] (or by Johnson [10], correcting signs and conjugates):

$$\frac{\sigma_{xx} + \sigma_{yy}}{2} = \frac{2P}{\pi b} \, \text{Im}\left(\frac{1}{z/b + \sqrt{(z/b)^2 - 1}}\right) ,$$

$$(1)$$

$$\frac{\sigma_{xx} - \sigma_{yy}}{2} - i\sigma_{xy} = -\frac{2P}{\pi b} \frac{y/b}{z/b + \sqrt{(z/b)^2 - 1}} \frac{1}{\sqrt{(z/b)^2 - 1}} .$$

Johnson [10] gives the residual stress due to the maximum shakedown load per unit contact length P_m and the depth y at which it occurs in terms of the contact half-width at maximum load b_m and the tensile yield strength σ_Y in terms of the contact pressure $(2P_m/\pi b_m)$ as,

$$\sigma_{xxr} = -0.134 \, (2P_m/\pi b_m) ,$$

$$y/b_m = 0.5 , \qquad\qquad (2)$$

$$\sigma_Y/\sqrt{3} = 0.25 \, (2P_m/\pi b_m) .$$

The contact half-width b is in turn related to the wheel diameter D and the modulus of elasticity E by the usual equation,

$$b = \sqrt{2 \ PD \ (1 - \nu^2)/\pi E} \ . \tag{3}$$

Since the crack turns out to be subjected mostly to shear, the extent of the plastic zone ahead of a crack under shear in such a stress field can be estimated from the Dugdale [12] - Barenblatt [13] yield zone. Under pure shear, with a shear yield strength of $\sigma_Y/\sqrt{3}$, with the crack faces subject to a coefficient of kinetic friction f_k, and with a normal pressure $-\sigma_{yy}$, the plastic zone extends along the crack line by

$$R/a = \sec \ (\pi(|\sigma_{xy}| - f_k|\sigma_{yy}|)\sqrt{3}\sigma_Y/2) - 1 \ . \tag{4}$$

The reduced shear strength due to differences in normal stress components may be estimated, neglecting through-thickness stress deviators, by

$$\sigma_{Yred}/\sqrt{3} = \sqrt{\sigma_Y^2/3 - (\sigma_{xx} - \sigma_{yy})^2/4} \ . \tag{5}$$

A small computer program was developed to evaluate the stress fields and the extent of the plastic zone, assuming various fractions of the shake-down load. The resulting history of shear and normal stress applied to the region of the crack are shown in Figure 2. The history is surprisingly involved. At first, the shear stress greatly exceeds the normal stress, so sliding occurs on the crack faces. The difference between the applied shear stress σ_{xy} and the surface shear stress σ_{xysfc}, here called the net tip shear σ_{xytip}, is available to cause stress intensities at the tip of the crack. If this is started to decrease immediately after reaching a maximum, there would be a backward sliding on the crack surface. Since this would require a reversal of the surface shear stress σ_{xysfc}, locking occurs instead. This occurs even though the applied shear stress is still increasing. The crack faces remain locked as the contact point rolls over the crack. They finally break loose when the backward stress due to the stress intensity plus the applied shear stress is enough to break the surfaces loose. Reversed sliding now occurs until the reversed stress intensity due to σ_{xytip} reaches the same value which it had in the forward direction. Again, the surface is locked. This time the decreasing normal stress soon leads to reversed (now forward) sliding, and both shear and normal components of stress decrease to zero as the point of contact passes away.

I had expected that the resulting crack tip stress and strain fields could be calculated from the displacement of an elastic linear or nonlinear (HRR) strain field, coupled with a slip line field analysis for the region very close to the crack tip. From Figure 2, however, the mean normal stress is 6 to 8 times larger than the net tip shear stress. This normal stress, increasing as the contact point approaches the crack, was expected to "set in" the forward displacement. Successive forward displacements should lead to, and be inhibited by, reversed residual shear stress components. It is interesting to note that these displacements would be opposite in direction to the deformations associated with exceeding the shakedown load, where observations of Crook [14] on uncracked material show a backward sliding of a layer near the surface when the point of rolling contact is moving forward relative to the substrate. A relatively complete and exact solution seems necessary to understand this in terms of plasticity, the

Bauschinger effect, and possible strain aging. In any event, the large normal stress components mean that a small scale yielding analysis may well not be appropriate. Furthermore, the complex interaction between residual stress, any possible Bauschinger effect, mean normal stress, and current stress increments, along with fluctuating loads and the macroscopic stresses due to bending of the rail, mean that a more detailed analysis is called for. An approximate numerical method is described in the next section. As an example, the stress history of Figure 2 is taken as boundary conditions.

NUMERICAL STUDIES

The Segmented Boundary Integral Method

The computer program used for this study is based on a boundary integral method that gives repeated plane elastic solutions for incremental, history-dependent boundary conditions. It allows the modelling of problems involving crack opening and closing, stick-slip friction, and any plasticity that can be simulated by shear displacement across discrete planes at a critical resolved shear stress. The body must be elastically homogeneous. It may be either infinite in extent or bounded by a multiply-connected polygonal boundary consisting of m (for margin) straight segments.

The boundary conditions are modelled by regarding the body as being contained in an infinite elastic solid and inserting displacement discontinuities between the body and its surroundings, as shown in Figure 3. For brevity, these displacement discontinuities are called "darts". The term comes from sewing, where it denotes the removal of a strip of material and re-joining the edges, as for example, to draw in a woman's dress at her waist. For cracks and slip surfaces the boundary tractions on opposite faces are equal and opposite, and the displacements are taken to be relative displacements between the faces, with crack opening as positive. The use of darts (relative displacements) rather than loads as kernels of the integral equation is especially convenient for modelling cracks and slip surfaces within a body. To improve the accuracy, not only average tractions and displacements, but also their gradients may be specified. Correspondingly, linearly varying ("gradient") darts are used to produce the desired boundary conditions.

Darts would give infinite forces on segments, if they ran from one end of a segment to the other. This follows from the fact that, as may be seen in Figure 3, they have edge dislocations at their ends. (Normal darts have a pair of climb dislocations; shear darts have a pair of glide dislocations The resulting $1/r$ stress singularities would integrate to logarithmically infinite forces of opposite sign on the two segments adjacent to any dislocation. To avoid these infinities the net dislocation of a node between segments was split and half was moved towards the other end of each segment by a fraction F of the segment length. At crack tips the entire dislocatio was moved back.

Known analytic solutions for dislocations are used to give average traction vectors t and displacement vectors u^{μ} on each affected segment μ, due to darts D^m on that or other dislocated segments m:

$$t^{\mu} = T^{\mu m} D^m ,$$

$$u^{\mu} = U^{\mu m} D^m ,$$

$$(6)$$

where summation over all normal and shear, displacement and gradient components and over all segments μ is understood.

The boundary conditions are specified as general linear relations between any or all of the traction and displacement vectors. The relations are given in terms of "parameters" P^k ($k = 1, \mu_{max}$) and "coefficients" $P^{k\mu}$ by

$$P^k = P^{k\mu}_{,u} u^\mu + P^{k\mu}_{,t} t^\mu . \tag{7}$$

For instance, static friction is assumed on a segment, if the prior shear stress was within the limits for incipient sliding. Then the boundary conditions on that segment are

$$u^\mu_n = 0 , \qquad u^\mu_s = (u_s)_{previous} . \tag{8}$$

The equation $k = \mu$ in (8) is reduced to this form by taking

$$P^\mu_1 = 0 , \; P^{\mu\mu}_{1,t} = 0 , \; P^{\mu\mu}_{1,u} = 1 ;$$

$$P^\mu_2 = (u^\mu_s)_{previous} , \; P^{\mu\mu}_{2,t} = 0 , \; P^{\mu\mu}_{2,u} = 1 ; \tag{9}$$

$$P^{k\mu}_\mu = 0 , \; \text{for } k \neq \mu .$$

Other friction and slip conditions are handled similarly. The decisions for crack opening or closing, for stick-slip friction, and for plastic sliding or locking are all made on the basis of the previous state, so small steps must be taken to avoid serious over-shoot.

When equations (6) for the tractions and displacements in terms of the darts are substituted into the boundary conditions (7), the loading parameters P^k are linearly related to the darts D^m:

$$P^k = (P^{k\mu}_{,t} T^{\mu m} + P^{k\mu}_{,u} U^{\mu m}) D^m . \tag{10}$$

Equation (10) is solved for the darts D^m. With the darts known, stress and displacement are calculated at any points desired by the user.

The program is written in FORTRAN IV, Level G, using complex numbers. It has been run on the IBM 370-168 computer in the MIT Information Processing Center in less than 5 seconds for a 25 segment problem and a single set of boundary conditions. The core requirement is roughly $257 + .2$ (segment capacity)^2 kbytes. The program has been used for elastic calculations of fatigue crack growth under general in-plane loading by Pustejovsky [15], where it was checked against known analytical and finite element solutions. A similar elastic test will be reported below in connection with the segments finally chosen.

Verification of the Boundary Integral Method for Plasticity Around Cracks

To verify the program for plasticity, it was used with the problem shown in Figure 4, consisting of a fully plastic ligament of unit half-length between two co-linear cracks each of total length 44.1. This saved modelling the external boundaries of an externally grooved specimen. Increasing the stress applied at infinity takes the ligament to its limit load and then applies a controlled extension.

The grid of Figure 4 was chosen to allow the development of a fully plastic flow field using a small number of segments. The lower half of the specimen moves down with a unit displacement, carrying the central tongue with it. Blocks adjacent to the central tongue move inwards, while the outer ones move diagonally upward and inward to replace the material flowing through the shear zones leading down to the crack tip. Thus the material flowing into the ligament comes from an extra crack *tip* opening displacement which is 1.5 times the crack *flank* opening displacement.

As shown here, the flow field is unique. If there were a slip line across the ligament itself, the field would not be unique due to the possibility of unequal flow from either side. The numerical method would still select the deformation field shown here, which is the first that became kinematically admissible. A small amount of strain hardening would give a unique flow field, as worked out by Neimark [16].

The mean normal stress across the ligament required to produce the slip in the various shear zones is $(10/3)\sigma_Y$, versus the value of $(1+\pi/2)(2/\sqrt{3})\sigma_Y = 2.9685 \, \sigma_Y$ of the exact solution for an isotropic material.

The computed displacement fields were taken to be the differences between displacements of four and eight times those at the limit load. The downward displacement increment of the material below the ligament relative to that above was normalized to unity. The analytical and *computed* displacement increments across the various slip lines and the crack face are compared in Figure 4 for the left hand side, which had the greater errors. The results are within 0.5% for all cases. The average traction across the ligament is within 2% of the expected value. The equivalent stress at the centers of the blocks is by no means as well calculated, differing from the yield strength by a factor of 1.5 either way. From other work with the computer program, this error is more likely to be due to the block sliding model adopted here than to any defect in the program itself. We therefore conclude that the computer program used here is self-consistent and gives good results within the limitations of the particular field of segments that is being considered.

The part of the program dealing with changes in frictional and stick-slip boundary conditions was checked by a complete set of some 25 combinations of initial conditions and changes, for example from an open crack to one sticking or sliding in either direction.

Choice of Grid for the Buried Crack

The grid was chosen to put slip segments where they were most needed. Preliminary runs showed that slip on radial planes out to unit radius occurred only directly in line with the crack. The desire to minimize computer cost then led to the field of segments shown in Figure 5. The missing slip plane at the top of the hexagon was left out not only to minimize the number of segments, but also because a preliminary run with a plastic zone just above the central plane, including that segment and radial ones out beyond it, had indicated no plastic strain would occur there. The fraction F by which dislocations were moved back from crack tips was taken to be 0.001. The crack and slip segments were extended to overlap by that amount so that the dislocation sites for adjacent segments would coincide exactly.

Verification of the Program for Elasticity Around the Crack

As a further check on the program, the first step of the loading for this crack was compared with independent calculations of stress and displacement from the usual Irwin-Williams elastic singularities. Additive constant stress terms can be chosen to match either the conditions on the crack or at infinity; the conditions along the crack were matched. In order to fix rigid-body motion, the displacements were fitted to the computer results at the first point in front of the crack and the rotation was fitted to the vertical displacement of the second point.

The results for the stress were all within 6% of the maximum stress component at any point. The maximum error in mean traction occurred directly in front of the crack, where it was higher than the local stress by almost exactly $\sqrt{2}$ as expected from the $1/\sqrt{r}$ singularity. The displacements were accurate to 2% on the inner ring but the error increased to 14% at sites with radius 0.375 ($\sqrt{r/a} = 0.22$), very likely because of higher order displacement terms.

Choice of Boundary Conditions

The crack was assumed to occur at the depth giving the maximum allowable load at shakedown, and the shakedown residual stress was assumed, both according to (2). The load was then decreased to correspond to small-scale yielding, taking the expected plastic zone size calculated from (4) and (5) to be just unity for a crack half-length of a = 7.875. This particular number came from taking six segments for the total crack length 2a and doubling adjacent segment lengths, progressing outward from the crack tip of interest, where the first segment was of length 0.25. This load resulted in the following maximum values for stress components regarded as being applied at infinity:

$$\sigma_{xxr} = -0.310 \, \sigma_Y \, ,$$

$$\sigma_{yymax} = -1.480 \, \sigma_Y \, , \qquad\qquad (11)$$

$$\sigma_{xymax} = 0.432 \, \sigma_Y \, .$$

Cost and Potential for Development

The stress cycle shown in Figure 2 was divided into 27 steps of varying size to reveal details with a small total number of steps. The program was run for four complete cycles at a total computer cost of $51. The cost could be reduced in a number of ways. Running consecutively at optimum rates would drop the cost to $34. Conditions at all segments and 11 other points were printed out at each step, giving a printing cost of 27% of the total. Two thirds of the cost is solution time, which varies as the cube of the number of segments. The full matrix, involving 17 segments and 4 degrees of freedom per segment was solved at each stage, even though the normal displacements were known to be zero for all the plastic segments, so the matrix could be correspondingly reduced in size. (Similarly for segments that are known to be in the locked, rather than sliding, mode). Instead of translational and gradient darts, triangular darts could be introduced at each node, halving the number of degrees of freedom for the plastic elements once again, and eliminating the question of displacing dislocation sites some arbitrary distance from the node to avoid infinite forces on segments. These savings would probably be used to get more details in the plastic zone, especially at higher levels of applied stress. Some studies of load history effects could be made. Thus the method has good promise as a research tool, and to test empirical equations for the growth of cracks in rails. Even with these savings, the program is not likely to be of use for routine predictions of life in individual rails until further reductions in computer cost are available, or more understanding has been found of just which are the critical events that must be calculated.

Results and Discussion

A number of specific observations were made from looking at the computed data.

1) A steady state was nearly attained after only three cycles, since the results of the fourth differed by at most 0.2% of the maximum stress or displacement. This is half the worst round-off error, as noted by asymmetry of the fully plastic problem.

2) The crack stayed closed throughout the process. Perhaps as a result of this continued closure, the out-of-plane sliding was no more than 10% of the sliding in the plane of the crack, and the shear displacement of the crack at its tip was very nearly identical to that of the first segment beyond the crack.

3) The displacements across slip lines radiating from the crack tip were calculated from the mean displacement and displacement gradient along each radial segment, and are shown in Figure 6 for the last cycle. Consider first the results for planes at 0° and 180° from the crack, here with normalized displacements scaled down by a factor of 10. The short reversals on the curves as plateaus are approached are due to the fact that if reversed sliding was found to occur after the tractions for forward sliding had been assumed, no correction was made. Instead, the displacements were assumed locked at their *previous* values and for the *next* iteration. The short reversals before a plateau of constant displacement should therefore be ignored. Finer step sizes or an iteration procedure would reduce or eliminate them.

4) Shear on cross slip planes was out of phase, and unequal amounts of slip occurred on planes across the crack tip from each other.

5) There is a very slight forward motion of the upper half of the crack in the direction of the rolling contact, but it was little more than the error in the elastic results.

6) The plastic zone extended primarily ahead of the crack, about as expected from the preliminary run. The extent of the plastic zone can be estimated quantitatively by extrapolating the slope of the displacements for the last segment from the mean value for that segment. This extrapolation indicated the plastic zone of forward sliding extended to 1.066 and that for reversed sliding to 1.064, surprisingly close to the value of 1 expected from the Dugdale-Barenblatt results, taking friction and normal stress into account according to (4) and (5). The stresses calculated at the point of the missing segment directly above the plastic zone indicated that shear would not be expected there except just before reversed sliding on the crack plane. There was a large difference of normal stress components, however, and check calculations should be made, allowing for the possibility of normal components of plastic strain at that point. The yield on radial planes above and below the crack was estimated by extrapolation to extend out no farther than r = 0.343. This justifies the extent of the grid chosen for possible slip, shown in Figure 5.

CRACK GROWTH CRITERION

With the displacement components at the crack tip known, the next question is to estimate the corresponding advance of the crack tip.

Shear from the Crack Tip on One Radial Plane

Slip on one radial plane from the crack tip can be thought of as being produced by a dislocation leaving the tip. The predominant mode of deformation for this loading is glide along the plane in front of the crack, shown in Figure 7a. (For a continuous flow field, one would consider the relative displacement across some finite angle $\delta\theta$.) Assume that the crack slides enough to accommodate the displacement introduced at the crack tip and that there is no rewelding. New surface will be introduced, as shown by the dashed line. The crack advance relative to the average displacement between the faces is just half the relative displacement across the glide plane. Assume there is enough compression to suppress any fracture by hole growth or cleavage in the tensile region below the tip of the crack, at A. Then the growth is purely in the glide direction. To describe any re-welding, define the "efficiency" of crack growth as the ratio of growth to half the component of the glide displacement (Burger's vector) on the crack plane:

$$\eta = da/(d|u_s|/2) \ . \tag{12}$$

With no re-welding, the crack growth efficiency would be $\eta = 1$. The amount of re-welding would depend on the pressure; a typical efficiency might be $\eta = 1/2$. On reversed sliding, the new material would be passing partially re-welded material, and the efficiency would drop further, say to 1/4. (Even on the initial deformation, the material nearest the new tip is passing an old crack face that has been rubbed for some distance and is likely to weld more easily. Thus the degree of coherence along the fresh crack is likely to vary, as indicated in Figure 7a).

Now consider a dislocation running off to the side, tending to produce crack opening, as shown in Figure 7b. There is no re-welding, and the crack growth efficiency relative to the average flank coordinate is $\eta = 1$. A dislocation of the same sign, moving downwards and to the left along the

same cross-slip plane through the crack tip would give the same final configuration as that of Figure 7b. A dislocation of opposite sign, however,
would bring the crack flanks together, as shown in Figure 7c. Because there
would be relatively little pressure on the glide plane as it slid closed,
the magnitude of the growth efficiency would still be approximately
$|\eta| = 1$. It should be noted, however, that the crack growth would actually
be negative during this half cycle ($\eta = -1$). This cancels out the growth
shown in Figure 7b.

If a dislocation of sign opposite to that shown in Figure 7b were to travel
upward to the right along the same slip plane it would require interfer-
ence at the crack tip. Such flow would more likely arise from dislocations
running in towards the tip from outside. Any that reach the tip are likely
to travel along planes somewhat behind it, as shown in Figure 7d. The
growth efficiency is $\eta = 1$ because there is no re-welding.

The above series of displacements has consisted of coplanar slip (i.e.,
slip on the crack plane) followed by cross-slip. Reversing the order, to
cross-slip followed by coplanar slip, gives similar results except that
producing crack opening first will increase the efficiency of crack growth
due to slip on the crack plane.

Shear on Opposing Planes Through the Crack Tip

The combination of dislocations of opposite sign running out from the
crack tip along the same plane, as shown in Figures 7b and 7c, amounts to
homogeneous shear on the cross-slip plane in front of the crack. Various
sequences of such cross-slip, combined with coplanar (crack-plane) slip,
are shown in Figure 8. Figure 8a shows the same process as Figure 7a.
Similarly, Figure 8b shows the combination of the processes in Figure 7b
and 7c. Cross-slip of the opposite sign is shown in Figure 8c. In the
processes of both Figures 8b and 8c, the efficiency would be very small
because only the discontinuities and oxide film at the very crack tip
would be spread out along the new cross-slip plane, tending to weaken it.
For later reference, the effects of both signs of slip on the other cross-
slip plane through the crack tip are shown in Figures 8d and 8e.

Now consider cross-slip *followed* by coplanar slip. Initial cross-slip
leaves the patterns shown in Figures 8b-e, but with no fresh cracking on
horizontal surfaces. The corresponding patterns after coplanar slip are
shown as Figures 8f-i in the second column of Figure 8. The reversal
of that for Figure 8b, shown in Figure 8f, gives the same crack advance
but the spur crack has been left behind. Reversing the sequence of Figure 8c
gives the same pattern as before, as shown in Figure 8g. Reversing the
sequence of Figure 8d gives a new result, however. Crack plane slip of the
opposite sign will tend to come in at the new crack tip and open up a hole,
as shown in Figure 8h. The crack growth efficiency will also be higher on
reversing the sequence. This lack of commutativity will be awkward to incor-
porate into an overall theory of crack growth. Finally, reversing the
sequence of Figure 8e gives the same pattern as before, shown in Figure 8i.

This discussion indicates some of the features that would be required in a
general criterion of fracture under combined compression and shear. Experiments so far (Jones and Chisholm [17]) deal with the coplanar slip deformation found to be most important here for these horizontal cracks in a rail,
but have not included the transverse compression. Monotonic shear with
pressure has been studied by Tipnis and Cook [18].

CONCLUSIONS

1) While the plasticity of a crack in a rail head is complicated, it can be approximated with reasonable economy.

2) For horizontal cracks in the elastic-plastic regime, the flow is primarily shear along the plane of the crack.

3) A change in crack direction would appear to require either a higher load or shear components of load.

4) A criterion for crack growth under sliding and compression is needed, and will have to take re-welding into account as a limiting case.

ACKNOWLEDGEMENT

The interest in and support of this work by the Transportation Systems Center, Department of Transportation, under contract TS-11653, is deeply appreciated. The help of Robert M. Russ with the computations, Erika M. L. Babcock with the typing, and Charles V. Mahlmann with the illustrations is also gratefully acknowledged.

REFERENCES

1. FEDERAL RAILROAD ADMINISTRATION, Office of Safety, Bulletin 141, Washington, D. C., 1972.
2. NATIONAL TRANSPORTATION SAFETY BOARD, Rept. RSS-74-1, Washington, D. C., 1974.
3. IRWIN, G. R., J. Applied Mech., 24, 1957, 361.
4. WILLIAMS, M. L., J. Applied Mech., 24, 1957, 109.
5. HUTCHINSON, J. W., J. Mech. Phys. Solids, 16, 1968, 13.
6. HUTCHINSON, J. W., J. Mech. Phys. Solids, 16, 1968, 337.
7. RICE, J. R. and ROSENGREN, G. F., J. Mech. Phys. Solids, 16, 1968, 1.
8. RICE, J. R., Fracture, 2, ed. H. Liebowitz, Academic Press, New York, 1968, 191.
9. McCLINTOCK, F. A., Fracture, 3, ed. H. Liebowitz, Academic Press, New York, 1971, 47.
10. JOHNSON, K. L., Proc. 4th U. S. Nat. Cong. Appl. Mech., Am. Soc. Mech. Eng., 2, 1962, 971.
11. RADZIMOVSKY, E. I., Univ. of Ill., Eng. Exp. Sta. Bull., 408, Urbana, 1953.
12. DUGDALE, D. S., J. Mech. Phys. Solids, 8, 1960, 100.
13. BARENBLATT, G. I., Advances in Appl. Mech., 1, eds. H. L. Dryden and T. von Karman, Academic Press, New York, N. Y., 1962.
14. CROOK, A. W., Proc. Inst. Mech. Eng., London, 171, 1957, 187.
15. PUSTEJOVSKY, M. A., Ph. D. thesis M. I. T., Cambridge, Massachusetts, 1976.
16. NEIMARK, J. E., J. Applied Mech., 35, 1968, 111.
17. JONES, D. L. and CHISHOLM, D. B., Fractography - Microscopic Cracking Processes, ASTM STP 600, eds. C. D. Brachem and W. R. Warke, 1976, 235.
18. TIPNIS, V. A. and COOK, N. H., J. Basic Eng., Trans. Am. Soc. Mech. Eng., 89D, 1967, 533.

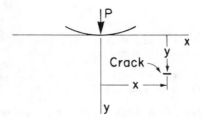

Figure 1 - Coordinates of crack relative to contact point

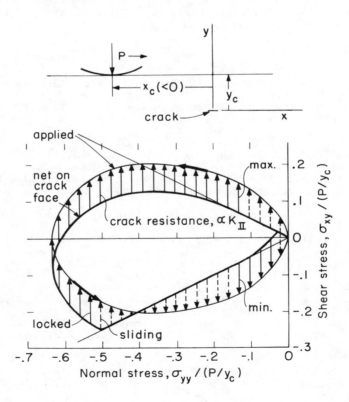

Figure 2 - Applied and crack face stresses on a buried crack

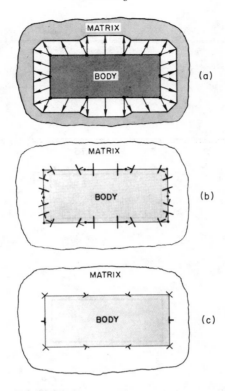

Figure 3 - Modelling biaxial tension on a body with translational displacement discontinuities or darts (a), giving dislocation dipoles on each segment (b), and resultant dislocations at nodes (c).

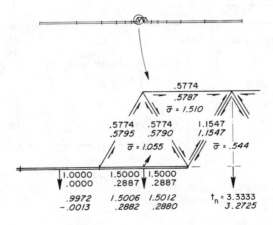

Figure 4 - Analytical and *computed* displacements and stresses for fully plastic test

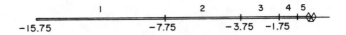

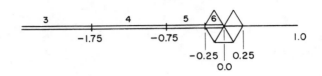

Figure 5 - Segments for modelling a crack with localized plastic flow

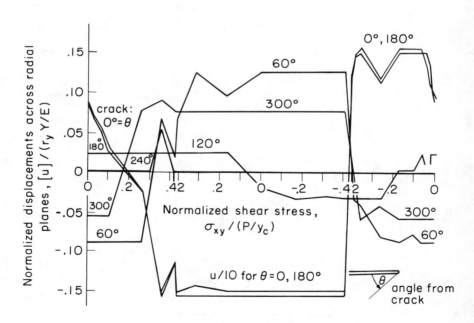

Figure 6 - Normalized displacement history across planes radiating from a crack tip under shear and pressure.

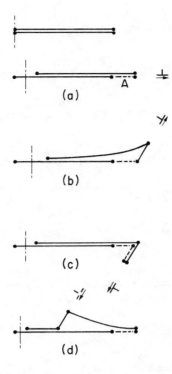

Figure 7 - Crack growth from slip on one radial plane
at a time

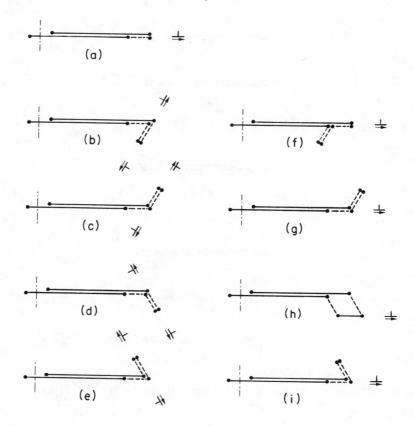

Figure 8 - Crack growth from co-planar slip (a) followed by cross-slip (b-e) and cross-slip followed by co-planar slip (f-i).

ON CRACK CLOSURE IN FATIGUE CRACK GROWTH[1]

A. J. McEvily[2]

INTRODUCTION

Crack closure is a term used to indicate that the opposing faces of a fatigue crack make contact before the minimum load in a cycle is reached, a phenomenon first noted by Elber [1],[2], some ten years ago. Since that time the subject has attracted considerable interest as a means of rationalizing the effects of mean stresses and overloads on the rate of fatigue crack growth. The closure effect itself is relatively easy to detect with appropriately placed strain or COD gauges, and is manifested by a continuous increase in specimen compliance on loading until the crack surfaces are fully separated. Thereafter with further loading in the linear elastic range the specimen compliance remains constant at a maximum value. Upon unloading from the maximum load in the cycle the reverse process takes place, usually with some hysteresis. Although the overall phenomenon is easily detected, the precise determination of the opening load itself does require some sophistication in measurement technique, especially after an overload. For R = 0 loading (R being the ratio of the minimum to maximum stress in a loading cycle), Elber has observed that the opening load is one-half of the maximum load with the ratio decreasing as the R value is increased. In Elber's approach to crack growth analysis, only the portion of the loading cycle above the opening load is considered to be effective in propagating the crack, and the effective stress intensity factor, K_{eff}, which is defined as the difference between K_{max} and K_{open} is used as a correlating parameter. However, there has been some uncertainty as to the nature of the closure process itself and also as to whether or not it explains the observed effect of mean stress on the rate of crack growth.

The controversy usually centres on whether the closure phenomenon is a plane stress effect which involves primarily only the near-surface region of the specimen or whether it is an effect present throughout a thick specimen, that is in both the plane strain as well as the plane stress regions. In addition, although it is clear that closure can occur at macroscopic distances behind the crack tip, especially in plane stress, closure of the tip itself is more difficult to ascertain, however, Bowles [3], using a plastic impregnation technique to replicate the crack tip region, has found that the tip remains blunt on unloading, but with some evidence of point-to-point contact present behind the crack tip. With respect to closure in the plane strain region, Lindley and Richards [4] tested a variety of steels and found that the effect vanished as the thickness increased. They concluded that it was a near surface effect associated with the greater stretch of material in the plane stress region. In more recent tests of 25 mm thick compact

---

[1] Professor McEvily kindly consented to present this Plenary Paper on Crack closure at short notice. The text is an edited transcript of his actual lecture.

[2] Department of Metallurgy, University of Connecticut, Storrs, Conn., U.S.A.

tension specimens of a high strength aluminium alloy, Paris and Hermann [5] found that closure occurred in plane strain with the ratio of the opening load to maximum load being about 0.5. However, this level of closure was found only in the near-threshold region. At higher values of the stress-intensity factor the ratio dropped to 0.23. The tendency for the closure level to decrease with increase in stress intensity was also noted by Bachmann and Munz in the case of the titanium alloy Ti-6A1-4V [6], and by Kikukawa *et al* for a variety of alloys [7]. In order to learn more of the influence of the surface region as well as that of K_{max} and overloads on closure the following experiments were carried out.

CLOSURE EXPERIMENTS [8]

A fatigue crack was grown at $R \simeq 0$ in a high strength aluminium alloy under decreasing stress intensity conditions, and the near-threshold region was approached. Strain gauges were then placed immediately ahead of the crack as well as across the crack to study crack opening behaviour. In these studies the ratio of opening load to maximum load was found to be about 0.5. The specimen thickness was then reduced by carefully machining 1.5 mm from each face of the initially 6 mm thick specimen. The opening load characteristics were then rechecked with strain gauges and it was found that the K_{op}/K_{max} ratio was still high, about 0.4, a finding which indicated that a significant closure effect was present in the plane strain portion of the specimen when tested in the near-threshold region. At higher stress intensity levels, however, the K_{op}/K_{max} ratio decreased in these tests to less than 0.1.

An important question which then arises is why does the K_{op}/K_{max} ratio decrease in plane strain specimens as the level of K_{max} increases. On this point the studies of several steels by Otsuka *et al* [9],[10], appear to be particularly helpful, for they have observed that the crack growth mode changes as the stress intensity level is increased. In the near-threshold region crack growth occurs in a shear mode on planes inclined to the direction of crack growth. This mode of growth is also known as Forsyth's Stage I or as Mode II of fracture mechanics. At higher stress intensities where a power law can be used to represent the dependency of the crack growth rate on the stress-intensity factor range, a shift to an opening mode of growth occurs which is accompanied by the appearance of fatigue striations on the fracture surface. In related work, Yoder *et al* [11], have observed that for Ti-6A1-4V the transition to power law dependency occurs when the plastic zone is equal to the average size of a Widmanstätten packet in this α-β alloy. Above the transition the fracture surface is featureless except for a regular array of fatigue striations. Below the transition striations are not observed, and the microstructural detail of the alloy is clearly revealed on the fracture surface, an indication that in this region the crack advanced by following easy paths through the complex microstructure and, based on Otsuka's observations, probably by a shear mode of propagation.

Since the threshold is generally above the opening load, crack growth in the near-threshold may occur by a combination of a sliding mode as well as an opening mode, with the presence of the sliding mode accounting for the observed high value of the K_{op}/K_{max} ratio in the near-threshold region. The higher the ratio the more important will be the contribution of the sliding mode, and, in fact, for certain aluminium alloys, Kikukawa *et al* have found the ratio to be equal to unity. In the wake of a shear crack a zig-zagged fracture surface may develop upon the material. However, in order for a crack to extend in this manner it may be necessary that fatigue crack growth alternate with tensile rupture to keep the average crack plane normal to the direction of principal tensile stress, and some evidence of mixed fatigue and tensile mode growth in the near-threshold region has been obtained by Pickard *et al* [12]. At higher stress intensities the opening mode becomes dominant and there is a corresponding decrease in the K_{op}/K_{max} level, at least for specimens of sufficient thickness to ensure that plane strain conditions prevail. In thin specimens it is possible that closure levels are much higher, a matter which should be checked in further experiments.

CLOSURE AND OVERLOADS [8]

Next, let us consider opening load behaviour above the threshold region as influenced by an overload. Paris and Hermann have made careful determinations of opening load behaviour after overloads of the order of 100% and found that two opening loads were observable. The first of these opening loads is less than that measured prior to the overload, whereas the second is higher and exhibits at first an increase in closure level followed by a gradual decrease as the crack advances through the overload plastic zone. The lower opening load is ascribed to a loss of contact behind the crack tip, and the second to a loss of contact at the crack tip itself. The application of the overload results in a considerable retardation in the subsequent rate of fatigue crack growth, and Paris and Hermann attribute this retardation to the low value of ΔK_{eff} which is present above the higher of the two opening loads. In this view the closure effect is not thickness dependent, but is considered to arise from a general residual stretch of the material behind the crack tip. However, the relative contribution of near surface regions and interior regions was not established.

To learn more of the contribution of the surface region to closure and retardation phenomenon after an overload we have carried out the following experiments. A fatigue crack was grown in a compact tension specimen of the aluminium alloy 6061-T6 and a 100% overload was applied. Constant amplitude cycling was resumed and the usual delayed retardation phenomenon was observed. When the crack had grown beyond the region of influence of the overload, a second 100% overload was applied, and 1.5 mm was machined from each face of the 6 mm thick specimen. Cyclic loading was then resumed at one-half of the previous load amplitude to compensate for the reduction of thickness. In this case, virtually no retardation was observed. A similar test was carried out with a second specimen which was machined after the first overload but not the second. Again after machining the retardation effect was virtually absent, whereas it was most pronounced after the second overload of the thinned-down specimen. These results demonstrate that in the linear-elastic range retardation is strongly related to the stretch of material and resultant closure in the near-surface region. These results also suggest that the first opening load observed by Paris and Hermann may be due to crack opening in the interior of the specimen, and the second may be due to opening at the surface. Since the experiments indicate that retardation is primarily due to a surface effect, it is to be expected that the magnitude of the retardation effect will decrease with increase in specimen thickness. It also seems likely that delayed retardation is related to the shape of the crack front and interactions between surface and interior regions since the crack progresses ahead more rapidly in the interior of the specimen rather than at the surface following an overload.

ACKNOWLEDGEMENT

The author is pleased to acknowledge the support of the U.S. Air Force
Office of Scientific Research under Grant No. 74-2703B. The capable
assistance of Mr. Richard Shover in carrying out the test programme is
also gratefully appreciated.

REFERENCES

1. ELBER, W., Eng. Fracture Mech., 2, 1970, 37.

2. ELBER, W., ASTM STP 486, 1971, 230.

3. BOWLES, Q., private communication.

4. LINDLEY, T.C., and RICHARDS, C.E., Mater. Sci. Engr., 14, 1974, 281.

5. PARIS, P.C., and HERMANN, L., presented at the Int. Cong. Appl. Mech.,
    Delft, 1973, to be published.

6. BACHMANN, V., and MUNZ, D., Int. J. Fract., 11, 1975, 713-716.

7. KIKUKAWA, M., JONO, M., and TANAKA, K., Proc. of the 2nd Int. Conf.
    on Mech. Behaviour of Materials, Boston, 1976.

8. McEVILY, A.J., Met. Sci., 11, 1977, to be published.

9. OTSUKA, A., MORI, K., and KAWAMURA, T., Jap. Soc. of Mech. Engrs.,
    1976.

10. OTSUKA, A., MORI, K., and MIYATA, T., Eng. Fract. Mech., 7, 1975, 429.

11. YODER, G.R., COOLEY, L.A., and CROOKER, T.W., Proc. of the 2nd Int.
    Conf. on Mech. Behaviour of Materials, Boston, 1976, 1010.

12. PICKARD, A.C., RITCHIE, R.O., and KNOTT, J.F., Metals Tech., 2, 1975,
    253.

SURVEY OF RECENT WORK ON THE EFFECT OF THE ATOMIC STRUCTURE'S
DISCRETENESS ON CLEAVAGE CRACK EXTENSION IN BRITTLE MATERIALS

E. Smith*

I. INTRODUCTION

The excellent pioneering work of Taylor [1], Orowan [2] and Polanyi [3] on
the one hand, and Griffith [4] on the other, has led to the acceptance
that a crystalline material's mechanical behaviour, and more particularly
its plastic deformation and fracture characteristics, are crucially dependent on the mobility of dislocations and cracks in the material. Furthermore, it is also accepted that in developing an understanding of these
processes, the simplest approach is to use a theoretical model in which
the real material is represented by an elastic continuum. If such a model
is used to describe a perfect dislocation, there is no resistance to the
dislocation's movement, because in the real situation the crystal structure
reverts to its original form following the passage of a perfect dislocation
through the crystal. However, as Griffith clearly appreciated, movement
of a cleavage crack tip is associated with a finite resistance even with
a continuum model, since the original crystal structure is not retained
after the passage of a crack tip. This is because atomic bonds break
irreversibly, a process that is incorporated into the continuum description
through the surface energy term γ; a crack tip is therefore analogous to an
imperfect dislocation, with the surface energy corresponding to the fault
energy associated with the imperfect dislocation's movement. Moreover,
using thermodynamic arguments for a perfectly elastic solid, Griffith [4]
showed that with prescribed values of γ and the applied stress σ, there
exists a critical crack size below which a crack should contract, and above
which it should extend unstably, eventually leading to complete failure of
a solid. Thus only a crack having the critical size can be in equilibrium,
which is of the unstable type, and the critical instability stress as a
function of unstable crack size is shown schematically in Figure 1.

Use of a continuum-type model, as in the preceding discussion, does not
really account for the atomic structure's discreteness, since the γ term
merely averages out the atomic bond rupture processes, and does not in-
corporate the interplay between the breaking of individual atomic bonds in
the vicinity of a crack tip. This neglect of the atomic structure's discreteness when using a continuum-type model was recognized over thirty
years ago by Peierls [5] in the case of a dislocation. Using a model that
allows for lattice discreteness within the crystal planes immediately
adjacent to a slip plane and their interaction via an appropriate interatomic force law, the remainder of the material obeying the classic laws
of linear infinitesimal elasticity theory, Peierls [5] and Nabarro [6]
showed that discreteness can have a marked effect on the mobility of a
straight dislocation, by virtue of the lattice providing periodic barriers
as the dislocation moves from one equilibrium position to the next; the
resistance to movement is greater the smaller the dislocation width, i.e.
a quantitive measure of the spread of distortion along the slip plane.

*Joint University of Manchester/UMIST Department of Metallurgy,
Grosvenor Street, Manchester M1 7HS, United Kingdom.

Subsequently, more detailed consideration has been given to the problem, and it is now generally accepted that there is a good correlation between a dislocation's mobility, its width, and also the type of atomic bonding. Thus, if the bonding is non-directional, as in an ideal metal which is less resistant to elastic shear than tension, a dislocation is wide and very mobile, whereas if the bonding is directional, as in a covalently bonded material which is more resistant to elastic shear than tension, a dislocation is narrow and its movement is difficult, being manifested in a high flow stress; these general conclusions are supported by experimental data, which has been admirably reviewed by Kelly [7].

Until quite recently, little consideration has been given to the way in which the discreteness of the atomic structure affects cleavage crack extension in a crystalline material. However, during the last few years, several investigations have led to the very important conclusion that stable crack tip configurations can exist in the sense that, within a range of crack tip stress intensifications, a crack can extend or contract in a stable manner as the stress is raised or lowered. This is because the atomic discreteness provides discrete barriers to a straight crack front as it moves through the crystal lattice, and is a behaviour that contrasts markedly with that of the Griffith-type continuum model in which the solid can sustain only cracks which are in unstable equilibrium. This phenomenon of 'lattice trapping' of a straight crack tip has been the focus of recent investigations, and this paper's purpose is to survey the main conclusions arising from this recent work, particularly with regard to the effect of the nature of the atomic bonding on the magnitude of the lattice trapping effect. It will be shown that lattice trapping is not likely to be responsible for brittle fracture energies, as determined via appropriate cleavage experiments, being very different to values of γ derived with the aid of an appropriate force law; the main significance is that lattice trapping allows a crack to propagate, within a range of crack tip stress intensifications, by a thermally activated process. This effect has been observed experimentally, as will be emphasized in the paper, but more importantly, the lattice trapping models provide a basis for understanding the effect of aggressive environments on cleavage crack extension, a problem that is of considerable technological significance.

II.   SURVEY OF RECENT THEORETICAL WORK

The first investigation on the effect of the atomic structure's discreteness on brittle cleavage crack extension (i.e. cleavage without any accompanying plastic deformation in the form of dislocation generation or mobility) was that of Goodier and Kanninen [8]. Using a numerical procedure based on a Peierls-Nabarro type approach which allows for discreteness within the atomic planes bounding the cleavage crack, the remaining material obeying the classic laws of linear infinitesimal elasticity, they investigated a model appropriate for a two-dimensional cubic lattice, and determined the critical crack extension stress for a range of idealized force laws describing the interaction between the atoms across the cleavage plane. They showed, that (a) linear cut-off and sinusoidal force laws are both characterized by the existence of stable cracks which are absent in the Griffith continuum-type model, (b) the critical crack extension stress exceeded that given by the Griffith relation, using a value of γ relevant to the force law. These results provided the first demonstration of lattice trapping and of discreteness having an effect and providing a a resistance to the mobility of a straight crack front.

Thomson, Hsieh and Rana [9] used an analytical procedure to investigate a Peierls-Nabarro type model, and by introducing a series of simplifying assumptions, particularly concerning the crack tip profile, were able to relate the magnitude of the discreteness effect or lattice trapping effect, as reflected in the ratio of the lower and upper critical crack extension stresses (see Figure 2), with a parameter which is representative of the crack front width (i.e. a measure of the spread of distortion along the cleavage plane). This conclusion is analogous to that for a dislocation where the stress, usually referred to as the Peierls-Nabarro stress, to move a dislocation increases as the dislocation width decreases, width in this case being defined as the spread of distortion along the slip plane. Smith [10] has also used a Peierls-Nabarro approach that to some extent circumvents the limitations of Thomson, Hsieh and Rana's investigation; using a simple Mode III model (Figure 3) to simulate the Mode I situation, he systematically related the magnitude of the lattice trapping effect to the crack width, and confirmed Thomson, Hsieh and Rana's conclusions. Smith's approach is applicable for the range of force laws

$$\frac{p}{\mu} = \frac{2(w/a)}{1 + 16m^4(w/a)^4} \tag{1}$$

where p is the stress acting across the cleavage plane, 2w is the relative displacement of adjacent atoms in the two planes bounding the crack, a is the distance between these planes, b is the atomic spacing in the direction of crack propagation (the lattice is assumed to be two-dimensional), μ is the shear modulus, and m is a parameter (Figure 4). Maximum and minimum values of the applied stress σ arise respectively when $\varepsilon = 0$ and $b/2$ (Figure 3), their magnitudes for a macroscopic crack of length 2c being

$$\sigma_{max} = \left[\frac{\mu^2 a}{2m^2 c} \coth\left(\frac{\pi a}{2b}\right) \right]^{1/2} \tag{2}$$

and

$$\sigma_{min} = \left[\frac{\mu^2 a}{2m^2 c} \tanh\left(\frac{\pi a}{2b}\right) \right]^{1/2} \tag{3}$$

whereupon

$$\frac{\sigma_{min}}{\sigma_{max}} = \tanh\left(\frac{\pi a}{2b}\right) \tag{4}$$

Stable equilibrium atomic configurations exist between the stress limits σ_{min} and σ_{max} and the ratio $\sigma_{min}/\sigma_{max}$ may be regarded as a measure of the lattice trapping effect, i.e. the effect is large when this ratio is small and small when the ratio approaches unity. Furthermore, the instability condition associated with a purely continuum model, and which corresponds to the Griffith relation, is

$$\sigma_{GRIFF} = (4\mu\gamma/\pi c)^{1/2} \tag{5}$$

where the surface energy γ relevant to the range of force laws (1) is

$$\gamma = \int_{0}^{\infty} p \, dw = \frac{\pi \mu a}{8m^2} \tag{6}$$

It must be remembered that expressions (2) - (5) refer to a Mode III model, and μ should be replaced by $\mu/(1 - v)$ when they are applied to the Mode I situation, v being Poisson's ratio. The crack front width is defined to be that distance from the crack tip along one of the bounding planes where the displacement is a half of the crack tip value. This width is calculated to be $15a/16$, whereupon expressions (2) and (4) show that the magnitude of the lattice trapping effect and the critical crack extension stress both increase as the crack front width decreases. These conclusions and indeed those of Thomson, Hsieh and Rana are applicable only for a specific range of force laws (see Figure 4 in the case of Smith's analysis); consequently such analyses are somewhat limited in their scope, particularly if the effect of force law and the type of atomic bonding on the lattice trapping effect are to be assessed. To attain this objective the approach must be broadened, and as a first step useful conclusions may be reached merely by using very simple physical reasoning procedures [11]. As already indicated, dislocation width can be correlated with the type of atomic bonding, a relation that may be seen by examining an idealized model of an edge dislocation. In Figure 5b where the material's elastic shear resistance is low and the planes normal to the slip plane are flexible, the dislocation is wide, while in Figure 5a where the elastic shear resistance is high and the planes normal to the slip plane are not so flexible, the dislocation is narrow. Similar simple physical arguments may be used for a crack, by removing the associated extra half-plane of atoms associated with the dislocations in Figure 5. Thus if the elastic shear resistance is low, the atomic planes parallel to the crack are flexible and the crack front width is small, while if the elastic shear resistance is high, these planes are fairly rigid and the crack front width is large. Taken together with the correlation of the lattice trapping effect with crack front width arising from the Peierls-Nabarro type analyses [9, 10], these simple arguments suggest that the trapping effect is greatest when a material's elastic shear resistance is low and least when this resistance is high.

It is obviously desirable to confirm this viewpoint using more rigorous methods, and with this in mind, Thomson, Hsieh and Rana [9] examined a very simple model (Figure 6), which is similar to the Frenkel-Kontorova model [12] of a dislocation; the planes bounding a crack are represented by chains of atoms that are linked by two types of spring: lateral bendable springs link the atoms within each chain, while the two chains are attached by transverse stretchable springs, both types of spring being linearly elastic. The transverse springs rupture upon the attainment of a critical displacement or tensile force, while the applied stress is represented by the force P applied to the terminal atoms in the chains, and the crack surface by atoms whose transverse bonds have ruptured; with this model a material's shear and tensile elastic resistances are simulated by respectively the bending and transverse stiffnesses of the two types of spring. Lattice trapping is observed with this model, its magnitude increasing with the ratio stretchable spring stiffness: bendable spring stiffness, in accord with the conclusion reached by simple physical arguments, since the bending stiffness may be associated with a material's elastic shear resistance.

However, with this model, Thomson, Hsieh and Rana did not take the intermediate step of quantitatively correlating the lattice trapping effect with crack front width. Accordingly, Smith [11] considered an even simpler

Frenkel-Kontorova type crack model, which enables the magnitude of the trapping effect, crack front width and the type of atomic bond, as reflected in a material's elastic characteristics, to be systematically related; furthermore, and most importantly, use of such a simple model enables its analysis to be readily extended, without introducing serious mathematical complications, so as to simulate different force laws describing the rupturing of the atomic planes bounding the cleavage crack. With Smith's simulation model (Figure 7), the atomic planes bounding a crack are again represented by two chains of atoms but the major difference as compared with the Thomson, Hsieh and Rana model is that the atoms are constrained to move in the direction of the chains rather than transverse to them; the applied stress is simulated by forces applied to the terminal atoms. Each chain is an aggregate of identical springs having an initial length b, these springs being linearly elastic with a tensile stiffness M. Furthermore, each atom in a chain interacts with its neighbour in the other chain, the two chains are a distance a apart, such that the force tending to restore an atom to its original position is $R(\bar{u}) = L\bar{u}/a$, where $\bar{u}$ is the relative displacement of neighbouring atoms in the two chains; this restoring force is assumed to be operative until the relative displacement attains some critical value beyond which the restoring force is zero, thereby simulating the 'linear bond-snapping' that is characteristic of the Thomson, Hsieh and Rana model [9]. Since the atoms move in the direction of the chains rather than transverse to them, their equilibrium positions are governed by a second order difference equation rather than a fourth order equation, and this simplifies the mathematical analysis in comparison with that of Thomson, Hsieh and Rana. The shear and tensile elastic resistances of the actual material are simulated by respectively the stiffness M of the springs within a chain and the slope L of the linear cut-off force law. Equilibrium of the atomic configuration is shown to be possible when $P_1 < P < P_u$ where

$$\frac{P_1}{\left(\frac{Lq}{2}\right)} = \sqrt{1 + \lambda} - 1 \qquad\qquad (7)$$

$$\frac{P_u}{\left(\frac{Lq}{2}\right)} = \sqrt{1 + \lambda} + 1 \qquad\qquad (8)$$

whereupon

$$\frac{P_1}{P_u} = \frac{\sqrt{1 + \lambda} - 1}{\sqrt{1 + \lambda} + 1} \qquad\qquad (9)$$

where $\lambda = 2Ma/Lb$ and $R(u) = 0$ when $u > qa/2$ with $u = \bar{u}/2$ being the displacement of an atom. It is instructive to relate these results with those obtained by assuming the chains to be continua. In this case it is readily shown that $P_c = qaM/b\sqrt{\lambda}$; Figure 8 shows how P_1, P_u and P_c are affected by the type of atomic bonding, as reflected in the magnitude of λ. Before entering a detailed discussion of these results, however, the critical stresses will be related to the crack front width. Again, defining the crack front width w as that distance from the crack tip at which the displacement is half the displacement at the crack tip, it is found that w is equal to $0.35\ b\sqrt{\lambda}$.

The magnitude of the lattice trapping effect is reflected in the ratio of P_1/P_u, and it therefore immediately follows from the relation $w = 0.35\ b\sqrt{\lambda}$

and relation (9) that if λ is small, w is small and the lattice trapping effect is large, whereas if λ is large, w is large and the lattice trapping effect is small. Consequently, with the real as distinct from the model situation, if the elastic shear resistance of a material is low (reflected in low values of M and λ) the crack front is narrow and the lattice trapping effect is large, while if the elastic shear resistance is high (reflected in high values of M and λ) the crack front is wide and the lattice trapping effect is small. These conclusions are entirely in accord with those arising from simple physical reasoning procedures and also with those obtained by Thomson, Hsieh and Rana for their more sophisticated Frenkel-Kontorova model; as indicated previously, these latter workers correlated a material's elastic shear resistance with the magnitude of the lattice trapping effect but did not develop a quantitative link with the crack front width.

The one-dimensional analyses reviewed to date have used a linear cut-off force law and have therefore simulated 'linear bond-snapping'; it is clearly desirable to extend the discussion to a wider range of force laws with the objective of correlating, in a fairly general quantitative manner, the effect of the non-linear characteristics of the force law with the magnitude of the lattice trapping effect. Smith's one dimensional model is ideal for such a study, in view of its mathematical simplicity. Thus consider the force law for which $R(u) = Lq$ for $qa/2 < u < (q + t)a/2$ while $R(u) = 2Lu/a$ for $0 < u < qa/2$ (Figure 9). Detailed analysis [13] for this force law shows that provided

$$\frac{t}{q} < \frac{2}{\lambda} \left[\sqrt{1 + \lambda} + 1 \right] \tag{10}$$

equilibrium of the system is possible when $P_1 < P < P_u$ where

$$\frac{P_e}{\left(\frac{Lq}{2}\right)} = \left(1 + \frac{t}{q}\right) \left(\sqrt{1 + \lambda} - 1\right) \tag{11}$$

and

$$\frac{P_u}{\left(\frac{Lq}{2}\right)} = \left(1 + \frac{t}{q}\right) \left(\sqrt{1 + \lambda} - 1\right) + 2 \tag{12}$$

whereupon

$$\frac{P_1}{P_u} = \frac{\left(1 + \frac{t}{q}\right) \left(\sqrt{1 + \lambda} - 1\right)}{\left(1 + \frac{t}{q}\right) \left(\sqrt{1 + \lambda} - 1\right) + 2} \tag{13}$$

and is a measure of the magnitude of the lattice trapping effect. This result clearly shows that the lattice trapping effect becomes more pronounced as the force law becomes sharper (i.e. t/q becomes smaller) and also as λ becomes smaller (i.e. as the ratio stretchable bond stiffness: bendable bond stiffness becomes larger in the actual material). Expression (13) is valid provided the force law is sufficiently sharp that t/q is less than the value given by expression (10); the upper limits of t/q for various values of $\lambda = 2Ma/Lb$ are shown in Table 1, which therefore gives an indication of the range of force laws for which expression (13) is valid. Suppose, for example, that $\lambda = 1$ when the broadest force law for which (13) is valid is that for which t/q = 4.83. At this limit P_1/P_u is equal to 0.54, decreasing to 0.17 when t/q = 0, a decrease that illustrates in dramatic form the effect of force law sharpness in providing a marked lattice trapping effect. When the inequality sign in (10) is reversed, the situation becomes more complicated, but a detailed consideration of the system's equilibrium again shows that the magnitude of the lattice trapping effect increases (i.e. P_1/P_u decreases) as the force law becomes sharper (i.e. t/q becomes smaller). Indeed, whereas both P_u and P_1 increase with t/q, assuming q remains constant, the difference between them is always Lq, as is also evident by examining the detailed results for small t/q values (i.e. those satisfying (10)).

Now consider the force law (Figure 10) for which $R(u) = [Lq/ta][(t+q)a-2u]$ for $qa/2 < u < (q+t)a/2$ while $R(u) = 2Lu/a$ if $0 < u < qa/2$. Detailed analysis for this force law shows that provided relation (10) is satisfied, equilibrium atomic configurations are possible when $P_1 < P < P_u$ where

$$\frac{P_1}{\left(\frac{Lq}{2}\right)} = \left(1 + \frac{t}{q}\right)(\sqrt{1 + \lambda} - 1) \tag{14}$$

and

$$\frac{P_u}{\left(\frac{Lq}{2}\right)} = (\sqrt{1 + \lambda} + 1) \tag{15}$$

whereupon

$$\frac{P_1}{P_u} = \frac{\left(1 + \frac{t}{q}\right)(\sqrt{1 + \lambda} - 1)}{(\sqrt{1 + \lambda} + 1)} \tag{16}$$

an expression which clearly shows that the lattice trapping effect is more pronounced the steeper the descent from the force law maximum, i.e. as t/q becomes smaller, and also as λ becomes smaller. Expression (16) is valid provided the descent from the force law maximum is sufficiently rapid that t/q is less than the value given by expression (10), and again the upper limits of t/q for various values of t/q are shown in Table 1, which therefore indicates the range of force laws for which (16) is valid. At the limits, P_u and P_1 are equal and there is no lattice trapping. However, this state of affairs is unique since with t/q values slightly in excess of the critical value, P and P_1 are different and there is a lattice trapping effect; the maximum force (P_u) situation is associated with an atom being subject to maximum restraint and another atom with a displacement on the decreasing part of the force law, while the minimum force (P_1) situation is associated with an atom just having lost its restraint with another atom having a displacement on the decreasing part of the force law. As t/q increases, there are other unique values of t/q for which $P_u = P_1$ and there is no lattice trapping; each of these unique situations corresponds to the existence of equilibrium atomic configurations with an atom in the force law maximum position, another having just lost its restraint, and with other atoms having displacements associated with the decreasing part of the force law.

All the models reviewed in this section to date, are idealized to the extent that they are of the Peierls-Nabarro type or are one-dimensional simulation models; such models allow for the atomic structure's discreteness only within the atomic planes that bound a cleavage crack. The simplicity of the models, particularly those of the one-dimensional type, has enabled the interplay between a variety of factors to be readily appreciated with a minimum of mathematical analysis. The conclusions reached from the simple models are supported by a limited number of investigations of more realistic models. As regards the effect of the shape of the non-linear part of the force law, Esterling [14] has examined the stability of a two-dimensional Mode I crack in a simple cubic lattice within a lattice statics approximation, investigating the effect of various idealized nearest neighbour force laws for the case where Poisson's ratio is zero; Esterling's analysis therefore extends a similar analysis due to Hsieh and Thomson [15], who specifically considered the force law appropriate to linear bond-snapping. Inspection of Esterling's results clearly shows that the lattice trapping effect is more marked the steeper is the descent of the force law from the maximum to the zero restraint position, and this accords with the predictions of the simple one-dimensional simulation model. Moreover, if the maximum is maintained at a constant value, the greatest effect is on the lower limiting crack tip stress intensification, a result which clearly agrees with the predictions of equations (14) and (15).

Sinclair has examined the behaviour of a two-dimensional Mode I (111) cleavage crack in silicon with a straight edge parallel to the [0$\bar{1}$1] direction, using an atomistic computer simulation model. Several non-central interatomic force laws were investigated, all being matched to the elastic constants and the cohesive energy, but varying in shape at long-range; a short-ranged force law gives a more pronounced trapping effect than a long-ranged law. This result agrees with the predictions from Smith's simple one-dimensional model, for if the general force law in Figure 10 is considered for two specific laws characterized by the same areas under the curves, t_q has the same value (say ε) for the two laws, if the area under the increasing linear portions is neglected. Consequently relation (13) becomes

$$\frac{P_1}{P_u} = \frac{\left(1 + \frac{t^2}{\varepsilon}\right)(\sqrt{1 + \lambda} - 1)}{\left(1 + \frac{t^2}{\varepsilon}\right)(\sqrt{1 + \lambda} - 1) + 2} \tag{17}$$

whereupon P_1/P_u decreases with t; thus a short-ranged force law (small t) shows a greater lattice trapping effect than a long-ranged law, which is precisely Sinclair's result.

III. DISCUSSION

As indicated in the Introduction, a particular aim of investigations concerned with the lattice trapping effect is to relate the critical crack extension stress, with discreteness taken into account, to the critical extension stress predicted by the classic Griffith theory, which is continuum based and incorporates the surface energy γ of the particular material under consideration. If this latter procedure is applied to the simple one-dimensional model, the simulated crack extends when the applied force P attains the critical value P_c given by

$$P_c = \sqrt{\frac{\lambda L}{a} \int R(u)\,du} \tag{18}$$

where R(u) is the force law, and the integration limits are zero and that value of u for which the restraining force becomes zero. For the force law in Figure 9, it immediately follows that:

$$\frac{P_c}{\left(\frac{Lq}{2}\right)} = \sqrt{\lambda\left(1 + \frac{2t}{q}\right)} \qquad (19)$$

when comparison with expressions (11) and (12) shows that $P_1 < P_c < P_u$ irrespective of the value of t/q; in other words, the Griffith force lies between the force limits within which equilibrium configurations are possible with the discrete atom model. The same conclusion is also valid for the force law in Figure 10, for then

$$\frac{P_c}{\left(\frac{Lq}{2}\right)} = \sqrt{\lambda\left(1 + \frac{t}{q}\right)} \qquad (20)$$

and the conclusion follows by comparison with expressions (14) and (15). Against this background it is worth looking very carefully at the results obtained by Esterling [14]. He showed that the Griffith stress was bounded by the upper and lower critical stresses for only a few of the wide variety of force laws studied; for most laws, the Griffith stress was less than the lower critical stress. The laws for which the Griffith stress is bounded are those where the descent from the force law maximum to the zero restraint position is particularly steep; laws characterized by a tail prior to zero restraint are associated with a Griffith stress which is less than the lower critical value. This suggests that if a force law with a similar tail is used with the simple one-dimensional model, a similar effect ought to be observed. This is indeed the case, since it is easily demonstrated that with the artificial force law shown in Figure 11, with $t/q < 2 \, [\sqrt{1 + \lambda} + 1]/\lambda$ (i.e. relation (10)), P_1 and P_u have the same magnitudes as for the dotted force law (i.e. the same law as that shown in Figure 10), and these are given by relations (14) and (15); furthermore, for a range of λ values there is a corresponding range of t/q values for which P_c is less than P_1. Thus when the force law has a pronounced tail prior to the zero restraint position, it is possible for P_c to be less than P_1, a result that accords with Esterling's behaviour pattern.

The results from the various investigations reviewed in the preceding section, strongly suggest that for a two-dimensional cleavage crack subject to Mode I loading conditions, the magnitude of the lattice trapping effect, as reflected in the ratio of the upper and lower crack tip stress intensification limits between which stable cleavage cracks can be sustained within a brittle solid, is greater:
(a) the larger is the ratio stretchable bond elastic stiffness: bendable bond elastic stiffness
(b) the narrower is the force law describing the tensile rupturing of atomic bonds across a cleavage plane
(c) the steeper is the descent of the force law from the maximum to the zero restraint position.

In viewing conclusions (a), (b) and (c), it is therefore quite clear that maximum lattice trapping of a straight crack front should be observed with a narrow force law, a steep descent from the maximum to the zero restraint position, and the larger the ratio stretchable bond elastic stiffness: bendable bond elastic stiffness, with the latter parameter likely to be having the dominant effect. However, if a particularly low bendable bond elastic stiffness is the cause of this ratio being large, it will at the same time promote dislocation activity. Whether or not dislocation generation and mobility does actually occur in the vicinity of a crack tip, is an aspect of the cleavage problem that is beyond the scope of the simple models described in the preceding section; however, it is a most important problem and has received detailed consideration (e.g. [17], [18]), for it is central to the classification [19, 20] of materials into ductile and brittle categories. As an example of a specific study, Gehlen, Hahn and Kanninen [21] have studied the configuration near the tip of a straight {100} cleavage crack in alpha iron using an atomistic computer simulation approach, and showed that the extension stress was appreciably in excess of the Griffith value, with dislocation nucleation being observed.

Because dislocation activity is easy in materials with a low elastic shear resistance (i.e. in the extreme case, those having an ideal metallic bond), it means that where brittle crystalline materials are concerned, very special circumstances indeed are likely to be required to give sufficiently strong lattice trapping of a straight crack front, for it to be reflected in the experimentally measured crack extension stress being markedly in excess of that predicted by the Griffith relation. Accordingly, use of the Griffith approach, based on a continuum-type model and a value of γ relevant to the force law describing the behaviour of the atoms along the cleavage plane, should suffice for most practical purposes.

However, as emphasized in several papers [15, 16, 22, 23], the main consequence of lattice trapping is that it is responsible for what is known as the 'creep mobility of cracks'. Thus, analogous to the behaviour of dislocations, it should be possible for a crack to propagate slowly, with the aid of thermal fluctuations, by the nucleation and movement of kinks along a crack front, rather than by the forward propagation of the entire crack front. Experimental evidence for thermally activated cleavage crack growth is provided by the observations of Wiederhorn, Hockey and Roberts [24] on {10$\bar{1}$0} cleavage cracks in sapphire tested in vacuum. Figure 12 shows the critical stress intensity factor for rapid cleavage fracture as a function of temperature, while Figure 13 shows the stress intensity factor - crack velocity variation at specific temperatures. Such observations clearly suggest that, for sapphire, cleavage cracks can propagate in vacuum by a thermally activated process, and this is powerful support for the existence of lattice trapping. Similar experimental results have been obtained [25] for some glasses, where the degree of local order in the crack tip vicinity is presumed to be sufficient for the processes discussed in this paper to become operative.

The behaviour of a kinked crack front in alpha iron has been investigated via an atomistic simulation approach by Kanninen and Gehlen [23], and in silicon by Sinclair [16], and also via a lattice statics approach by Esterling [14] in a general material for a variety of idealized nearest neighbour force laws. As with a straight crack front, there is a range of crack tip stress intensification within which a crack kink is lattice trapped, but this range is appreciably narrower than the corresponding range for a straight crack front. More importantly, Esterling found that the widths of the ranges in which a crack kink and a straight crack front are lattice trapped increase and decrease together, with the kink limit stresses always lying between the straight crack limit stresses. Thus, although this paper has concentrated on the lattice trapping of a straight crack front in terms of the force law and the nature of the atomic bonding, the conclusions should nevertheless be equally applicable to the trapping of an irregular crack front, and therefore very relevant to thermally activated cleavage crack extension. Interest in this phenomenon is reflected in the recent series [15, 16, 22] of analyses concerned with crack kink kinetics; such analyses have followed similar lines to those used more than a decade earlier for dislocation kinks. A major reason for this interest is that an understanding of thermally activated cleavage crack growth provides a basis [22] for explaining the effects of aggressive environments on cleavage crack extension in brittle materials, a problem that is of considerable technological importance. It is the author's opinion that this research area will receive extensive study, both theoretical and experimental, in the next few years.

ACKNOWLEDGEMENTS

The author is grateful for discussions on this topic with many colleagues during the last few years, especially Dr. R. Thomson, Dr. B. R. Lawn and the Battelle (Columbus) Metal Science Group.

REFERENCES

1. TAYLOR, G. I., Proc. Roy. Soc., A145, 1934, 362.
2. OROWAN, E., Z. Phys., 89, 1934, 634.
3. POLANYI, M., Z. Phys., 89, 1934, 660.
4. GRIFFITH, A. A., Phil. Trans. Roy. Soc., A221, 1920, 163.
5. PEIERLS, R. E., Proc. Phys. Soc. (London), 52, 1940, 34.
6. NABARRO, F. R. N., Proc. Phys. Soc. (London), 59, 1947, 256.
7. KELLY, A., "Strong Solids", Clarendon Press, Oxford, 1966.
8. GOODIER, J. N., "Fracture", Ed. H. Liebowitz, Academic Press, New York, 2, 1969, 1.
9. THOMSON, R., HSIEH, C. and RANA, V., J. Appl. Phys., 42, 1971, 3154.
10. SMITH, E., J. Appl. Phys., 45, 1974, 2039.
11. SMITH, E., Materials Science and Engineering, 17, 1975, 125.
12. FRENKEL, J. and KONTOROVA, T., Physik. Z. Sowjet-union, 13, 1938, 1.
13. SMITH, E., Materials Science and Engineering, to be published.
14. ESTERLING, D. M., J. Appl. Phys., 47, 1976, 486.
15. HSIEH, C. and THOMSON, R., J. Appl. Phys., 44, 1973, 2051.
16. SINCLAIR, J. E., Phil. Mag., 31, 1975, 647.
17. KELLY, A., TYSON, W. R. and COTTRELL, A. H., Phil. Mag., 15, 1967, 567.
18. RICE, J. R. and THOMSON, R., Phil. Mag., 29, 1974, 73.
19. SMITH, E., Materials Science and Engineering, 15, 1974, 3.
20. SMITH, E., Paper No. 5 presented at Institute of Physics/Metals Society Conference, Churchill College, Cambridge, 1975.
21. GEHLEN, P. C., HAHN, G. T. and KANNINEN, M. F., Scripta Met., 6, 1972, 1087.
22. LAWN, B. R. and WILSHAW, T. R., "Fracture of Brittle Solids", Cambridge University Press, 1974.
23. KANNINEN, M. F. and GEHLEN, P. C., "Interatomic Potentials and Simulation of Lattice Defects", Eds. P. C. Gehlen et al., Plenum Press, New York, 1972, 713.
24. WIEDERHORN, S. M., HOCKEY, B. J. and ROBERTS, D. E., Phil. Mag., 28, 1973, 783.
25. WIEDERHORN, S. M., JOHNSON, H., DINEES, A. M. and HEUER, A. H., Jnl. American Ceramic Society, 57, 1974, 336.

Table 1 - The Limits of t/q for which expression (13)
is valid, for various values of λ = 2Ma/Lb

λ	t/q
0	∞
0.50	8.89
1	4.83
2	2.73
3	2.00
8	1.00

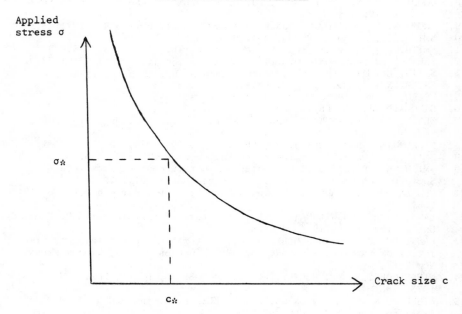

Figure 1 - The relation between the critical applied stress σ required to
extend a crack as a function of its length, as predicted via
the Griffith continuum-type model; the relation follows a σ∝1/√c
form. For a given applied stress σ_*, a crack of length c > c_*
extends, while a crack of length c < c_* contracts.

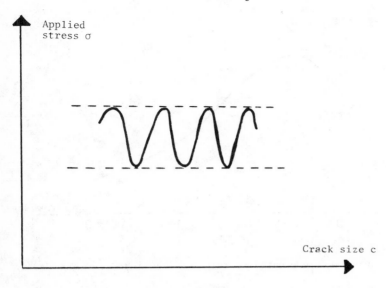

Figure 2 - The schematic relation that exists between the applied stress and stable crack size, when atomic discreteness is taken into account; it is important to note that the distances between the peaks and troughs are of atomic dimensions, which are of course very small in comparison with the macroscopic crack size.

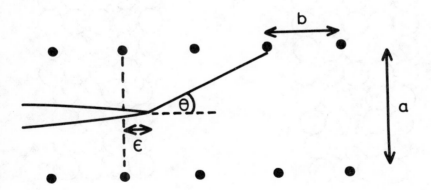

Figure 3 - Discrete atoms in the vicinity of a crack tip in Smith's Mode III Peierls-Nabarro type model. The atoms in the upper plane are displaced with respect to those in the lower plane as the crack tip moves from left to right.

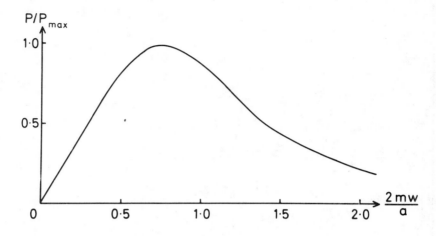

Figure 4 - Force law represented by relation (1) with $P_{max} = (3\mu/4m)(1/3)^{1/4}$

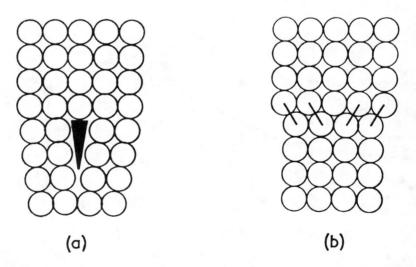

Figure 5 - Idealized models: (a) a narrow edge dislocation, (b) a wide edge dislocation

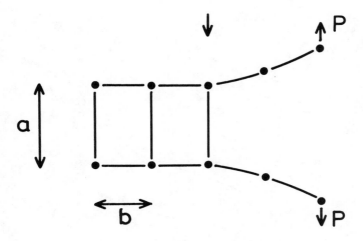

Figure 6 - Thomson, Hsieh and Rana's one-dimensional model [9] of the cleavage process

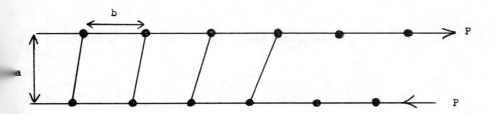

Figure 7 - Smith's one-dimensional model [11] of the cleavage process

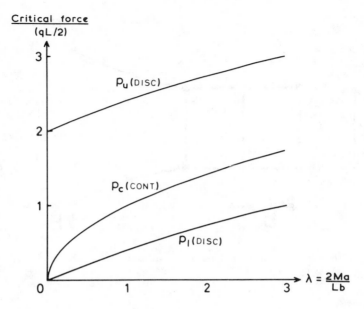

Figure 8 - Upper and lower bounds P_u and P_l for the crack extension
force in Smith's one-dimensional discrete atom model, compared
with the value P_c for a continuum model; the results are for a
linear cut-off force law.

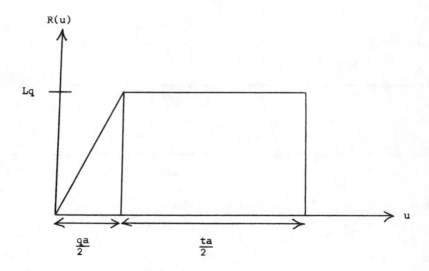

Figure 9 - The force law leading to expressions (11) - (13); u is the
displacement of an atom, and R(u) is the restraining force
due to its interaction with an adjacent atom in the neighbouring chain

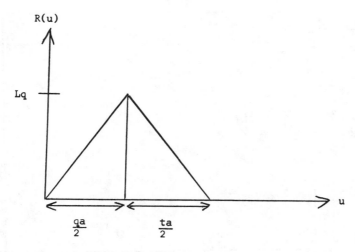

Figure 10 - The force law leading to expressions (14) - (16); u is the displacement of an atom, and R(u) is the restraining force due to its interaction with an adjacent atom in the neighbouring chain

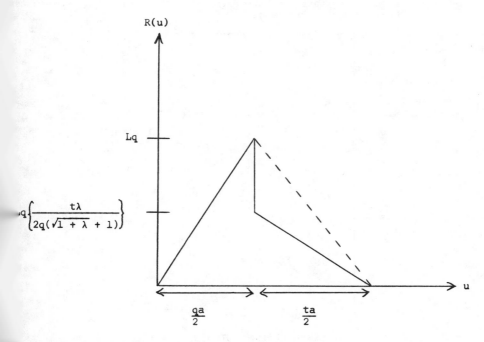

Figure 11 - The force law (full lines) used to show that it is possible for P_c to be less than P_1 with the simple one-dimensional model

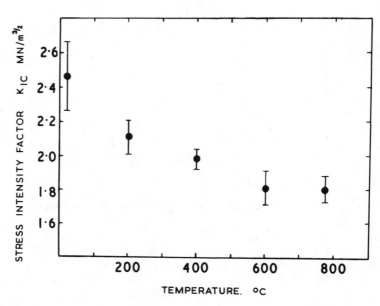

Figure 12 - The critical stress intensity factor for rapid cleavage fracture
of sapphire as a function of temperature; the brackets give the
standard deviation of the data which was obtained by Wiederhorn
Hockey and Roberts [24].

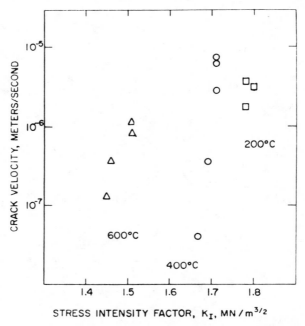

Figure 13 - The stress intensity factor - crack velocity variation for
sapphire at specific temperatures [24].

GRAIN SIZE: THE FABRIC OF (BRITTLE) FRACTURE
OF POLYCRYSTALS

R. W. Armstrong[1]

MICROSTRUCTURE AND FRACTURE

More than two hundred and fifty years ago, Réaumur [1] reported that the
quality of a steel material was established by the fineness of its microstructure. In recording this observation, Réaumur noted that the distribution of grains within the material was able to be revealed by fracturing
a piece of it and examining the surfaces of separation. There is an important connection between these statements. No doubt, the correlation of
the quality of a material with the smallness of the scale of its microstructure seemed reasonably correct, as far as concerned the ambient mechanical properties of steel materials, at least, because the fracture strength
of steel is itself found to be greater as the average size of the grains is
smaller. This is particularly true for the (brittle) cleavage fracture
strength of iron and steel materials as reported nearly twenty-five years
ago in the careful experiments of Petch [2]. A recent assessment of results for the brittle fracture strength dependence on average grain diameter
for a number of iron and steel materials has been made by Madhava [3]. The
data are shown in Figure 1 according to the Petch analysis which predicts
that

$$\sigma_f = \sigma_{of} + k_f d^{-1/2} .$$ (1)

The references for these data are given in Table 1 including the reference
to the original results of Petch.

The total data in Figure 1 do show that the fracture strength increases as
the grain size is refined even though there are considerable differences
between the results obtained in the various studies. The lowest fracture
stresses in Figure 1 are reported for the intercrystalline fracture of three
iron materials. Nevertheless, these low values of fracture stress are shown
to be larger than the stress values calculated from the Sack equation [4]
describing the Griffith fracture condition for a circular crack of size
equal to the average grain diameter (see Table 1). The reason for all of
the measured fracture stresses in Figure 1 being larger than is predicted
by the Griffith fracture condition is normally taken to be the requirement
for crystalline materials that plastic flow must necessarily accompany the
most brittle fracture process. This plastic flow, which may occur to a
vanishingly small degree for extremely brittle failures, affects the
fracture process in two ways. For a crack-free material, an amount of
deformation must occur initially to produce an internal concentration of
stress of sufficient intensity to cause a micro-cleavage crack to form;
hence, the presence of σ_{of} in equation (1) for the movement of dislocations
and the influence of different values of σ_{of} on the measurements in Figure 1.
The growth of the cleavage crack, certainly past one grain diameter, may

---

[1]On leave, at the U.S. National Science Foundation, from the University
of Maryland.

require an additional amount of plastic flow at the crack tip; hence, the internal concentration of stress measured in equation (1) by k_f may be greater even than is specified by the Griffith condition, i.e. $k_f \geq [\pi E \gamma_s / (1 - \nu^2)]^{1/2}$ from equation (1) and Table 1. The largest fracture stresses in Figure 1 are reported for the condition whereby fracture was preceded by plastic yielding. These results are consistent with the presumed influence of plastic flow on raising the fracture strength of materials.

The striking appearance of brittle cleavage fracture surfaces which is caused by the varying orientations of individual micro-cleavage cracks within the polycrystal grains and by the differing areas of the grains covered by these micro-cracks is revealed by the two reasonably similar scanning electron micrographs of Figure 2a,b. These fracture examples are taken from separate steel materials which have failed under quite different circumstances. Figure 2a shows the fracture surface obtained for crack-free (plain carbon) 1010 steel which was tested by Madhava [3] in tension at 4.2°K. Figure 2b is from the research study of Stonesifer [5] involving A533 B steel material tested at 77°K in the form of a pre-cracked compact specimen designed for fracture mechanics evaluation of the toughness of the material in plane strain deformation. In this latter case, the fracture stress is characterized by the equation, after Irwin [6] and Orowan [7]:

$$\sigma_f = K_{Ic} (\pi a_e)^{-1/2} . \tag{2}$$

Stonesifer found that the prior austenite grain size, which in Figure 2b is only slightly larger than the ferrite grain size for this A533 B specimen, had a significant infleunce on determining the nature of the fracture surface morphology and, correspondingly, on the fracture toughness of the material.

The similar appearances of the fracture surfaces which are shown in Figure 2 for these two engineering materials are roughly indicative of the same degree of brittleness in them even though this brittleness was ahcieved by means of different external testing conditions. The fracture stress for the 1010 specimen at 4.2°K was measured to be 930 MPa and this is greater than the 164 MPa which should apply at 77°K for this A533 B material subject to the condition of it containing a surface crack nearly 13 mm deep. The ($\epsilon = .002$) yield stress of the A533 B material at 77°K in a crack-free condition was found to be 1020 MPa and this compares with 840 MPa for the yielding of the 1010 specimen in compression at 4.2°K. The brittle fracture stress measurement for the 1010 steel material is plotted in Figure 1 according to its ferrite grain diameter of 0.03 mm. If the fracture stress determined for the A533 B material according to equation (2) is plotted in Figure 1 by substituting the appropriate $a_e^{-1/2}$ value for $d^{-1/2}$, say $a_e^{-1/2} = 0.28$ mm$^{-1/2}$, then, the value of K_{Ic} is required to be clearly greater than any k_f value measured for the fracture stress-grain size experiments. The value of $K_{Ic} = 33$ MPa·m$^{1/2}$ for this steel as compared with the value of $k_f = 3.3$ MPa·m$^{1/2}$ which is determined for the grain size dependence of the fracture stress of 1010 steel in Figure 1. This typically large value of K_{Ic} is attributed to the controlling influence of plastic deformation on determining the unstable growth of a macroscopic crack in even the most brittle circumstance of any fracture mechanics experiment.

Another view of the relationship between the microstructure of a material and the nature of cleavage cracking within it is shown in Figure 3, this time from a study by Prasad [8] involving polycrystalline zinc material which has been deformed in compression at 4.2°K. The cleavage cracks are revealed as white bands in this micrograph which has been produced with polarized light from a surface film put on to the specimen in a post-deformation anodization treatment. The technique has been found to be useful for revealing the relationship between the pattern of grains and the orientation or extent of cleavage cracking within hexagonal close packed metals. Extensive deformation twinning is observed within the microstructure of this zinc material, and, as for the cleavage cracks, the twin bands are related to the crystallographic orientations of the grains.

Figure 3 is important for one reason because the formation of these clea-vage cracks in an externally applied compressive stress field gives empha-sis to the local tensile character of the internal concentrations of stress within the slip bands which have produced the cracks. More significantly, perhaps, the figure is useful for describing the importance of information which, sometimes, either is totally absent or, at least, is not able to be observed easily in studies of cracking processes within the microstructures of materials. The deformation by slip which has produced the cracking and twinning events of Figure 3 is not able to be observed with this metallo-graphic technique. The absence of critical information about the role of slip deformation - the primary agent - in effecting cleavage is true, also, for most studies of fracture surfaces such as are shown in Figures 2a, b. A close examination of Figure 2a is required to detect evidence of the extensive deformation by twinning which has occurred in this steel material. Thus, the connection between the microstructure of materials and their fracture properties, which has been described in historical times by Réaumur, has not been carried forward today as far as scientists and engineers desire because of this problem of less than complete information being obtainable with current observational techniques. Modern theoretical models for understanding fracture processes are normally described on the level of atomic dimensions via the dislocation events which are operative.

THE BRITTLENESS OF CRACK-FREE POLYCRYSTALS

The dislocation theory for the ductile-brittle transition behavior of steel and related materials, such as the refractory metals, has been de-veloped by Cottrell [9] and by Petch [10]. Their analyses represent the starting point for our current understanding of the brittle fracture pro-cess within essentially crack-free polycrystalline materials. Both analyses are based on the result computed by Stroh [11] for the stability of a crack formed from an idealized (slip band) dislocation pile-up, as follows:

$$\sigma[n'b] = c'\gamma_s \quad , \tag{3}$$

where σ is the applied stress, n' is the number of dislocations, and c' is a numerical constant. Cottrell evaluated $n'b$ and σ in terms of several equations:

$$n'b = (\sigma_y - \sigma_{oy})d/2\mu \tag{4a}$$

$$\sigma = c''\sigma_y \tag{4b}$$

$$\sigma_y = \sigma_{oy} + k_y d^{-1/2} \quad , \tag{4c}$$

where σ_y is the yield stress of the material and σ_{oy} is the grain size in-dependent friction stress component of the yield stress. Beyond equations (4a)-(4c), Cottrell proposed that the onset of brittle fracture was con-trolled by the break-out of cleavage past one grain diameter so that γ_s

should be replaced by a surface energy term including the plastic work associated with the unstable growth of the cleavage crack, say, γ_p. The combination of terms in equations (3) and (4) gives the implicit specification of a ductile-brittle transition in the well-known Cottrell form:

$$k_y[\sigma_{oy}d^{1/2} + k_y] = C\mu\gamma_p \; . \tag{5}$$

Equation (5) is particularly useful for indicating the effect on the potential brittleness of a material of those factors determining the yield stress according to the Hall [12] - Petch [2] grain size dependence in equation (4c). Brittleness is promoted: (a) by a large k_y value, measuring the slip band stress concentration for propagating plastic yielding across grain boundaries; (b) by a large σ_{oy} value, for the average friction stress to move the dislocations within slip bands; and (c) by a large grain size, the length over which an appreciable internal concentration of stress may be generated. At low temperature, σ_{oy} is increased appreciably due to the reduced availability of thermal fluctuations for assisting the movement of the large number of dislocations involved in the general plastic yielding of a material, hence, its brittleness is enhanced.

Petch obtained in his analysis [10] an explicit value of the ductile-brittle transition temperature which is measured for the fracturing energy absorbed in Charpy impact tests. This was accomplished by expressing the temperature dependence of σ_{oy} in one way as

$$\sigma_{oy} = Be^{-\beta T} \tag{6}$$

where B and β are experimental constants [13]; and, also, by assuming for this ductility transition temperature that the value of k_y was increased to $k_f{}^*$, the slope of the true ductile fracture stress dependence on the reciprocal square root of the grain size [14]. By substitution for σ_{oy} and k_y in equation (5), for the transition condition at $T=T_D$,

$$T_D = \beta^{-1}[\ell nB - \ell n \; (C\mu\gamma_p k_f{}^{*-1} - k_f{}^*) - \ell n \; d^{-1/2}] \; . \tag{7}$$

Heslop and Petch [15] showed that the predicted grain size dependence of T_D was in agreement with experimental results measured for mild steel material and, also, they showed that an increase in T_D could be accounted for if the yield stress of a material were raised by the addition of a temperature independent contribution, say, $\Delta\sigma_{oy}{}^*$, as might occur due to solid solution strengthening, precipitation hardening, or even neutron irradiation of the material.

The Cottrell and Petch analyses for the ductile-brittle transition behavior of materials have been related by Armstrong [16] to the experimental condition of this transition being specified since the beginning of this century by the yield stress of the material becoming equal to the tensile brittle fracture stress, i.e.

$$\sigma_y = \sigma_f \; . \tag{8}$$

Because both stresses show a Hall-Petch stress-grain size dependence, these dependences can be substituted directly into equation (8) and the parameters rearranged in the form of equation (5) developed by Cottrell as

$$k_y[\; \sigma_{oy}d^{1/2} + k_y] = k_y \; [\sigma_{of}d^{1/2} + k_f] \; . \tag{9}$$

Thus, on this basis, the value of $C\mu\gamma_p$ in equation (5) is obtained as

$$C\mu\gamma_p = k_y[\sigma_{of}d^{1/2} + k_f] \ . \qquad (10)$$

This experimental condition which applies at the ductile-brittle transition requires, then, that the value of γ_p is not an independent parameter; it is larger itself either as the value of k_y is increased, as the grain size is increased, or as the fracture stress parameters, σ_{of} and k_f, are increased.

The dislocation theory leads to the expected inequality $k_f > k_y$ for separate model calculations of hypothetical brittle cracking versus plastic yielding processes. Therefore, the balance reflected in equations (8) and (9) is achieved for a given grain size by increasing the value of σ_{oy} over that of σ_{of} to account, when multiplied by $d^{1/2}$, for the difference between k_f and k_y. The condition that $\sigma_{of} < \sigma_{oy}$ means that the increase in γ_p with increase in grain size is smaller than applies for the collection of terms on the left side of equation (5) - and for the idealized case of $\sigma_{of} = 0$, which should apply for a pre-existing crack, the value of γ_p is independent of the grain size.

An advantage of describing the ductile-brittle transition in terms of equal values of the yield stress and the tensile fracture stress, as given by equation (8), is that these stresses are reasonably well-defined experimentally, despite the apparent scatter in Figure 1, and the theoretical considerations for understanding the transition behavior are mainly those involved in understanding the controlling factors for these individual stresses. The yield and tensile fracture stresses are related to those stresses which are operative in the Charpy impact testing procedure by accounting for the effect of the strain rate on the yield stress through reducing the parameter β in equation (6) and by accounting for the effect of the notch on raising the yield stress through a multiplying factor α in equation (8). In this manner the Charpy impact transition temperature, T_D, is evaluated in an analogous way to that given by Petch as

$$T_D = \beta^{-1}[\ln(\alpha B) - \ln\{(k_f - \alpha k_y) + (\sigma_{of} - \alpha\Delta\sigma_{oy}*)d^{1/2}\} - \ell n d^{-1/2}] \ . \qquad (11)$$

Equation (11) has been applied by Armstrong [17] to evaluating the T_D dependence on ferrite grain size for mild steel. Figure 4 shows a comparison of this calculated result with separate experimental measurements which have been reported for the ferrite grain size dependence of a number of iron and steel materials [18-21] and for the prior austenite grain size dependence of T_D measured for A533 B steel [5] and for an iron-nickel alloy material [22]. The agreement of results for the ferrite grain size dependence appears to be satisfactory. The indication for the different average values of the T_D results for the A533 B steel and the iron-nickel alloy - but otherwise an austenite grain size dependence which is similar to the ferrite result - is mainly attributed by Armstrong and Stonesifer [23] to a large value of $\Delta\sigma_{oy}*$ for A533 B steel and to an effective small (martensitic) ferrite grain size for the iron-nickel alloy.

The stress-grain size analysis for the ductile-brittle transition behavior of iron and steel materials, as described for equation (8), has been carried over by Armstrong [24] to describe the onset of brittleness in hexagonal close-packed (hcp) metals. The situation is shown schematically in Figure 5. The brittle fracture stress of these materials, too, is found experimentally not to change in any significant way with the temperature or the strain rate. However, for decreasing temperature or increasing strain rate, the yield stress of an hcp metal such as zinc, say, as measured in compression, is increased due to increases in both the σ_{oy} and k_y parameters of equation (4c). Thus, the temperature enters into equation (8) in two ways also. The k_y temperature dependence is important generally and the dislocation theory for it has been described by Armstrong as

$$k_y = c''' \, m[m^*\mu b\tau_c]^{1/2} \quad , \tag{12}$$

for which m is the Taylor orientation factor for the distribution of slip systems required to maintain continuity of the material during straining, m* is a less-restrictive Sachs orientation factor, and τ_c is the concentrated stress required for propagation of deformation across the grain boundary during general yielding of the material.

Prasad, Madhava and Armstrong [25] have shown for the polycrystalline zinc material revealed in Figure 3 that the temperature dependence of k_y followed the prediction from previous metallographic observations of τ_c being controlled by slip on the $\{11\bar{2}3\}$ $\langle11\bar{2}2\rangle$ systems. The friction stress σ_{oy} for this material followed the temperature dependence corresponding to slip on $\{0001\}$ $\langle11\bar{2}0\rangle$ systems, as expected. The extensive twinning in Figure 3 does not enter, therefore, into either the σ_{oy} or k_y values which are measured over a range of temperatures in this case. For zinc, as was previously indicated for beryllium and magnesium materials, the temperature dependence of k_y was found to be very much larger than that for σ_{oy} and so the σ_{oy} temperature might be neglected in obtaining an approximate relationship for T_D from equation (8). The study verified the result that a relatively small grain sizes the value to T_D should follow the predicted dependence:

$$T_D = \beta_c^{-1} \ln[c^*m^2m^*\mu bB_c k_f^{-2}] + [2(\sigma_{oy}+\Delta\sigma_h-\sigma_{of})\beta_c^{-1} k_f^{-1}]d^{1/2} \tag{13}$$

where β_c and B_c apply for τ_c according to equation (6) and $\Delta\sigma_h$ is a work hardening contribution added to σ_y so as to specify the value of T_D for the flow stress at some small value of strain becoming equal to the tensile fracture stress. In a separate study of other polycrystalline zinc materials, Pszonka [26] has shown that the grain size dependence of equation (13) applies very well for the brittle fracture transition observed in tension.

THE BRITTLE FRACTURE OF PRE-CRACKED MATERIALS

The K_{Ic} characterization of brittleness brought on by the presence of a macroscopic crack within a material, as prescribed by equation (2), naturally excludes any direct consideration of grain size on fracture toughness. Nevertheless the K_{Ic} parameter is able to be related to the preceding description of the onset of brittleness in crack-free materials via the Dugdale [27] or Bilby-Cottrell-Swinden [28] models of continuum crack growth with an associated plastic zone at the crack tip. Keer and Mura [29] have compared the theoretical bases for calculations of this type when described in terms of continuous distributions of dislocations or in terms of discrete dislocations. Yokobori [30] has considered the energetics of fracture involving such models. Most recently, Bilby [31] has reviewed the connection between cracks and dislocations for this fracture description in terms of the equations

$$(a_s-a)/a = \sec(\pi\sigma_f/2\sigma_y)-1 \tag{14a}$$

$$\Phi_a/a = (4\sigma_y/\pi M)\ell n(a_s/a) \tag{14b}$$

where $a_s \equiv (a+s)$; the critical displacement parameter $\Phi_a = (2\gamma_p/\sigma_y)$; and, the modulus factor $M = \mu/2(1-\nu)$ or $\mu(1+\nu)/2$ for plane strain versus plane stress deformation. Armstrong [32] has pointed out that equation (14a) is able to be usefully expanded by series approximation so as to change it into a form that is directly comparable with equation (2). The result has been obtained by Stonesifer [5] that

$$\sigma_f = (8^{1/2}\sigma_y/\pi)\ [s/a_a]^{1/2}\ . \tag{15}$$

A comparison of equations (2) and (15) shows for $a_s \tilde{~} a_e \tilde{~} a$, a linear dependence of σ_f on $a^{-1/2}$ requires a constant plastic zone size to characterize the fracture toughness. The comparison gives the result

$$K_{Ic} = (8s/\pi)^{1/2}\sigma_y\ . \tag{16}$$

The same result is obtained from equation (14b) by using the condition that $K_{Ic}^2 = 4M\gamma_p$.

A direct utility of equation (15) is to demonstrate the effect of the plastic zone size s on modifying the linear $a^{-1/2}$ dependence of σ_f as (σ_f/σ_y) increases, even to the extent of approaching 1.0. This consideration has been demonstrated in one way for results to be expected for PMMA (polymethylmethacrylate) material by Armstrong [32]. The same consideration is shown in Figure 6 for the dependence of σ_f on $a^{-1/2}$ which is computed from very accurate measurements of K_{Ic} at different a values for tempered 4340 steel as determined by Jones and Brown [33]. Also shown in Figure 6 is the σ dependence on $a^{-1/2}$ which was obtained for PMMA material in the study of Williams and Ewing [34]. The greater toughness of steel material relative to PMMA is established in the figure by the increased slope of the steel result which corresponds to a larger value of the plastic zone size s.

The influence of the polycrystal grain size on the fracture toughness of a material as expressed in K_{Ic} should follow a Hall-Petch dependence because of the yield stress factor in equation (16). This dependence has been verified by Stonesifer and Armstrong [35] for the ferrite grain size dependence of a number of steel results [36,37] and for prior austenite grain size dependence of A533 B steel material. Figure 7 indicates this Hall-Petch dependence may apply for the fracture toughness of several aluminum alloys in sheet form as reported by Hahn and Rosenfield [38] from a study of Thompson, Zinkham and Price [39]. A Hall-Petch dependence has been established for the yield strength of recrystallized 7075 aluminum alloy by Waldman, Sulinski and Markus [40]. In this latter study, the grain diameter was measured perpendicular to the rolling direction in a longitudinal direction. The k_y value for the material of 0.16 MPa$\cdot$m$^{1/2}$ compares favorably with the range of values $0.12 < k_y < 0.19$ MPa$\cdot$m$^{1/2}$ reported in an analysis of previous studies by Armstrong [41]. Thompson and Zinkham [42] have indicated for the results in Figure 7, plus others, that the microstructures for these alloys ranged from being coarse equiaxed grains to being highly elongated and almost completely unrecrystallized grains - though the grain size through the thickness of the sheet was able to be correlated with the fracture toughness of the alloys. A quantitative evaluation of the fracture toughness dependence on grain size in Figure 7 does indicate that the plastic zone size should have to increase as the average grain size is reduced for these materials. A similar result is indicated for the high temperature fracture toughness measurements for

A533 B steel material which have been obtained by Stonesifer [5]. One consideration for this result is that the post-yield fracture mechanics description proposed by Heald, Spink and Worthington [43] might apply in these cases so that the value of σ_f should be employed in place of σ_y in equation (16) and, correspondingly, the grain size dependence of K_{Ic} should be greater, as measured, for example, by k_f or k_f^*.

A special application of equations (15) and (16) may be to describing the fracture toughness dependence on grain size for polycrystalline ceramic materials. This occurs because these materials fracture brittlely due to the presence of cleavage cracks within the microstructure even at stress levels near to those required for general plastic yielding. Armstrong [41] has suggested that the curvature due to the dual effect of the plastic zone size in equation (15) may be responsible for the bimodal fracture stress dependence on grain size frequently reported for these materials. Bradt, Dulberg and Tressler [44] have interpreted the effect of surface finishing on the strength - grain size dependence of MgO material to be explicable in terms of an equation similar to equation (15). The measurement of an increased fracture energy as the grain size is reduced has been reported for $A\ell_2O_3$ by Simpson [45]. An increased fracture energy has been measured by Ahlquist [46] as the yield strength is increased for alkali halide materials.

Finally, it should be noted that despite the differences which have been described for the models of fracture on the macroscale versus the microscale there is one more connection to be made between the various equations describing the operative mechanics of the cracking processes. This is best illustrated, perhaps, by combining equation (16) with $K_{Ic}^2 = 4M\gamma_p$ to give

$$\sigma_y^2 s = (\pi/2)M\gamma_p \ . \tag{17}$$

This result for the fracturing energy involved in the failure of a brittle material is to be compared with equation (5) from Cottrell for the fracture transition as rewritten in the form

$$\sigma_y^2 d \ [1 - (\sigma_{oy}/\sigma_y)] = C\mu \ \gamma_p \ . \tag{18}$$

Putting aside comparative questions about the strain rate and triaxiality of the stress state, the equations are nicely matched by considering for the onset of brittleness in a crack-free material that the slip length was entered into equation (3) as a measure of the dislocation displacement involved in forming a cleavage crack. The presence of σ_{oy} is due to the finite friction resistance which is attributed to the average shear strength of individual crystals. For the limiting case of brittleness to be obtained for a crack containing material, then, it might be imagined according to equation (17) that the value of s could be reduced to a dimension of the order of the polycrystal grain diameter as described for equation (18). The decrease in K_{Ic} for steel as the temperature decreases is well-known. The explanation for it is produced by the dramatic decrease in s which must occur to override the appreciable increase in σ_y accompanying the decrease in temperature. A lower limiting size for the micro-cleavage events signally the onset of brittle fracture at the crack tip of pre-cracked specimens has been estimated by Ritchie, Knott and Rice [47] as occurring at a distance of two average grain diameters ahead of the crack tip. Stonesifer and Cullen [48] have obtained striking confirmation of this estimate for the fracture toughness testing of A533 B steel at 77°K - hence the connection of equations (17) and (18) seems reasonable on the basis of microstructural observations, also. The implication that the most brittle of cleavage fractures are controlled by events occurring locally within the microstructure of the material, on the scale of several grains at most, must contribute something to the observed scatter of experimental results in Figure 1.

ACKNOWLEDGEMENTS

This research has been supported by the U.S. Office of Naval Research and by the Department of Mechanical Engineering at the University of Maryland.

REFERENCES

1. RÉAUMUR, R. A. F. de, L'Art de Convertir le Fer Forge en Acier, Paris, 19722, see SMITH, C.S., Sources for the History of the Science of Steel: 1532-1786, Society for the History of Technology and the MIT Press, Mass., 1968, p. 125.
2. PETCH, N. J., J. Iron and Steel Inst., 1974, 153, 25.
3. MADHAVA, N. M., Ph.D. Dissertation, Univ. of Maryland, 1975.
4. SACK, R. A., Proc. Phys. Soc. London, 58, 1946, 729.
5. STONESIFER, F. R., Ph.D. Dissertation, Univ. of Maryland, 1975.
6. IRWIN, G. R., Trans. ASM, 40, 1948, 147.
7. OROWAN, E., Rept. Prog. Phys., 12, 1948-49, 214.
8. PRASAD, Y.V.R.K., Post-doctoral Research, University of Maryland, 1973-74.
9. COTTRELL, A. H., Trans. TMS-AIME, 212, 1958, 192.
10. PETCH, N. J., Phil. Mag. 3, 1958, 1089,
11. STROH, A. N., Advanc. Phys. 6, 1957, 418.
12. HALL, E. O., Proc. Phys. Soc. London, B64, 1951, 797.
13. HESLOP, J. and PETCH, N. J., Phil. Mag. 1, 1956, 866.
14. PETCH, N. J., Phil. Mag. 1, 1956, 186.
15. HESLOP, J. and PETCH, N. J., 1958, 1128.
16. ARMSTRONG, R. W., Phil. Mag., 9, 1964, 1063.
17. ARMSTRONG, R. W., Brown University Report E38, 1967; Met. Trans. 1, 1970, 1169.
18. LESLIE, W. C., SOBER, R. J., BABCOCK, S. G. and GREEN, S. J., Trans. ASM, 62, 1969, 690.
19. BUCHER, J. H. and GOODENOW, R. H., Met. Trans. 1, 1970, 2344.
20. GUPTA, I., Met. Trans. 2, 1971, 323.
21. LESLIE, W. C., Met. Trans. 2, 1971, 1989.
22. SASAKI, G. and YOKOTA, J. M., Met. Trans. 6A, 1975, 586.
23. ARMSTRONG, R. W. and STONESIFER, F. R., in preparation, 1977.
24. ARMSTRONG, R. W., Acta Met. 16, 1968, 347.
25. PRASAD, Y.V.R.K., MADHAVE, N. M. and ARMSTRONG, R. W., Grain Boundaries in Engineering Materials, Fourth Bolton Landing Conference, 1974, p. 67.
26. PSZONKA, A., Scripta Met., 8, 1974, 81.
27. DUGDALE, D. S., J. Mech. Phys. Sol. 8, 1960, 100.
28. BILBY, B. A., COTTRELL, A. H. and SWINDEN, K. H., Proc. Roy. Soc. London, A272, 1963, 304.
29. KEER, L. M. and MURA, T., Proc. First Intern. Conf. on Fracture, 1, Sendai, 1965, 99.
30. YOKOBORI, T., Intern. J. Fract. Mech. 4, 1968, 179.
31. BILBY, B. A., Dritte Intern. Tagung über den Bruch, 11, Munich, 1973, 1.

32. ARMSTRONG, R. W., Dritte Intern. Tagung uber den Bruch, 3, 1973, 421.
33. JONES, M. H. and BROWN, W. F., Jr., Rev. Develop. Plane Strain Fract. Tough. Test., ASTM STP 463, 1970, 63.
34. WILLIAMS, J. G. and EWING, R. D., Fracture 1969, Chapman and Hall, 11/1.
35. STONESIFER, F. R. and ARMSTRONG, R. W., Fracture 1977, ICF4.
36. YOKOBORI, T., KAMEI, A. and KOGAWA, T., Dritte Intern. Tagung uber den Bruch, 1, 1973, 431.
37. CURRY, D. A. and KNOTT, J. F., Met. Sci. 10, 1976, 1.
38. HAHN, G. T. and ROSENFIELD, A. R., Met. Trans. 6A, 1975, 653.
39. THOMPSON, D. S., ZINKHAM, R. E. and PRICE, C. W., Tech. Rept. AFML, TR-74-129, Part 1, 1974, 174.
40. WALDMAN, J., SULINSKI, H. and MARKUS, H., Met. Teans. 5, 1974, 573.
41. ARMSTRONG, R. W., Canad. Met. Quart. 13, 1974, 187.
42. THOMPSON, D. S. and ZINKMAN, R. E., Eng. Fract. Mech. 7, 1975, 389.
43. HEALD, P. T., SPINK, G. M. and WORTHINGTON, P. J., Mat. Sci. and Eng. 10, 1972, 129.
44. BRADT, R. C., DULBERG, J. L. and TRESSLER, R. E., Acta Met., 24, 1976, 529.
45. SIMPSON, L. A., J. Amer. Ceram. Soc. 56, 1973, 7 and 610.
46. AHLQUIST, C. N., Acta Met., 22, 1973, 1133.
47. RITCHIE, R. O., KNOTT, J. F. and RICE, J. R., J. Mech. Phys. Sol. 21, 1973, 395.
48. STONESIFER, F. R. and CULLEN, W. H., Met. Trans. 7A, 1976, 1803.

Table 1 - Brittle Fracture Stress Investigations for Iron and Steel Materials

SYMBOL	Material and Remarks	Testing Temperature °K	Reference
O	Mild Steel, 0.1 w/o C.	4.2	N. M. Madhava, Ph.D. Thesis, Md., 1975.
O	Fe-3% Si, 0.037 w/o C.	77	D. Hull, Acta Met., 9, (1961), 191.
□	Manganese Steel, 0.15 w/o C, 1.8 w/o Mn.	60 to 140	F. de Kazinczy and W.A. Backofen, Trans. ASM, 53, (1961), 55.
▷	Mild Steel, 0.155w/o C.	77	N.J. Petch, Progress in Metal Physics, 5, 1954, p. 1; J. Iron and Steel Institute, 174, (1953),25.
▽	Ingot Iron, 0.036 w/o C.	77	Ibid.
◁	Spectrographic Iron, 0.07 w/o C.	77	Ibid.
⊕	Iron, 0.03 w/o C.	77	J.R. Low, Progress in Material Science, 12,1963, p.1
⟶	Rimmed Steel, 0.07 w/o C.	78	J.R. Low, "Relation of Properties to Microstructure", ASM, Metals Park, Ohio, 1954, p. 163.
◇	Rimmed Steel Decarburized; Intercrystalline Fracture	78	Ibid.
◊	Mild Steel, 0.2 w/o C.	77	F.S. Deronja and M. Gensamer, Trans. ASM, 51, (1959), 666.
⟷	Mild Steel, 0.2 w/o C.	4.2	Ibid.
⌂	Steel M, 1.3 w/o Mn, 0.16 w/o C.	77	G.T. Hahn, M. Cohen and B.L. Averbach, J. Iron and Steel Institute, 200, (1962), 634.
▷	Steel E, 0.22 w/o C, 0.36 w/o Mn.	77	G.T. Hahn, M. Cohen, B.L. Averbach and W.S. Owen, "Fracture", 1959, (New York, John Wiley), p. 91; J. Iron and Steel Institute, 200, (1962), 634.
▽	Steel F2, 0.039 w/o C.	77	G.T. Hahn, B.L. Averbach, W.S. Owen and M. Cohen, "Fracture", 1959, (New York, John Wiley), p. 91.
◁	Steel X-52, 0.26 w/o C, 1.15 w/o Mn	77	A.R. Rosenfield and G.T. Hahn, Trans. ASM, 59, (1966), 962.
△	Armco Iron, 0.022 w/o C.	128 to 183	E.A. Almond, D.H. Timbras and J.D. Embury, "Fracture 1969", (London, Chapman and Hall), p. 253.
⊞	EN 2 Steel, 0.15 w/o C; Yielding Preceded Fracture.	77	D. Hull and I.L. Mogford, Phil. Mag., 3, (1958),1213.
■	High Purity Quenched Iron, 0.0025 w/o C; Intercrystalline Fracture	77	C.E. Richards, C.N. Reid and R.E. Smallman, Trans. Japan Inst. Metals, 1968, Vol. 9 Supplement, p.961
▼	Decarburised NPL Iron, 0.018 w/o C; Intercrystalline Fracture	77	E.A. Almond, D.H.Timbres and J.D. Embury, Phil. Mag., 23, (1971), 971.
▲	Decarburised Armco Iron; Intercrystalline Fracture.	218	Ibid.

$\sigma = 351.5 + 3.30d^{-1/2}$ MPa ; d in m.

$\sigma = \left[\pi E v/(1-v^2) \right]^{1/2} d^{-1/2}$

A.R. Rosenfield and G.T. Hahn, Trans. ASM, 59, (1966), 962.

R.A. Sack, Proc. Phy. Soc. (London)58, (1946),729.

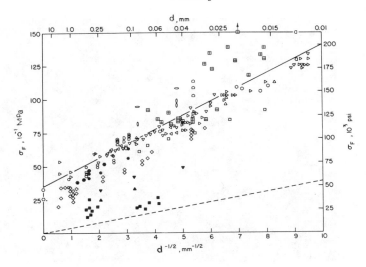

Figure 1 - Hall-Petch dependence for the brittle fracture stress
of iron and steel materials, after Madhava (1975).

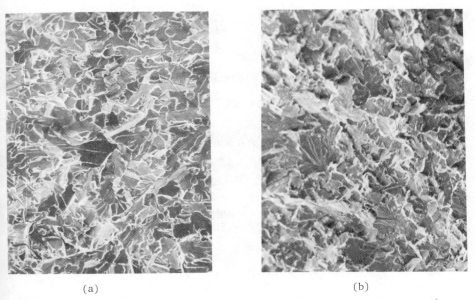

(a)   (b)

Figure 2 - Scanning electron micrographs of polycrystal cleavage surfaces
for two steel materials:
(a) tensile fracture at 4.2°K, after Madhava (1975); and
(b) fracture mechanics test at 77°K, after Stonesifer (1975).

Figure 3 - Cleavage cracks within polycrystalline zinc compressed at
4.2°K, after Prasad (1974).

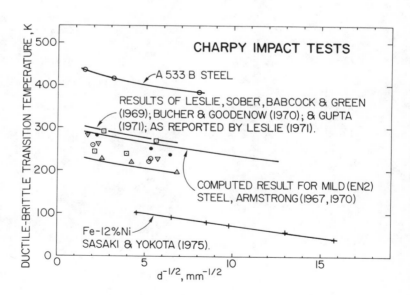

Figure 4 - Ductile-brittle transition temperature dependence on grain
size for Charpy impact testing of steel materials, after
Stonesifer and Armstrong (1977).

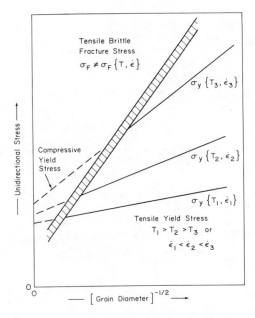

Figure 5 - The tensile ductile-brittle transition behavior of hcp
materials according to a Hall-Petch analysis, after
Armstrong (1968).

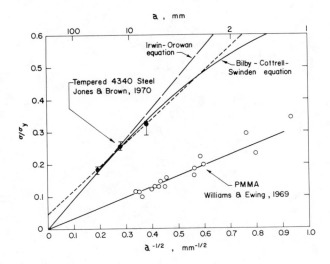

Figure 6 - The fracture stress dependence on crack size for steel,
after Jones and Brown (1970), and for PMMA, after Williams
and Ewing (1969).

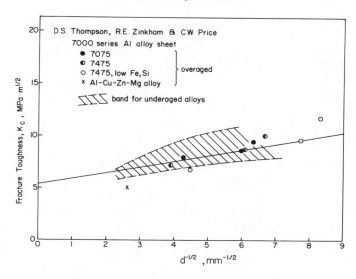

Figure 7 - The fracture toughness dependence on polycrystal grain size for variously treated aluminum materials; results of Thompson, Zinkham and Price (1974).

STRESS CORROSION CRACK GROWTH FOR SHORT CRACKS WITH
PARTICULAR REFERENCE TO THE ZIRCALOY-2/IODINE SYSTEM

B. Tomkins and J. H. Gittus*

INTRODUCTION

A study of the behaviour of short cracks (< 1 mm) in various fracture
fields involving slow stable crack extension e.g. (creep, fatigue) is of
increasing interest. In the development of fracture studies, it provides
a link between crack nucleation processes and long crack fracture mechanics
analyses. In practice, the life of many high quality components (e.g.
nuclear fuel cladding, gas turbine blades) is dependent on the behaviour
of small cracks and defects. Now short crack growth often involves high
stresses with accompaning plastic straining and the problems of elastic-
plastic or fully plastic fracture analyses are well known. However recent
developments such as the use of the J contour integral and earlier concepts
such as crack tip opening displacement have improved our ability to
examine the short crack problem particularly in relation to fatigue [1, 2].
A current problem in water reactor fuel cladding has precipitated a study
of the behaviour of short cracks in a stress corrosion situation. The
particular cladding alloy, Zircaloy 2 (a zirconium, 1.5% tin alloy), is
susceptible at reactor operating temperature ($\sim 300°C$) to failure during
a rapid uprating of the fuel power in the presence of gaseous iodine
fission products [3]. As the maximum cladding thickness is of order
0.6 mm, failure involves only initiation and short crack growth.

This short paper examines one possibility for the mechanics of the onset
of stress corrosion crack extension from a small crack and its subsequent
growth. The results are applied to the behaviour of cracked Zircaloy
fuel cladding during a power ramp transient.

CRACK TIP OPENING FOR SHORT CRACKS

The crack tip opening displacement (δ) of long cracks in a predominantly
elastic field (small scale yielding) is well characterised. For the
idealised B-C-S crack model where crack tip plasticity is confined to
discrete slip lines radiating from the crack tip it is possible to
estimate δ up to general yield (Y), as,

$$\delta = \frac{2AYa}{\pi E} \ell n \ [\sec \ (\beta\sigma)] \qquad\qquad (1)$$

for an edge crack (of length a) in a semi infinite plate subjected to an
applied tensile stress σ. A $\sim$ 1 for a plane strain crack with plastic
relaxation by two 45° slip lines at the tip, and $\beta = \pi/2Y$. For a work
hardening material, it is possible to extend this model for stresses

*Reactor Fuel Element Laboratories, UKAEA, Springfields Works, Salwick,
 Preston, Lancashire, England.

exceeding yield if Y is increased to the ultimate stress T and a variable modulus $d\sigma/d\varepsilon$ is used for E [4, 5]. The resulting δ has an elastic and plastic term,

$$\delta = \frac{Aa}{\beta_1 E} \ell n \ [\sec \ (\beta_1\sigma)] + Aa\varepsilon_p \left[\frac{\beta_1\sigma}{(n+1)} + \frac{(\beta_1\sigma)^3}{3(3n+1)} + \ldots \right]$$

$$\approx \frac{Aa}{\beta_1 E} \ell n \ [\sec \ (\beta_1\sigma)] + \frac{Aa\varepsilon_p}{(n+1)} \tan \ (\beta_1\sigma) \tag{2}$$

for small work hardening exponent (n, < 0.1). $\beta_1 = \pi/2T$. For higher n values and $(\sigma/T) > 0.8$, a good approximation is,

$$\delta = Aa \ \ell n \ [\sec \ (\beta_1\sigma)] \left[\frac{1}{\beta_1 E} + \frac{2\varepsilon_p}{(n+1)} \right] \tag{3}$$

Equations (2) and (3) show that any plastic strain (ε_p) contributes approximately twice as much to crack opening as would an equal elastic strain.

Now mechanistically, the achievement of a given crack opening whether by elastic or plastic straining involves the same processes of very localised flow at the crack tip [2]. Hence the crack tip state for a given opening should be the same in terms of dislocation structure and density, and if a critical crack tip situation is involved in a fracture process, it should apply to a whole range of short as well as long crack sizes. In this regard it is becoming increasingly clear that the Stage I/Stage II transition in fatigue crack growth occurs in both low stress and high strain fatigue when the correct crack opening condition is achieved. In stress corrosion crack growth, K_{1scc} is such a crack state parameter. In terms of the equivalent crack opening.

$$\delta_{1scc} = \frac{K_{1scc}^2}{4EY} \tag{4}$$

$$= \frac{Aa}{\beta_1 E} \ell n \ [\sec \ (\beta_1\sigma^*)] + Aa\varepsilon_p^* \left[\frac{(\beta_1\sigma^*)}{1+n} + \ldots \right] \tag{5}$$

from equation (2) where σ^* and ε_p^* are the critical applied stress and plastic strain for stress corrosion crack extension from a short crack. The need for a significant degree of plastic straining is consistent with the well known conditions of a high stress (σ^*), of order Y, for the maintenance of stress corrosion failure in smooth specimens.

This argument does not indicate why a particular value of crack opening should be required but that once it has been achieved, crack extension can occur. In corrosion fatigue studies, there is some evidence that the threshold for enhanced growth, the equivalent to K_{1scc}, is the Stage I/Stage II transition i.e. a mechanistic change at the crack tip in the accommodation of a given crack opening [6]. In stress corrosion, δ_{1scc} could be related to a wider variety of structural sizes (e.g. grain size) or microstructural events (e.g. localised microcracking which occurs at $K \sim 15 \ MNm^{-3/2}$ in high strength steels).

THE GROWTH OF SHORT STRESS CORROSION CRACKS

Because stress corrosion crack extension above K_{1scc} is stable up to the much higher instability fracture toughness condition (K_{1c}), it must be possible to exceed or at least maintain the critical crack opening condition δ_{1scc}. In the long crack low stress situation this is easily achieved because the crack extension force increases with increasing crack length, even for constant end displacement.

However, for a short crack, the maintenance of δ_{1scc} is dependent on continued plastic straining. From the second half of equation (2) it can be seen that the plastic strain needed decreases in proportion to the increase in crack length for a constant value of σ^*.

An upper limit on plastic strain needed for growth is given by the condition of continuous crack extension where,

$$\Delta a = \Delta\delta/2 \tag{6}$$

Now, $$\Delta\delta \approx \alpha a \; \Delta\varepsilon_p \tag{7}$$

Hence for crack extension from a_o to a_f, as ε_p increases from ε_{po} to ε_{pf},

$$(\varepsilon_{pf} - \varepsilon_{po}) = \frac{2}{\alpha} \ln\left(\frac{a_f}{a_o}\right) \tag{8}$$

Now, $$\alpha \approx \delta_{1scc}/\varepsilon_{po} \; a_o \tag{9}$$

and hence, $$\varepsilon_{pf} = \frac{2a_o\varepsilon_{po}}{\delta_{1scc}} \ln\left(\frac{a_f}{a_o}\right) \tag{10}$$

for $\varepsilon_{pf} \gg \varepsilon_{po}$ and $a_f \gg a_o$.

Such crack extension would result in large ductility and crack openings comparable with crack length. This is not typical of most stress corrosion cracking failures and certainly not of those which are dangerous in service i.e. low ductility failures. This indicates a discontinuous crack advance process, in which the crack grows by several times the critical crack opening, once the initiation condition has been achieved. A further plastic strain increment is then needed to attain this condition again. A reasonable assumption for such a process is that the increment of crack advance is proportional to the critical crack opening (δ_{1scc}). This is consistent with the fact that for a crack, the important linear dimensions are proportional to crack length, including δ and the plastic zone size. Then, equation (6) becomes,

$$\Delta a = m \; \delta_{1scc} \tag{11}$$

The process is shown schematically in Figure 1. After i steps, the crack length a_i is given by,

$$a_i = a_o + i\,m\,\delta_{1scc} \approx a_o\,(1 + i\,m\,\alpha\,\varepsilon_{po}) \tag{12}$$

and the increment of plastic strain ε_{pi} needed for the ith step is,

$$\varepsilon_{pi} = \varepsilon_{po}\,a_o/a_i \tag{13}$$

For failure after r steps, the failure strain ε_{pf} is given from equations (12) and (3) as,

$$\varepsilon_{pf} = \sum_1^r \varepsilon_{pi} = \varepsilon_{po}\left[1 + \frac{1}{1 + m\alpha\varepsilon_{po}} + \frac{1}{1 + 2\,m\alpha\varepsilon_{po}} + \cdots\right] \tag{14}$$

Figure 2 shows how $\varepsilon_{pf}/\varepsilon_{po}$ varies with m for a material where $\delta_{1scc}/a_o = 0.5$, $a_o = 10$ μm (a typical initiated crack size) and $a_f = 400$ μm. It can be seen that the ductility is reduced to 2 ε_{po} if m is 5.5.

APPLICATION TO ZIRCALOY-2 CRACKING IN IODINE VAPOUR

The failure of Zircaloy-2 water reactor fuel cladding by stress corrosion in the presence of fission product iodine vapour is a good example of a displacement limited failure involving only short crack growth. Cladding thickness is of the order 0.6 mm and failures are known to occur following rapid power upratings (time period $\sim$ 30 mins) involving cladding stresses of order Y ($\sim$ 480 MN/m^2). The cladding is forced out by the oxide fuel during the uprating to a total hoop strain of order 1% (i.e. a plastic strain of order 0.6%). Immediately after the uprating stress relaxation will result in a further plastic strain increment of approximately 0.25%. Assuming an initial crack of 10 μm, can such strain increments overcome the initiation condition given by δ_{1scc}? At present it is difficult to know what the effective value of σ/T is as $\sigma \rightarrow Y$, but it will $\rightarrow$ 1.0. Therefore, although equation (2) cannot be used without knowing σ/T, it is clear that a bound on the problem is penetration of the flow zones across the section. In this case, the simpler equation

$$\delta \approx 2\varepsilon_p \cdot t \tag{15}$$

can be used where t is the clad thickness. When combined with equation (4), this gives

$$\varepsilon_p^* = \frac{K_{1scc}^2}{8\,ETt} \tag{16}$$

and taking current values for K_{1scc} (32 MNm$^{-3/2}$ - ref (7)) and T (550 MN/m^2), $\delta_{1scc} = 4.2$ μm and $\varepsilon_p^* = 0.35\%$. This plastic strain is well within the values expected during the uprating.

However, for failure to occur a further increment of strain is needed. Figure 2 indicates that an initiated crack could be pushed to 2/3t by an additional 0.5% plastic strain if m = 4, i.e. the crack advance increment averages 16.8 μm. This is 2-3 times the typical grain size for Zircaloy-2 and is therefore consistent with the cleavage mechanism observed.

SUMMARY

This initial study of the crack opening displacement of short cracks in a high stress field has indicated that it provides a means of comparing low stress, long crack extension with short crack behaviour. Early calculations in relation to the Zircaloy-2 cladding problem indicate that its failure is consistent with a K_{1scc} concept.

ACKNOWLEDGEMENT

The authors acknowledge the value of discussions with their colleague Dr. H. Hughes.

REFERENCES

1. DOWLING, N. E., in "Cracks and Fracture", ASTM STP 601, 1976, 19.
2. TOMKINS, B., Phil. Mag., 18, 1968, 1041.
3. COX, B. and WOOD, J. C., in "Corrosion Problems in Energy Conversion", (ed. C. S. Tedman, The Electrochemical Soc.), 1974, 275.
4. HEALD, P. T., SPINK, G. M. and WORTHINGTON, P. J., Materials Science, 10, 1972, 129.
5. TOMKINS, B., Trans. ASME - J. Eng. Matls. and Tech., 97, No. 4, 1975, 289.
6. TOMKINS, B., Proc. Conf. on the Influence of Environment on Fatigue, I. Mech. E., May 1977.
7. WOOD, J. C., J. Nucl. Mat., 45, 1973, 105.

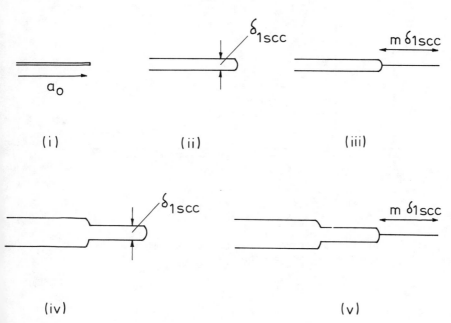

Figure 1 - The discontinuous crack advance process for stress corrosion crack growth

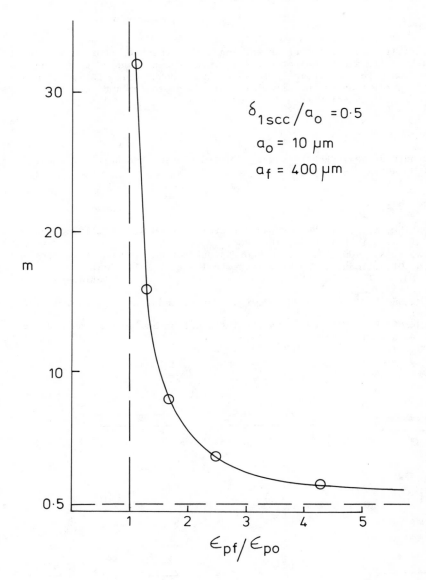

Figure 2 - Ratio of fracture strain to initiation strain as a function of crack advance coefficient

CREEP-FATIGUE INTERACTION FAILURE IN TYPE 316 STAINLESS STEEL

J. Wareing and B. Tomkins*

INTRODUCTION

Most fatigue failures in metals and alloys at both ambient and elevated temperature occur by the initiation and subsequent propagation of a dominant surface nucleated crack. For smooth specimens and relatively undefected components in the limited life region above the endurance limit, initiation is often rapid in terms of final endurance although the initiated crack (length a_0) is very small (of order 10 μm). Endurance N_f is then an integration of crack growth rate, da/dN, from a_0 to a final crack size a_f at which unstable ductile or brittle fracture occurs. Now, a_0 and a_f are both structure sensitive but on a different structural scale, a_0 is defined materially by dislocation substructure and is also influenced by surface environment e.g. microscale corrosion effects. a_f however is related to the material macro-defect structure, e.g. second phase particles, inclusions, which dominate fracture. The third parameter determining N_f, da/dN, may or may not be structure sensitive but is dependent on material mechanical properties (σ_y, n) and also applied stress-strain field (σ, ε_p) and environment.

Now endurance at elevated temperature, is much more sensitive to the types of cycle imposed than at ambient temperature. Variations in strain rate and the introduction of dwell periods can have order of magnitude effects on endurance. This has led to difficulties in attempting to draft design codes for elevated temperature power plant e.g. ASME Code Case 1592, on a similar basis to low temperature codes such as ASME III in relation to fatigue endurance.

The present paper shows briefly how cycle shape affects the three parameters which determine endurance at elevated temperature for one material, type 316 stainless steel. The principles involved, however, apply to a wider range of alloys.

FATIGUE CRACK GROWTH AT ELEVATED TEMPERATURE FOR CONSTANT STRAIN RATE CYCLES

It is now well established that fatigue cracks propagate incrementally by shear decohesion at the crack tip during the crack opening process [1, 2]. Hence the means by which cracks accommodate the plastic opening demanded, sometimes termed plastic blunting, is the means by which they advance. If one examines the equations for crack tip opening (δ) under elastic and fully plastic applied stress-strain fields a general equation can be derived with the form,

*Reactor Fuel Element Laboratories, UKAEA, Springfields Works, Salwick, Preston, Lancashire, England.

$$\delta \approx \frac{Aa}{\beta E} \ln [\sec(\beta\sigma)] + \frac{Aa\,\varepsilon_p}{(n+1)} \tan (\beta\sigma) \tag{1}$$

for small values of work hardening exponent n [3]. A is a constant (~ 1 for plane strain crack opening), $\beta = \pi/2T$ (where T is the material flow stress) and σ and ε_p are the applied stress and plastic strain. For small values of σ/T, equation (1) reduces to,

$$\delta \approx \frac{\beta\sigma^2 a}{2E} + \frac{\beta\sigma\varepsilon_p a}{(n+1)} \tag{2}$$

Now the general equation for fatigue crack growth rate [3] has elastic and plastic components similar to equation (2), viz.

$$\frac{da}{dN} = \frac{A_1 \Delta K^m}{E f(T)} + A_2 \left(\frac{\Delta\sigma}{2T}\right)^2 \frac{\Delta\varepsilon_p a}{(2n + 1)} \tag{3}$$

A comparison between equations (2) and (3) shows some similarity in form between elastic and plastic components but da/dN is usually only a fraction of the CTOD, δ.

Now in the plastic straining region of fatigue failure, where most creep-fatigue studies have been made and cycle wave shape effects investigated [4], it is the plastic component of equation (3) which dominates crack growth. For a given applied strain range, the crack growth rate is sensitive to the ratio ($\Delta\sigma/T$). In earlier work on a 20 Cr/25 Ni stainless steel, Wareing et al [5] showed that this ratio was increased by increasing temperature and decreasing strain rate at elevated temperature. The result in endurance terms was a considerable decrease, reflecting the crack growth rate effect. The increase in strain rate sensitivity of T at elevated temperature is a result of stress relaxation creep processes which operate at moderate and low strain rates.

Figure 1 shows the effect of cycle strain rate on the fatigue endurance of type 316 stainless steel tested at a plastic strain rate of 1.43% at 625°C. It can be seen that the endurance is approximately halved when the strain rate drops below 10^{-4}/s. Fractographic evidence showed that the striation spacing, which at this strain level corresponds to the crack growth increment, to crack depth ratio increased by a corresponding factor of two as the cyclic strain rate decreased. This is also shown in Figure 1. It is worth noting also that at the strain rates below the transition, the crack path was increasingly intergranular but striations of the expected size were found on grain boundary facets. This indicates that at low strain rates grain boundaries represent a weaker although more constrained crack path with a lower effective flow stress, T. There was no evidence that intergranular failure occurred ahead of the current crack front. Woodford and Coffin [6] have noticed similar crack path behaviour in the A286 alloy at 593°C.

In this type of cycle involving a constant strain rate, the growth rate and hence endurance is determined by the material flow properties. An upper limit on growth rate occurs when $\Delta\sigma/2 \to T$ and work hardening is no longer effective in limiting crack opening and hence extension - the work hardening equivalent to general yield in perfectly plastic material. The crack growth rate is then given by the plastic displacement applied to the cracked section (W - a) and,

$$\frac{da}{dN} = \frac{\Delta\epsilon_p W}{(W - a)} (W - a) \tag{4}$$

which is independent of crack length. Integrating between a_a and a_f gives a lower bound on endurance as,

$$N_f = \frac{(a_f - a_o)}{\Delta\epsilon_p W} \tag{5}$$

All the work to date on type 316 stainless steel for the symmetrical constant strain rate cycle indicates that endurance variations are related primarily to crack growth rate variations and a_f and a_o are essentially constant.

CYCLES INVOLVING VARIABLE STRAIN RATES

In early work on type 304 stainless steel at 650°C, Berling and Conway [7] discovered that the introduction of some imbalance into a cycle, e.g. by means of a dwell period at maximum tensile strain, could lead to a reduction in endurance for an otherwise high strain rate cycle. They also found that the reduction in endurance was primarily due to the formation of grain boundary cracks or cavities in the bulk material into which the crack was propagating. These observations have since been confirmed by many experimenters on various high temperature alloys, particularly austenitic stainless and ferritic steels. The bulk cavitation seems to be a result of tensile creep deformation during stress relaxation in the dwell period which is reversed by rapid plastic straining elsewhere in the cycle. Several stress relaxation periods are needed to generate significant cavitation [5].

Fractography and metallographic sections have shown that the overall failure process is complex but follows a three phase pattern shown in Figure 2, which is a schematic section through a failure. The first phase is simple fatigue with insufficient cavitation damage developed to affect either growth rate or crack path. Striation observations in this phase have confirmed that the crack growth rate is just that expected for the constant higher strain rate of the main tensile plastic strain increment. In other words, no decrease in endurance would occur for the imbalanced loop if all growth were of the phase I type. However, after a time phase I is succeeded by phase II growth where cavitation ahead of the crack front predetermines the crack path as a grain boundary path and influences the growth rate. No direct measurements have been made of phase II growth rates. Finally, phase III intervenes to give premature unstable fracture of cavitated material. In terms of the three parameters mentioned as those determining endurance, in this type of failure a_o is unaffected, a_f decreased and da/dN increased, but only during phase II.

In earlier papers, Tomkins [3] and Wareing [8] examined a simple lower bound endurance based on a maximum crack growth rate given by equation (4) and a fixed reduced a_f of 0.8W. For this bound it was assumed that no phase I growth existed and that all growth was by phase II through material which was cavitated enough to satisfy the displacement criterion given by equation (4). This does represent a lower bound if a correct value for a_f is chosen but it is most pessimistic at lower strain levels. It also lacks some reality in that for the effective flow stress, $(T = T(1 - (p/L)^2)$ where p/L is the ratio of cavity size : spacing) to be reduced significantly p/L must be large which will only apply after a considerable accumulated creep strain.

Recently, Raj, Tomkins and Wareing [9] have proposed a bound based on phase I growth and a final crack size a related to cavity spacing. This bound is based on the fractographic evidence that phase I growth rates are not significantly affected by small amounts of bulk cavitation. This is consistent with the small effect on T of a modest p/L ratio. It is also observed that phase I terminates when the crack opening displacement, δ, equals the cavity spacing L. Once this criterion has been satisfied, cavity linkage can occur readily along the flow bands radiating at ±45° from the crack tip where displacement falls relatively slowly [2, 10]. Thus phase II growth is rapid and the effective a_f is given by the crack length at the end of phase I. This is given by equation (1) with δ = L.

Now in practice L is related in type 316 stainless steel to grain boundary carbide particle spacing, as these particles provide the main cavity nucleation sites. Two batches of type 316 steel were tested with an un-balanced cycle involving a tensile dwell period. These batches had an average particle spacing of 8 μm (Batch I) and 2 μm (Batch II). Figure 3 shows how the endurances of these two batches varied and how they related to predictions of endurance based on phase I growth and the δ = L criterion for a_f. Also included in Figure 3 are some data of Brinkman and Korth [11], which showed particularly severe hold time effects. The particle spacing for the material used in the latter tests is estimated at 1 μm on the basis of this criterion.

It is thought that continued examination of failure patterns in such un-balanced cycle, creep-fatigue situations will enable more rational rules to be developed for design based on true failure criteria.

SUMMARY

A study of some elevated temperature cycles on type 316 stainless steel has shown that an understanding of the three main parameters involved in failure development can lead to rational explanation of endurance vari-ations and provide a more realistic basis for engineering design in cyclic situations.

ACKNOWLEDGEMENTS

The authors are indebted to T. Topping and H. G. Vaughan for experimental assistance.

REFERENCES

1. LAIRD, C., Spec. Tech. Publ. 415, Amer. Soc. Test. Mater., 1967, 131.
2. TOMKINS, B., Phil. Mag., 18, 1968, 1041.
3. TOMKINS, B., J. Eng. Matls. and Tech., 97, 1975, 289.
4. COFFIN, L. F. Jr., Proc. Inst. Mech. Engrs., 188 9/74, 1974, 169.
5. WAREING, J., TOMKINS, B. and SUMNER, G., Spec. TEch. Publ. 520, Amer. Soc. Test. Mater., 1973, 123.
6. WOODFORD, D. A. and COFFIN, L. F. Jr., Fourth Bolton Landing Confer-ence, June 9-12, 1974 (to be published in the Proceedings).

7. BERLING, J. T. and CONWAY, J. B., Proc. of the First Int. Conf. on Pressure Vess. Tech. Part II, Delft, Holland, ASME, New York, 1970, 1233.
8. WAREING, J., Paper (76-84-E) Met. Trans. to be published.
9. RAJ, R., TOMKINS, B. and WAREING, J., unpublished work.
10. LANKFORD, J. and KUSENBERGER, F. N., Phil. Mag., 26, 1972, 1485.
11. BRINKMAN, C. R., KORTH, G. E. and HOBBINS, R. R., Nucl. Tech., 16, 1972, 299.

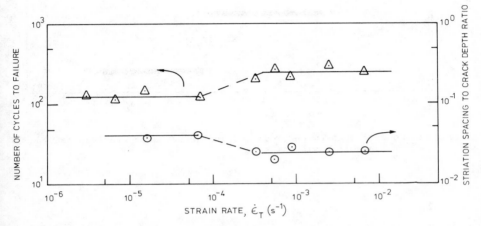

Figure 1   Relationship between fatigue endurance, striation spacing to crack depth ratio and cyclic strain rate for type 316 stainless steel at 625°C and plastic strain range 1.43%.

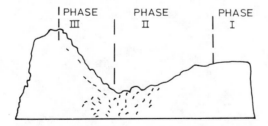

Figure 2   Schematic representation of the three phases of crack growth which occur during creep-fatigue failure.

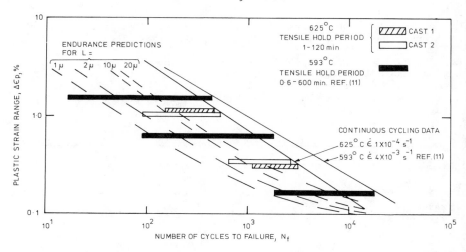

Figure 3   Comparison between predicted and actual creep-fatigue endurances
of type 316 stainless steel

MACROSCOPIC CREEP CRACK GROWTH

V. Vitek*

1. INTRODUCTION

Large engineering components often contain flaws or develop macroscopic
cracks during service. These defects may start to grow at high temperature
and thus cause a catastrophic failure even though the nominal strain in
the component may be small. This type of creep crack growth has to be distinguished from creep fracture which occurs as the final stage of tertiary
creep. In the latter case cavities and/or microcracks are formed more or
less homogeneously throughout the whole body and final rupture takes place
by linking up of these defects. On the other hand the creep crack growth
considered in this paper is due to creep deformation concentrated only near
the tip of a macroscopic crack and fracture occurs by creep assisted propagation of this crack. Extensive experimental studies of both these phenomena
have been carried out in the last years. However, whilst the mechanisms of
creep fracture have been extensively studied theoretically (e.g., [1] - [6]),
only some empirical correlations describing macroscopic creep crack growth
have been suggested [7 - 15]. In the present paper these correlations are
first discussed and their applicability assessed. A recently developed
fracture mechanics theory of creep crack growth based on the time dependent
Dugdale-Bilby-Cottrell-Swinden type model is then reviewed and suggestions
for further experimental studies of creep crack growth are made.

2. EMPIRICAL LAWS DESCRIBING CREEP CRACK GROWTH

The experimental studies show that the process of creep crack growth usually
consists of an incubation period and growth period [8, 9, 16, 17]. Depending
on the material and testing conditions the incubation period during which
creep damage develops ahead of the crack, may represent a substantial part
of the specimen life. Nevertheless, the incubation stage has been studied
so far very little and most of the authors concentrated on the growth period
with the aim to establish a correlation between the rate of crack growth,
da/dt, and various macroscopic parameters. The following three correlations
have been most commonly pursued:

(i) The rate of crack growth is assumed to be determined at a given
temperature by the stress intensity factor, K, [7 - 10] according to the
equation

$$\frac{da}{dt} = A \ K^m \ ,$$

(1)

where A is a constant and m an integer. However, this type of correlation
is usually obtained only for rather limited ranges of applied stresses and
K values and extrapolation to, for example, a wider range of K values are

---

* Materials Division, Central Electricity Research Laboratories, Kelvin
  Avenue, Leatherhead, Surrey, United Kingdom

doubtful. There is, of course, no fundamental reason for equation (1).

(ii) The rate of crack growth has been correlated in [11], [12] and [13] with a path independent integral C* obtained from the J integral (which is commonly used in the deformation fracture mechanics) by replacing strains and displacements by their time derivatives. This integral is not the time derivative of J and no physical meaning can be assigned to this quantity. Hence, though C* may perhaps be a useful engineering correlating parameter, it cannot provide any deeper insight into the mechanics of creep fracture.

(iii) Net section stress has been proposed in [14] and [15] to be a suitable correlating parameter. However, if this is so it means that the principle effect of the crack is to modify the stress field in the whole ligament, e.g., by decreasing its cross-section, but not to concentrate the creep strain near the crack tip. This situation does not correspond to the true creep crack growth and it is better described as creep rupture process which occurs under rather complicated stress conditions. The apparent crack growth then corresponds to the necking of the ligament.

In the following we shall consider only the case when crack growth occurs due to localised creep deformation. Hence, the rate of crack growth will principally be determined by the stress intensity factor K, although not necessarily given by equation (1). Since the damage is localised, a fracture mechanics type approach may be developed.

3. FRACTURE MECHANICS TYPE THEORY OF CREEP CRACK GROWTH

The principle problems are (a) to set up macroscopic criteria governing the crack growth and (b) to take into account time dependent stress relaxation ahead of the crack. In the deformation fracture mechanics with a well defined yield stress a very successful model of plastic zone was developed by Dugdale [18] and Bilby, Cottrell and Swinden [19], hereafter called the DBCS model. Recently, a time dependent analogy of this model has been developed by Vitek [20, 21]. The plastic zone has been represented by an array of edge dislocations coplanar with the crack the Burgers vectors of which are perpendicular to the plane of the crack, similarly as in [19]. However, unlike in the DBCS model no fixed friction stress acting upon the dislocations was assumed but the strain rate, $\dot{\varepsilon}$, ahead of the crack was determined by the equation

$$\dot{\varepsilon} = A(\tau/G)^n ,\qquad (2)$$

where τ is the local tensile stress, A is a constant, n is an integer and G is the shear modulus. It was assumed that equation (2) is the same as the creep law describing the creep behaviour of an uncracked specimen subject to the same tensile stress τ. The time dependent development of the dislocation density, B(x,t), describing the plastic zone, takes place according to the equation

$$\frac{\partial B(x,t)}{\partial t} = - h \frac{\partial \dot{\varepsilon}(x,t)}{\partial x}\qquad (3)$$

where t is the time and h a gauge length independent of the applied stress, time and crack length, which has to be introduced if the concept of strain is to be used in the framework of a one-dimensional model.

The numerical solutions of equation (3) for a given applied tensile stress, σ, and different values of n have been presented in [20] and [21]. With increasing time a region of uniform stress develops ahead of the crack. This region extends with time and the stress decreases. Thus at any time the stress distribution is similar to that assumed in the DBCS model. Furthermore, the dislocation distributions found by solving equation (3) are very similar to those given in (19) if the uniform stress ahead of the crack is taken as an apparent friction stress. Owing to this result all the analytical formulae of the DBCS model can be readily used replacing the friction stress, σ_1, by the apparent friction stress σ_1^F which is a function of both time, t, and applied tensile stress σ. In particular the crack opening displacement can be written as

$$\Phi = \frac{4(1-\nu)}{\pi} \, a \left(\frac{\sigma_1^F}{G}\right) \ln \, \sec \frac{1}{2} \pi \frac{\sigma}{\sigma_1^F} \, , \tag{4}$$

where ν is Poisson's ratio. The dependence

$$\sigma_1^F = \sigma_1^F(t,\sigma) \, , \tag{5}$$

is, of course, obtained as a result of the numerical calculation and approximate analytical expressions for equation (5) have been given in [20] and [21]. However, the time does not figure out in equation (5) explicitly but always as the dimensionless quantity

$$T = A \frac{h}{a} 10^{4-2n}t \, . \tag{6}$$

It has been suggested by Wells [22] and Cottrell [23] that attainment of a critical crack opening displacement (COD), Φ_c, is a condition for the onset of crack propagation. It has been shown in [21] that this condition is also applicable to the initiation of creep crack growth. The incubation time is then the time needed for accumulating Φ_c.

If the critical COD, Φ_c, is known the incubation time, t_i, can be calculated by inverting equations (5) and (4). The result of this calculation is, however, the dimensionless time T_i, related to t_i by equation (6), and to calculate t_i the gauge length h has to be chosen. Hence, the present theory possesses two semi-empirical parameters, the critical COD, Φ_c, and the gauge length, h, the determination of which will be discussed later.

In order to describe the crack propagation we assume that both the COD and h are constant during this process. The crack propagation can then be regarded as a sequence of repeated initiations and the following three possibilities have been studied in detail [24, 25]:

(i) The crack always travels a fixed length, d, when the critical COD is reached, eliminating the whole of plastic zone during this extension. This may correspond to cavity growth on an inclusion or to formation of a microcrack on the most favourably oriented grain boundary and subsequent linking of these defects with the main crack. d can then be identified with the average separation of inclusions or with average grain size.

(ii) The crack always extends by a distance equal to the length of the plastic zone eliminating the plastic zone during this extension. This may correspond to the case when cavities and/or microcracks form throughout the whole of the plastic zone.

(iii) The crack extends by the amount equal to the plastic opening at the crack tip, i.e., Φ_c, and only this part of the plastic zone which was contained inside this length is eliminated. This corresponds to continuous crack growth by plastic tearing.

In all three cases the rate of crack growth was found [24, 25] to be strongly dependent on K but for a given K a substantially weaker, but not negligible, dependence on σ always exists. The rate of crack growth can conveniently be expressed as

$$\frac{da}{dt} = B \ K^{\alpha} , \tag{7}$$

where both B and α are functions of the applied stress σ and their numerical values have been given in [24]. Furthermore α is different in three different ranges of values K and generally decreases with increasing K. Hence, although equation (7) is similar to equation (1) it is not the same because of the above mentioned functional dependences of B and α.

In order to apply the present theory we have to determine the two semi-empirical parameters Φ_c and h. This can be done on the basis of the following measurements:

(a) The incubation time, t_i, measured as a function of the initial crack length a_0 and applied stress σ.
(b) The rate of crack growth as a function of the current crack length, a, and of σ.

If the mode of the crack propagation has been identified a combination of measurements (a) and (b) for a limited range of a_0 and σ provides sufficient data for determination of Φ_c and h [24] when using equations (4), (5) and (7).

4. CONCLUSIONS

In the case of true creep crack growth the rate of crack propagation is principally governed by the stress intensity factor, i.e., the crack growth is due to the accumulation of the creep damage near the crack tip. However, the rate of the crack growth is not generally described by equation (1). On the other hand when the rate of crack growth can be correlated with the net section stress the main effect of the crack is to modify the stress field and thus the creep strain rate, in the whole specimen. The latter case should then be treated using a creep rupture theory.

The fracture mechanics type theory of the creep crack growth the principle results of which have been reviewed in this paper, enables both the calculation of the incubation period and of the rate of crack growth if the creep behaviour of the material described by equation (2) is known.

However, the theory contains two semi-empirical parameters - the critical COD, Φ_c, and the gauge length, h. These have to be determined experimentally as described above. Once these two parameters have been determined from measurements for a limited range of applied stress, σ, and crack length, a, the theory may be used for extrapolation to different ranges of σ and a. In particular, the measurements are usually done for relatively high stress levels so that they can be carried out in a reasonably short time. However, the application is often needed at much lower loads and the theory may be used to carry out this extrapolation provided the same mechanism of creep crack growth operates at both high and low loads.

However, most experimental data on creep crack growth cannot be analysed using the present theory. For example when K appears to control creep crack growth the measurements are usually presented as plots of da/dt vs. K but variation of this dependence with σ and the incubation period have usually not been studied. In order to use the present theory the following information is needed and has to be obtained from experiments:

(i)   Creep behaviour of uncracked specimens giving the relation (2).

(ii) Mode of creep crack growth as defined in paragraph 3.

(iii) The incubation time as a function of the initial crack length a_0 and the applied stress σ.

(iv) Rate of creep crack growth as a function of the current crack length a and of the applied stress σ but not only of the stress intensity factor K.

ACKNOWLEDGEMENTS

This work was carried out at Central Electricity Research Laboratories and is published by permission of the Central Electricity Generating Board.

REFERENCES

1. HULL, D. and RIMMER, D. E., Phil. Mag., 4, 1959, 673.
2. SPEIGHT, M. V. and HARRIS, J. E., Met. Sci. J., 1, 1967, 83.
3. RAJ, R. and ASHBY, M. F., Acta Met., 23, 1975, 653.
4. HEALD, P. T. and WILLIAMS, J. A., Phil. Mag., 22, 1970, 1095.
5. KELLY, D. A., Metal Science, 10, 1976, 57.
6. HANCOCK, J. W., Metal Science, 10, 1976, 319.
7. SIVERNS, M. J. and PRICE, A. T., Nature, 228, 1970, 760.
8. ROBSON, K., Int. Conf. on Properties of Creep Resistant Steels, Düsseldorf, 1972, Paper 4.5.
9. NEATE, G. J. and SIVERNS, M. J., Int. Conf. on Creep and Fatigue in Elevated Temperature Applications, 1973/1974, Philadelphia/Sheffield, Paper C234.
10. FLOREEN, S., Met. Trans., 64, 1975, 1741.
11. KENYON, J. L., WEBSTER, G. A., RADON, J. C. and TURNER, C. E., Int. Conf. on Creep and Fatigue in Elevated Temperature Application, 1973/1974, Philadelphia/Sheffield, Paper C156.
12. WEBSTER, G. A., Proc. Conf. on Mechanics and Physics of Fracture, 1975, Cambridge, Paper 8.
13. LANDES, J. D. and BEGLEY, J. A., ASTM Special Technical Publication, 590, 1976, 128.

14. NICHOLSON, R. D. and FORMBY, C. L., Int. J. Fracture, $\underline{11}$, 1975, 595.
15. NICHOLSON, R. D., Mat. Sci. Eng., $\underline{22}$, 1976, 1.
16. TAIRA, S. and OHTANI, R., Int. Conf. on Creep and Fatigue in Elevated Temperature Applications, 1973/1974, Philadelphia/Sheffield, Paper C213.
17. HAIGH, J. R., Mat. Sci. Eng., $\underline{20}$, 1975, 213.
18. DUGDALE, D. S., J. Mech. Phys. Solids, $\underline{8}$, 1960, 100.
19. BILBY, B. A., COTTRELL, A. H. and SWINDEN, K. H., Proc. Roy. Soc., $\underline{A272}$, 1963, 304.
20. VITEK, V., "Computer Simulation for Materials Applications", Nuclear Metallurgy, $\underline{20}$, 1976, 909.
21. VITEK, V., Int. J. Fracture 13, 1977, 39.
22. WELLS, A. A., Proc. Crack Propagation Symp., $\underline{1}$, 196, 201, Cranfield.
23. COTTRELL, A. H., Iron and Steel Inst. Special Report, No. 69, 1961, 231.
24. VITEK, V., Int. J. Fracture, 1977, to be published.
25. VITEK, V., Int. Conf. on Fracture Mechanics and Technology, 1977, Hong Kong.

VARIATIONAL BOUNDS AND QUALITATIVE METHODS
IN FRACTURE MECHANICS

R. V. Goldstein* and V. M. Entov*

INTRODUCTION

Fracture mechanics has given rise to the problem of calculating values of
stress intensity factor, K, in elasticity, i.e. the value of K at square
root type singularities of stress near a crack contour. This problem,
not simple by itself for complicated shapes of crack contour, becomes
almost irresolvable if variations in K have to be investigated for the
sequence of contours assumed by the crack as it develops (as is necessary
in an analysis of the kinetics of crack growth). Hence, a problem
naturally arises regarding the estimation of the values of K from which
it would be possible to derive sufficient conditions for fracture (or non-
failure) of a solid. The problem here is to find upper and lower bounds
for the maximum and minimum values of K for a given contour or set of
contours.

The present paper reports recent results obtained in this area, and also
some other bounds of integral characteristics of the solutions of
elasticity which are closely associated with these upper and lower bounds.

1. BOUNDS OF STRESS INTENSITY FACTORS FOR A PLANE OPENING-MODE CRACK
   IN AN ELASTIC SPACE

Consider an elastic space with a crack occupying the domain G in the plane
$x_3 = 0$ bounded by a piecewise smooth curve Γ. It may be assumed that the
external load is represented by wedging tractions symmetrical with respect
to the plane $x_3 = 0$:

$$\sigma_{33} = -q(x_1, x_2), \quad x_3 = 0 \quad (x_1, x_2) \epsilon G \tag{1.1}$$

Only normal stresses act in the plane $x_3 = 0$, while the vertical component
of displacement on this plane outside the crack is zero:

$$\tau_{13} = \tau_{23} = 0 \ (x_3 = 0), \ w(x_1, x_2) = 0 \quad (x_1, x_2)\overline{\epsilon G}$$

In the plane $x_3 = 0$ along the smooth parts of crack contour the stress
σ_{33} and displacement w have singularities of the type:

$$\sigma_{33} = \frac{N}{\sqrt{5}} \ , \quad w = \frac{4(1-\nu^2)}{E} N\sqrt{S} \tag{1.2}$$

where S is the distance along the normal from the crack contour;
$N(t)$, $(x_1(t), x_2(t))\epsilon\Gamma$ is the stress intensity factor at a given point

*Institute of Mechanical Problems, U.S.S.R. Academy of Science,
 1250 Moscow, U.S.S.R.

on the contour corresponding to the value of the parameter t.

The following *theorem of comparison* is essential for further discussion.

Suppose that there are two crack domains G and (G'<G) with contours Γ and Γ' having a common region Γ" = Γ Γ'. Assume that Γ" consists of a certain number of smooth arcs and (or) isolated points of contact of the contours Γ and Γ'. Further assume that the corresponding loads $q(x_1,x_2)$ and $q'(x_1,x_2)$ satisfy the conditions:

$$q(x_1,x_2) \geq 0 \ (x_1,x_2)\epsilon G/G', \quad q(x_1,x_2) \geq q'(x_1,x_2), \ (x_1,x_2)\epsilon G \quad (1.3)$$

Then, at those points of Γ", where the normals to the contours coincide, the stress intensity factor N for the crack occupying the domain G is not less than the stress intensity factor N' for the crack occupying the domain G':

$$N'(M) \leq N(M) \ , \ M\epsilon\Gamma" \tag{1.4}$$

The proof of this theorem and examples of its application are given in [1].

Of the corollaries to this theorem, we may mention the following assertion, which is valid within the framework of quasi-statical growth of cracks. Let there be two contours Γ_0 and Γ_0', of which the first envelops the second at the initial moment, and a system of wedging loads q and q' such that $q \geq q'$ at any instant and at every point. In this case a crack developing from the contour Γ_0 will always remain inside the bounds of a crack developing from the contour Γ_0. Hence, we have two simple, but important, conclusions:

(a) If the crack bounded at the initial instant by the enveloping contour is not critical for a given history of loading (i.e. it does not lead to failure), then the crack bounded at the initial instant by the enveloped contour is also not critical.

(b) If the enveloped crack is critical, then the enveloping crack is also.

Because of these assertions, the need to analyze complicated "irregular" curves is eliminated, and thus it is possible to restrict consideration to relatively simple crack contours. Indeed, these assertions are quite natural, and it may be assumed that they hold true for opening-mode cracks, not only in the whole elastic space, but also in bodies of other shapes. Unfortunately, this statement has so far not been proved, and we have therefore to limit ourselves to far less general statements.

It is not difficult to perceive that in all these arguments we have made use only of the property of *positiveness* rather than any particular solution of the elastic problem: positive (wedging) loads give rise to positive displacements of points on the crack surface and positive (stretching) stresses outside the crack in its plane. Thus, in order to widen the field of applicability of these assertions, it suffices to establish the positiveness of the corresponding class of elastic problems. It is natural to expect that the opening-mode crack problem will be a positive one not only for an infinite body, but also for bodies with boundaries sufficiently remote from the crack. This assertion is proved below for the particular case of a layer with a crack in the midplane.

2. OPENING-MODE CRACK IN A THICK LAYER

Suppose that the crack under consideration occupies a plane domain G of diameter d in the mid-plane $x_3 = 0$ of an elastic layer of thickness 2h. Assume that there are no tractions on the layer faces and the crack surface is acted upon by normal forces:

$$\sigma_{33} = -q(x_1, x_2) \quad , \quad (x_1, x_2) \epsilon G \tag{2.1}$$

The problem can be reduced to the following pseudo-differential equation (see Appendix A):

$$Ar = -\frac{2(1-\nu)}{\mu} q \quad , \quad (x_1, x_2) \epsilon G \tag{2.2}$$

where A is a pseudo-differential operator with symbol $A(\xi)$

$$Ar = \frac{1}{4\pi^2} \iint_{-\infty}^{\infty} A(\xi) \tilde{r}(\xi) e^{-i(x,\xi)} d\xi, \quad A(\xi) = -2\pi |\xi| + 2\pi K(h|\xi|) \tag{2.3}$$

$$K(h|\xi|) \equiv \frac{1}{h} K(h|\xi|) = \frac{2|\xi| \left(1 + 2h|\xi| + 2h^2 |\xi|^2 - e^{-2h|\xi|}\right)}{4h|\xi| + e^{2h|\xi|} - e^{-2h|\xi|}}$$

Here $\xi = (\xi_1, \xi_2)$ stands for the parameter of the Fourier transformation in (x_1, x_2) $\tilde{r}(\xi)$ is the Fourier transform of the displacement of the points on the surface in the direction of x_3, and ν, μ are the Poisson's ratio and shear modulus of the material, respectively.

Let $R_h(x_1, x_2, u_1, u_2) = R_h(x,u)$ denote the resolving operator of Eq. (2.2), so that

$$r(x_1, x_2) = \underset{G}{} R_h(x,u) q(u) du \tag{2.4}$$

For an infinite medium (h = ∞) we have $K(h|\xi|) = 0$, and the left-hand side of Eq. (2.2) contains only one term. Hence, by the assertion proved in Section 1, we find that the posi-function q(x) in the right-hand side corresponds to the positive solution of Eq. (2.2) regarded as an equation in r(x). Thus, $R_\infty(x,u) > 0$, i.e. the operator R_∞ is positive. In order to show that the operator R_h is positive for sufficiently large h, we shall construct the solution of Eq. (2.2) by iteration, assuming that

$$r_0 = R_\infty q \quad , \quad r_K = R_\infty(q + q_K) \quad , \quad K \geq 1$$

$$q_K = \frac{\mu}{2(1-\nu)} \frac{1}{(2\pi)^2} \iint_{-\infty}^{\infty} (2\pi) K(h|\xi| \tilde{r}_{K-1}(\xi) e^{-i(x,\xi)} d\xi \tag{2.5}$$

The quantitites q_K may be looked upon as fictitious additional loads which have to be applied to the surface of a crack in an infinite space so that crack opening may coincide with crack opening in this layer to the same approximation.

Clearly, positiveness of R_h will be proved if it can be shown that $(q+q_K)$ is positive for any K.

Let $q' = \lim_{K \to \infty} q_K$. From the convergence of successive approximations (see Appendix B) it follows that for sufficiently large values of the ratio h/d, the correction term in Eq. (2.5) will be close to

$$q_1 = \frac{\mu}{2(1-\nu)} \frac{1}{(2\pi)^2} \iint_{-\infty}^{\infty} (2\pi)K(h|\xi|)\overset{\sim}{r_0}(\xi)e^{-i(x,\xi)}d\xi \qquad (2.6)$$

In turn, using the smallness of h/d ,   this expression can be asymptotically represented by

$$q_1 \simeq q_1^o = \frac{\mu}{2(1-\nu)} \frac{1}{(2\pi)^2 h^3} \iint_{-\infty}^{\infty} (2\pi)K(n)\overset{\sim}{r_0}(0)dn \qquad (2.7)$$

This relationship defines in the domain   G   a constant additional load acting in the same direction as that of the initial loads applied to the crack surface. The fact that this additional load, irrespective of the initial load q distribution, does not vanish anywhere in the domain G ,   shows by itself that for sufficiently large h/d , the quantity   q'   and the sum (q_i + q')   are strictly positive in the domain G. By estimating the contribution of successive terms in the expansion in powers of d/h (see Appendix C), we find that this assertion remains valid at least for d/h<0.7 which proves that the operator R_h is positive for d/h<0.7 (i.e. for a sufficiently thick layer) and that all those assertions enunciated in Section 1 and [1] can also be applied for a thick layer with a crack. At the same time, it has been demonstrated that the opening and stress intensity factor for a crack in a thick layer may be asymptotically represented (with an error of the order of $0(d^3/h^3)$) in the form:

$$r(x) = r_0 + r_{10} \quad , \quad N = N_0 + N_{10} \qquad (2.8)$$

where r_0 is the opening of a crack under the action of given loads q in a body occupying the whole space; r_{10} is the opening of the same crack under the action of a constant load q_1^2 defined by (2.7). In the range of validity of the bounds (d/h<0.7), crack opening and stress intensity factor at each point diminish on passing from the layer to the space when the load and crack configuration remain unaltered.

3. EXTREMAL CRACK CONTOURS

3.1 The following approach may be of use in constructing the bounds of stress intensity factors.

Consider an elastic body whose plane of symmetry $x_3 = 0$ contains an opening mode crack wedged by symmetrical normal loads $q(x_1,x_2)$. Take an arbitrary closed domain G_0 (of area S_0) in the plane $x_3 = 0$. Consider the set of all domains G containing G_0 as a subdomain and having an area S, S>S_0. To each such domain we may attribute an energy W defined as the potential energy of the body having a crack occupying this domain in the given field of loads. The quantity W is a functional of the contour Γ of the domain G. Further assume that among all these contours, there exists one (or several) contour on which W attains its maximum. Such a contour is hereafter called the extremal contour. The extremal contour consists of two parts: Γ" coinciding with a part of the boundary of the domain G_0, and Γ' lying outside G_0 (the part Γ" may, in general, be absent). The main property of the extremal contour is that the stress intensity factor N is constant on the part Γ'. Indeed, along with Γ', consider an arc Γ* close to Γ' such that its end points coincide with the end points of Γ' Then Irwin's formula (see [2]) shows that the variation in energy on passing from the contour Γ"+Γ' to the contour Γ"+Γ* is

$$\sigma W = \frac{\pi(1-\nu)}{\mu} \int_{\Gamma'} N^2 (\vec{n}\sigma\vec{r}) d\ell \qquad (3.1)$$

where $(\vec{n}\sigma\vec{r})$ is the distance between the arcs Γ* and Γ' along the normal to Γ'. On the other hand, the corresponding change in the area is

$$\sigma S = \int_{\Gamma'} (\vec{n}\sigma\vec{r}) d\ell \qquad (3.2)$$

For the attainment of conditional extremum of W we should have

$$N^2 - \lambda = 0 \quad , \text{ where } \lambda = \text{const on } \Gamma' \qquad (3.2)$$

Thus, the invariance of N over Γ' is the necessary condition that the contour Γ"+Γ' may be extremal.

If the arc Γ" does not vanish, then the stress intensity factor on it is not greater than N* over the remaining part of the extremal contour. In the contrary case, indeed, by deforming the contour Γ' such that the area decreases so that the area inside Γ'+Γ" remains constant, we find that the energy increases, the area remaining constant. Using asymptotic expressions for the stresses near corner points of the contour, we find that there cannot exist angular points on the free part Γ' of the contour, at any case for N≠0 on the free part of the contour.

Suppose that we can find a sequence of embedded extremal contours Γ(S) corresponding to increasing the parameter S from S_0, and thus determine the corresponding stress intensity factors N(S) on the "free" part Γ' of the contour. Introduce a loading parmeter P, such that

$$q = Pq_0(x_1, x_2) \qquad (3.4)$$

Then for each $S > S_0$, there exists such a value P for which N = N* = const.

$$P = \frac{N^*}{N_0(S)} \quad , \quad N_0 = N(S, q_0) \qquad (3.5)$$

If N* is identified with the critical value of stress intensity factor (N*=K/π , K is the cohesion modulus of the material) at which limit equilibrium is attained at the crack edge, then the extremal contour of area is W(S). We have obviously

$$N^2 = \frac{\mu}{\pi(1-\nu)} \frac{dW}{dS} \qquad (3.6)$$

Since the true opening of a crack with a given contour can be found from the condition of minimum potential energy of an elastic body, the problem of finding the extremal crack contour can be reduced, in a general case, to the problem of finding the minimum, i.e. to finding, for a given area S, a contour enveloping the domain such that $S_0 < S$, for which $W(S)$ attains the value

$$W_0(S) = \max_{\text{mes}G=S} \quad \min_{G} W$$

3.2 We shall now give an example of an extremal crack contour. Consider an infinite medium with a crack occupying the domain G_0 in the plane $x_3=0$. Assume that the extremal loads are normal stresses acting on the crack surface:

$$\sigma_{33} = P(1+\varepsilon x_1^2) \tag{3.7}$$

where ε is small. If $\varepsilon = 0$, in (3.7), then a circular contour Γ will be the extremal contour without common points with the initial contour Γ_0, since the stress intensity factor does not vary over the contour of a penny-shaped crack under uniform wedging loads. Therefore, by virtue of the smallness of ε and symmetry of load (3.7) with respect to the axes x_1, x_2, it is natural to search for an extremal contour among elliptical contours close to circular ones. Let the ellipse be defined by the parametric equations:

$$x_1 = a_1\cos\theta \quad , \quad x_2 = B_1\sin\theta$$
$$a_1 = a(1+\sigma_1) \quad , \quad B_1 = a(1+\sigma_2) \tag{3.}$$

where σ_1 and σ_2 are small quantities to be determined.

The stress intensity factor N on an elliptical crack contour under the load (3.7) is given by the formula [4]:

$$N = -\mu\left(\frac{2}{a_1 B_1}\right)^{\frac{1}{2}}\left(a_1^2\sin^2\theta + B_1^2\cos^2\theta\right)^{\frac{1}{4}}\left[\frac{4C_{00}}{a_1 B_1} - \frac{16C_{02}\sin^2\theta}{a_1 B_1^3} - \frac{16C_{20}\cos^2\theta}{a_1 B_1^3}\right] \tag{3.}$$

where the constants C_{00}, C_{02}, and C_{20} are determined from the solution of the system of linear algebraic equations:

$$||M||\begin{pmatrix} C_{00} \\ C_{20} \\ C_{02} \end{pmatrix} = \frac{1}{2\mu}\begin{pmatrix} 1 \\ \varepsilon \\ 0 \end{pmatrix}$$

$$||M|| = \begin{Vmatrix} \dfrac{4E(k)}{a_1 B_1^2} \,, & \dfrac{8[k^{12}K(k)-(1-2k^2)E(k)]}{a_1^3 B_1^2 k^2} \,, & \dfrac{8[(1+k^2)E(k)-k^{12}K(k)]}{a_1 B_1^4 k^2} \\[3mm] 0 \,, & \dfrac{8L(k,k^1)}{a_1^5 B_1^2 k^2} & \dfrac{8U(k,k^1)}{a_1^3 B_1^4 k^4} \\[3mm] 0 \,, & \dfrac{8U(k,k^1)}{a_1^3 B_1^4 k^4} & \dfrac{8Q(k,k^1)}{a_1^3 B_1^4 k^4} \end{Vmatrix} \qquad (3.10)$$

$$L = (3-8k^2+2/k^2)E(k)-2k^{12}(2+11k^2)K(k)$$

$$U = k^{12}(2-k^2)K(k)-2(k^{12}+k^4)E(k)$$

$$Q = 2(3k^2-1)K(k)+(3k^2+\frac{2-10k^2}{k^{12}})E(k)$$

Here $K(k)$ and $E(k)$ are total elliptical integrals of the first and second kind respectively; $k^1 = 1 = B_1/a_1$, $k^{12} = B_1^2/a_1^2$. From (3.10), we find

$$C_{00} = \frac{a^3 P}{4\pi\mu} \,, \quad C_{02} = \frac{3a^7 P}{460\pi\mu}\varepsilon \,, \quad C_{z_0} = -\frac{29a^7 P}{7380\pi\mu}\varepsilon \qquad (3.11)$$

accurate to the order of ε. By virtue of (3.11) and the smallness of σ_1, σ_2, from (3.9) we obtain

$$N = \frac{P\sqrt{2a}}{\pi}\left\{\left(1-\sigma_1-\sigma_2\right)-\sin^2\theta\left(\frac{\sigma_2}{2}+\frac{3a^2\varepsilon}{230}\right)+\cos^2\theta\left(-\frac{\sigma_1}{2}+\frac{29a^2\varepsilon}{690}\right)\right\} \qquad (3.12)$$

Hence, it is seen that for

$$\sigma_1 = \varepsilon a^2\frac{29}{345} \,, \quad \sigma_2 = -\varepsilon a^2\frac{3}{115} \qquad (3.13)$$

the stress intensity factor does not change over the contour. Thus, ellipses with semi-axes given by (3.8) and (3.13) will be the extremal contours for a load of the type (3.7).

This family of extremal contours may be used in estimation of the limiting equilibrium of a crack of arbitrary configuration in a field of loads of the type (3.7). We shall construct a family of ellipses with semi-axes a_1 and B_1 given by Eqs. (3.8) and (3.13) respectively for different α. From these ellipses, take that ellipse Γ^* to which the given critical stress intensity factor $N = N^*$ corresponds. Then all the cracks enveloped by the ellipse Γ^* will be non-critical; all the cracks enveloping Γ^* will be critical. In a similar manner, as noted in 3.1, we may use the family of extremal contours constructed for any (symmetric with respect to the plane of crack) load field.

4. ENERGETIC BOUNDS OF AN INHOMOGENEOUS BODY

4.1 The following theorems on the variation of deformation energy with the elastic constants of a body will be of use in further discussion.

Consider an isotropic, and in general, inhomogeneous elastic body. A body is regarded as inhomogeneous in the sense that its elastic constants are different at different points: $a = \lambda(x_i)$; $\mu = \mu(x_i)$; or $E = E(x_i)$; $\nu = \nu(x_i)$. Let $S = S_0 + S_1 + \ldots S_i$ be its boundary, where S_0 is the outer boundary, while $S_1, \ldots, S_i$ are the inner boundaries (cavities, cracks, etc.) Assume that the external loads act only on the part S' of the surface S, while the displacements on the remaining portion S'' are fixed:

$$\vec{\sigma}_n(\vec{x}) = f(\vec{x}) \quad , \quad \vec{x} = (x_1, x_2, x_3) \epsilon S' \tag{4.1}$$

$$\vec{U}(\vec{x}) = g(\vec{x}) \quad , \quad \vec{x} \epsilon S'' \tag{4.2}$$

and that the body is in equilibrium under the action of these loads. The following theorem holds true:

Theorem 4.1: On increasing (decreasing) the elastic constants λ and (or) μ or the Young's modulus E of the material at any region of the body, the quantity

$$Q = \iint_{S''} \vec{f}(\vec{x}) \vec{U} d\sigma - \iint_{S''} \vec{\sigma}_n(\vec{x}) \vec{g}(\vec{x}) d\sigma \tag{4.3}$$

does not diminish (increase).

Proof:

Consider the potential energy of a system in equilibrium [5]:

$$W(\vec{U}) = \int_V A d\tau - \int_{S''} \vec{\sigma}_n(\vec{x}) \vec{U}_s(\vec{x}) d\sigma \tag{4.4}$$

where $\vec{U}$ is the displacement field corresponding to the equilibrium state of the body, and integration is taken over the volume of the body V, $\vec{U}_s$ is the displacement of points on the surface S, and A is the elastic energy density.

$$A = \frac{1}{2}[\lambda(\varepsilon_{11} + \varepsilon_{22} + \varepsilon_{33})^2 + 2\mu(\varepsilon_{11}^2 + \varepsilon_{22}^2 + \varepsilon_{33}^2 + 2\varepsilon_{12}^2 + 2\varepsilon_{23}^2 + 2\varepsilon_{31}^2)] \tag{4.5}$$

Diminish the elastic constants λ and (or) μ in some domain V^1 (with the boundary S^1) to a value λ_1 and (or) μ_1 without altering the displacement field. Evidently here the first positive term in (4.4) will diminish, therefore,

$$W(\vec{U}) \geq W_1(\vec{U}) \tag{4.6}$$

where $W_1(\vec{U})$ is the potential energy of the deformed body that corresponds to the displacement field $\vec{U}$. Note that the states with displacement field $\vec{U}$ are admissible [5] for a deformed elastic body, i.e. the displacements $\vec{U}$ take the same values on S'' as the displacements $\vec{U}_1$ corresponding to the equilibrium state under the previous loads (4.1). By virtue of the minimum potential energy principle, we have

$$W_1(\vec{U}) \geq W_1(\vec{U}_1) \tag{4.7}$$

For the equilibrium states of the initial and deformed bodies, by the Clapeyron formula [5], we have

$$W(\vec{U}) = \frac{1}{2}\left[-\iint_{S''} \vec{f}(x)\vec{U}_s(x)d\sigma + \iint_{S''}\vec{\sigma}_n(x)\vec{g}(x)d\sigma\right]$$

$$W_1(\vec{U}_1) = \frac{1}{2}\left[-\iint_{S''} \vec{f}(x)\vec{U}_{19}(x)d\sigma + \iint_{S''}\vec{\sigma}_{n_1}(x)\vec{g}(x)d\sigma\right] \tag{4.8}$$

where $\vec{\sigma}_n(\vec{x})$ and $\vec{\sigma}_{n_1}(\vec{x})$ are the tractions developed on that part of the surfaces of the initial and deformed bodies on which displacements are given.

Due to (4.8), from (4.4) and (4.6), it follows that

$$Q = \iint_{S''} \vec{f}_n(\vec{x})\vec{U}_s(\vec{x})d\sigma - \iint_{S''}\vec{\sigma}n(\vec{x})\vec{g}(\vec{x})d\sigma \leq$$

$$\iint_{S''}\vec{f}_n(\vec{x})\vec{U}_{15}(x)d\sigma - \iint_{S''}\vec{\sigma}_{n_1}(x)\vec{g}(x)d\sigma = Q_1 \tag{4.9}$$

Evidently, on interchanging the initial and deformed bodies, we find that Q does not increase with the increasing constants λ and (or) μ. Thus, the theorem has been proved.

Corollary 1

On the whole surface, if the external loads (S = S'') are given or zero displacements $(\vec{g}(\vec{x}) = 0)$ are defined on a part of the surface S'', then from (4.9) it follows that

$$\iint_{S'}(\vec{f})(\vec{x})\vec{U}_s(\vec{x})d\sigma \leq \iint_{S'}\vec{f}\vec{U}_{15}d\sigma \tag{4.10}$$

i.e. the work done by the external forces does not diminish with the decreasing of elastic constants λ and (or) μ in an arbitrary domain of the body.

Corollary 2

If there are no loads on a part S of the boundary, i.e. $\vec{f} = 0$, or displacements are given on the whole surface (S = S''), then from (4.9), we have

$$\iint_{S''}\vec{\sigma}_n(x)\vec{g}(x)d\sigma \geq \iint_{S'}\vec{\sigma}_{n_1}(x)\vec{g}(x)d\sigma \tag{4.11}$$

i.e. the integral of stresses with weight $\vec{g}(\vec{x})$ over that part of the surface where the displacements are defined does not increase with decreasing of the elastic constants λ and (or) μ in an arbitrary part of the body.

From this corollary it immediately follows that an increase (decrease) in Young's modulus gives rise to an increase (or decrease) in Q, as the constants λ and μ are proportional to Young's modulus.

Notes: 1: A few cases are mentioned for which Theorem 4.1 holds true and its proof is practically unchanged.

1. Moments but not forces are applied to the surface of the body.

2. The body consists of several parts in contact and there is no friction at the contact surface.

3. The body is composite and total adhesion exists at the contact surfaces. If partial adhesion or a slipping condition is substituted for total adhesion, then Q increases.

2: As the problem (4.1), (4.2) is linear, its fields of stresses and displacements can be represented in the form of the sum $\sigma_{ij}=\sigma_{ij}^{f}+\sigma_{ij}^{g}$, $U_{i}=U_{i}^{f}+U_{i}^{g}$ where the superscripts f and g denote the solutions of the elasticity problem under the following boundary conditions:

a) $\vec{\sigma}_{n}(\vec{x})=\vec{f}(\vec{x})$, $\vec{x}\varepsilon S''$, $\vec{\sigma}_{n}(\vec{x})=0$, $\vec{x}\varepsilon S''$

b) $\vec{\sigma}_{n}(\vec{x})=0$, $\vec{x}\varepsilon S''$, $\vec{U}(\vec{x})=\vec{g}(\vec{x})$, $\vec{x}\varepsilon S''$

respectively. With the help of this partition, the expression (4.3) for Q may be rewritten as follows:

$$Q = \iint_{S'}\vec{f}(\vec{x})\vec{U}^{f}(\vec{x})\,d\sigma - \iint_{S''}\vec{\sigma}_{n}^{g}(\vec{x})\vec{g}(\vec{x})\,d\sigma$$

where account has been taken of the fact that, by virtue of the reciprocity theorem, we have

$$\iint_{S'}\vec{f}(\vec{x})\vec{U}^{g}(\vec{x})\,d\sigma = \iint_{S''}\vec{\sigma}_{n}^{f}(\vec{x})\vec{g}(\vec{x})\,d\sigma$$

Thus, the quantity Q is the difference between the work done by external forces in the problem (a) and the quantity which may be called the work of pre-straining which has to be applied so that the displacement defined in the problem (b) may be attained on S''.

4.2 In the case of a crack subjected to uniform internal pressure P, the work done by external forces is, evidently, equal to the product of pressure P and the volume of opened crack (or increment in volume due to deformation if the crack was opened initially). Therefore, from the theorems proved above, it follows that a decrease in the rigidity of a body in some of its parts leads to an increase in the increment of the volume of internal crack in the body under a given internal pressure, whereas an increase in rigidity leads to a decrease in the volume increment*.

---

*Hereafter, the term "volume of crack" is used everywhere, and it is implici: that the crack is closed in the absence of stresses.

For an infinite body, in the case of uniformly distributed loads acting normal to the crack plane, circles are obviously the extremal free contours and the energy of a body with a crack equals half the product of stress and crack volume. Hence, it follows that the volume of a crack occupying an arbitrary two-dimensional domain G of area S, opened by internal pressure P does not exceed the volume of a circular crack of the same area. This inequality

$$V \leq \frac{P}{E} \frac{16(1-\nu^2)}{3\pi^{3/2}} S^{3/2}$$

is equivalent to the inequality known for the case of capacitance of a plane domain [6].

5. APPLICATION OF ENERGY BOUNDS TO PROBLEMS IN FRACTURE MECHANICS

In certain cases, the energy bounds given in Section 4 are directly applicable to problems of fracture mechanics, i.e. to the estimation of stress intensity factors and assessment of the possibility of failure of a cracked body. We shall give some examples.

5.1 Consider an axisymmetric piecewise homogeneous body having the symmetry plane $x_3=0$ normal to the symmetry axis ($x_1=x_2=0$). In the plane $x_3=0$ let there be a penny-shaped crack of radius R with centre at the origin. Let uniform wedging loads act on the crack surface. Let $d_1,\ldots,d_i$ denote the characteristic dimensions* of the body, and $\nu_1,\ldots,\nu_K$ and $E_1,\ldots,E_K$ be the elastic constants of its homogeneous parts. As the problem is linear, from dimensional considerations it follows that the total potential energy of a body with a crack is

$$W = \frac{P^2 R^3}{E_1} \Phi \left(\frac{d_1}{R},\ldots,\frac{d_K}{E_1}, \frac{E_2}{E_1},\ldots,\frac{E_K}{E_1}, \nu_1,\ldots,\nu_K \right)$$

$$= \frac{P^2 R^3}{E_1} \Phi \left(\xi_m, \frac{E_n}{E_1}, \nu_t \right) \tag{5.1}$$

m = 1,...,i
n = 1,...,K-i
t = 1,...,K

where the Young's modulus of that part containing the crack edge is denoted by E_1.

Differentiating (5.1) with respect to R, we obtain

$$\frac{\partial W}{\partial R} = \frac{3P^2 R^2}{E_1} \Phi \left(\xi_m, \frac{E_n}{E_1}, \nu_t \right) - \frac{P^2 R}{E_1} \sum_{m=1}^{i} \frac{\partial \Phi}{\partial \xi_m} dm \tag{5.2}$$

---

*Here in considering the boundaries of an axisymmetric body or cavity it is assumed that the characteristic dimensions $d_1,\ldots d_i$ are so chosen that they do not diminish under all possible axisymmetric expansions of the body or cavity.

We shall distinguish two particular cases:

$$1. \quad \partial\Phi/\partial\xi_m > 0, \quad m = 1,\ldots,i \tag{5.3}$$

$$2. \quad \partial\Phi/\partial\xi_m < 0, \quad m = 1,\ldots,i \tag{5.4}$$

Note that the signs of $(\partial\Phi/\partial\xi_m)$ for many practically interesting cases can be directly determined using Theorem 4.1.

In the first case, obviously we have

$$\frac{\partial W}{\partial R} \leq \frac{3P^2 R^2}{E_1} \Phi \frac{3W}{R} \tag{5.5}$$

and in the second case

$$\frac{\partial W}{\partial R} \geq \frac{3P^2 R^2}{E_1} \Phi \frac{3W}{R} \tag{5.6}$$

According to Irwin's formula

$$N^2 = \frac{E_1}{2(1-\nu_1^2)\pi} \frac{1}{2\pi R} \frac{\partial W}{\partial R} \tag{5.7}$$

Hence, by virtue of (5.5) and (5.6) we obtain

$$N^2 \leq \frac{3E_1}{4(1-\nu_1^2)\pi^2} \frac{W}{R^2}, \quad \frac{\partial\Phi}{\partial\xi_m} > 0$$

$$\tag{5.8}$$

$$N^2 \geq \frac{3E_1}{4(1-\nu_1^2)\pi^2} \frac{W}{R^2}, \quad \frac{\partial\Phi}{\partial\xi_m} < 0$$

Thus, the upper bounds of stress intensity factor for the first case and the lower bounds for the second case can be derived either from the energy W, or what is more important, from the upper and lower bounds of W respectively, which may be derived from Theorem 1* We shall illustrate these with the help of examples:

1. Consider a space with an axisymmetric cavity surrounded by a crack of radius R in the plane $x_3 = 0$, assuming that a pressure P acts in the cavity and on the crack. Let d_1 denote the diameter of the cavity in the plane of the crack. From Theorem 4.1 it follows that W increases with the increasing cavity size for a fixed R.

---

*The closeness of the estimated values of $\partial W/\partial R$ in (5.5) and (5.6) to the true values depends on the ratio between the retained and discarded terms in (5.2). Inequalities (5.5) and (5.6) become equalities for a penny-shaped crack in a homogeneous space. It is not difficult to show that the terms discarded from (5.2) will be small compared to the retained terms if $R/d_m \ll 1$ or $R/d_m \gg 1$. In these cases, the bounds of stress intensity factor may be expected to be very accurate. In other cases, the bounds are only a rough approximation.

Therefore, in this case we have $\partial\Phi/\partial\xi_m > 0$ and (5.8) can be used to estimate the stress intensity factor. Note that by virtue of Theorem 1, we have

$$\Phi(\xi_1,\ldots,\xi_m) \leq \Phi(1,\xi_2,\ldots,\xi_m) \quad \text{where} \quad W(R,1,\xi_2,\ldots,\xi_m,P,E) = \frac{P^2 R^3}{E}\, \Phi(1,\xi_2,\ldots,\xi_m)$$

is the energy of an elastic space with a cavity obtained by any axisymmetric expansion of the initial cavity in such a way that the cavity intersection contour with the $x_3 = 0$ coincides with the crack contour. Thus, from (5.8) we have

$$N^2 = \frac{3E}{4(1-\nu^2)\pi^2 R^2}\, W(R,1,\xi_2,\ldots,\xi_m,P,E) \tag{5.10}$$

In particular, for a spherical cavity, from (5.10) we have

$$N^2 \leq \frac{3E}{4(1-\nu^2)\pi^2 R^2}\, W(R,1,P,E) = \frac{3P^2 R}{4\pi(1-\nu)} \tag{5.11}$$

where $W(R,1,P,E)$ is the energy of a space with a spherical cavity of radius R. For $\nu = 0.25$, we find that $N \leq 0.56 P\sqrt{R}$, and for $\nu = 0.5$, we have $N \leq 0.7 P\sqrt{R}$. Note that for a penny-shaped crack of radius r_0,

$$N = \frac{\sqrt{2r_0}}{\pi}\, P \approx 0.45\, \sqrt{r_0}\ .$$ Thus, a crack of radius R which protrudes out of a spherical cavity is less hazardous than a penny-shaped crack of radius $r_{ef} = \frac{3\pi}{8(1-\pi)} R$ (for example, $r_{ef} = \frac{1}{2}\pi R$ when $\nu = 0.25$).

2. Under the assumptions made at the beginning of 1, consider a homogeneous body of finite dimensions with a penny-shaped crack. For this case, Theorem 1 gives that $\partial\Phi/\partial\xi_m < 0$, and thus the lower bound of N can be deduced from (5.9). Consider a sphere with centre at the origin and diameter equal to the diameter of the initial body and a right circular cylinder, its generator being parallel to the symmetry axis. The cylinder envelops the sphere. From Theorem 1 it is evident that

$$W \geq W_s\left(R,\frac{d_1}{R},P,E,\nu\right) \geq W_c\left(R,\frac{d_1}{R},P,E,\nu\right) \tag{5.12}$$

where W_s and W_c are the energies of the sphere and the cylinder with the same crack, respectively. From (5.9), therefore it follows that

$$N^2 \geq \frac{3E}{4\pi^2(1-\nu^2)} \frac{W_s}{R^2} \geq \frac{3E}{4\pi^2(1-\nu^2)} \frac{W_c}{R^2} \tag{5.13}$$

The quantitites W_s and W_c can be calculated from the solutions of the axisymmetric problem for a crack in a sphere [7] and a cylinder [8], respectively. Using (5.13) we shall estimate the stress intensity factor N_s on the contour of a penny-shaped crack in a sphere under no load.

Table 1 lists the exact values of N_s/N_∞ taken from [7], and estimated values corresponding to the middle and right-hand side terms in (5.13) denoted by X_{Bs} and X_{Bc}, and the values calculated from the data reported

Table 1

R/d_1	0.5	0.7	0.8
N_s/N_∞ [7]	1.156	1.32	1.47
X_{Bs} [7]	1.08	1.17	1.25
X_{Bc} [8]	1.03	1.12	1.21

in [7,8]*.

From the table it is seen that the estimated values are quite close to the true values. For $R/d_1 < 0.5$, the agreement, as would be expected, is even better.

   3.   In an infinite body with elastic constants E_1 and ν_1, let there be a spherical inclusion of radius ρ with centre at the origin in perfect adhesion with the medium; let the elastic constants of the inclusion be E_2, ν_1. Assume that a circular crack, with constant pressure P acting on its surface, envelopes the inclusion in the plane $x_3=0$ and occupies the domain $\rho \leq r \leq R$.

If the inclusion is more rigid than the matrix $E_2 > E_1$, then we have $\partial\Phi/\partial\xi < 0$ $(\xi = \rho/R)$, and the stress intensity factor on the contour of the crack, N_{R_1}, is bounded from below by (5.9). Clearly, the quantity

$$W\left(\frac{\rho}{R}, \frac{E_2}{E_1}, P, \nu_1\right) \geq W\left(\frac{\rho}{R}, \frac{E_2}{E_2}, P, \nu_1\right)$$ in (5.9) is the potential energy of a homogeneous body with an annular crack and elastic constants equal to the constants of the inclusion. Therefore, from (5.9) we have

$$N_{R_1}^2 \geq \frac{3E_1}{4\pi^2(1-\nu_1^2)R^2} W\left(\frac{\rho}{R}, \frac{E_2}{E_2}, P, \nu_1\right) \tag{5.14}$$

If the inclusion is less rigid than the matrix, $E_2 < E_1$, then the stress intensity factor N_{R_2} is found similarly from (5.8) as

$$N_{R_2}^2 \leq \frac{3E_1}{4\pi^2(1-\nu_1^2)R^2} W\left(\frac{\rho}{R}, \frac{E_2}{E_2}, P, \nu_1\right) \tag{5.15}$$

The quantity $W\left(\frac{\rho}{R}, \frac{E_2}{E_2}, P, \nu_1\right)$ can be calculated from the data reported in

---

*We may mention that Fig. 1 in [8] does not exactly specify the scale for the vertical axis. The corresponding reduction coefficients can easily be established with the help of formulae (3.10), (3.11) and the data listed in Table 1 published in [8].

[9], which gives the numerical results of solution of the problem of a flat annular crack in a homogeneous elastic space. For example, for $v_1 = v_2 = 0.3$ and $R = 1$, $\rho = 0.7$, according to [9]:

$$N^2_{R_1} \geq \frac{E_1}{E_2} P^2 0.027 \quad , \quad N^2_{R_2} \leq \frac{E_1}{E_2} P^2 0.027 \tag{5.16}$$

4. Let the piecewise homogeneous body described in Section 1 be an infinite body with an inclusion in perfect adhesion with the body. A penny-shaped crack is wholly contained within the inclusion. The elastic constants of the inclusion are E_1, v_1, and of the surrounding medium are E_2, v_2. If $E_1 > E_2$, then from (5.9) we may estimate the stress intensity factor from below. Using Theorem 1 we may write the following sequence of inequalities:

$$W \geq W_s \left(R, \frac{\rho}{R}, \frac{E_2}{E_1}, v_1 \right) \geq W_c \left(R, \frac{\rho}{R}, \frac{E_2}{E_1}, v_1 \right) \geq W_\infty \left(R, \infty, E_1, v_1 \right) \tag{5.17}$$

where $W_s (R, \rho/R) E_2/E_1, v_1)$ is the total potential energy of the body with spherical inclusion of radius equal to the diameter of the initial $W_c \left(R, \frac{\rho}{R}, \frac{E_2}{E_1}, v_1 \right)$ is the potential energy of a body containing a cylindrical inclusion, its generator being parallel to the axis of symmetry of the body and W_∞ is the energy of space with constants E_1, v_1 and containing the same penny-shaped crack. Substituting (5.17) into (5.9), for $E_1 > E_2$ we obtain

$$N^2 \geq \frac{3E_1}{4\pi^2 (1-v_1^2) R^2} \frac{W_s}{} \geq \frac{3E_2}{4\pi^2 (1-v_1^2) R_c^2} \frac{W_c}{} \geq \frac{3E_1}{4\pi^2 (1-v_1^2) R^2} \frac{W_\infty}{} = N^2_\infty \tag{5.18}$$

Here N_∞ is the exact value of stress intensity factor for a penny-shaped crack in a space with constants E_1, v_1. Similarly, for an inclusion less rigid than the matrix $E_1 < E_2$, by virtue of (5.8), we obtain a system of inequalities of opposite sense:

$$N^2 \geq \frac{3E_1}{4\pi^2 (1-v_1^2) R^2} \frac{W_s}{} < \frac{3E_1}{4\pi^2 (1-v_1^2) R^2} \frac{W_c}{} \leq \frac{3E_1}{4\pi^2 (1-v_1^2) R^2} \frac{W_\infty}{} = N^2_\infty \tag{5.19}$$

We may mention here one qualitative implication that follows from (5.18) and (5.19):

In a space with a penny-shaped crack, if the rigidity is decreased (increased) outside some axisymmetric domain containing a crack, the stress intensity factor on the crack contour decreases (does not increase).

5. Consider a cylinder containing a penny-shaped crack in perfect adhesion with a rigid medium and under conditions of sliding contact at the boundary. Then

$$N^2_{c(r,c)} \leq \frac{3E_1}{4\pi^2 (1-v_1^2) R^2} \frac{W_{c(r,c)}}{} \leq \frac{3E_1}{4\pi^2 (1-v_1^2) R^2} \frac{W_{c(s,c)}}{} \tag{5.20}$$

107

where $N_{c(r,c)}$, $W_{c(r,c)}$ are the stress intensity factor and total energy for the case of perfect adhesion of the cylinder to the medium and $W_{c(s,c)}$ is the total energy for the case of sliding contact at the interface. The values of $N_{c(r,c)}$ are reported in [10], while the values of $W_{c(s,c)}$ in [11]. In particular, for $R/d_1 = 0.6$, $N_{c(r,c)}/N_\infty = 0.87$, and $(N_{c(r,c)}/N_\infty)_B = 1.14$. It is clear that the agreement between the estimated and true values becomes worse as R/d_1 approaches unity. This is natural because in a problem with sliding contact, the value of N will increase without bounds as $R/d_1 \to 1$, while in the case of perfect adhesion there will be no such effect. Nonetheless, it should be borne in mind that the elasticity problems for a piecewise homogeneous body under sliding contact at the boundary are solved in far simpler manner than the corresponding problems under perfect adhesion at the boundaries. Therefore, the bounds of stress intensity factor for $R/d_1 \to 1$ in perfect adhesion problems can be obtained using the energy determined from the solution of the simpler problem with sliding at the boundaries.

5.2 If a crack grows being in the state of limit equilibrium, or at least there exists such a value of stress intensity factor N_* that the crack growth can be disregarded for $N < N_*$, then it is possible to derive the necessary conditions for a crack to be non critical directly from the energy bounds. Indeed, by Irwin's formula, to N_* there corresponds some value

$$T_* = \frac{\pi(1-\nu^2)}{E} N_*^2 \qquad (5.21)$$

which is the minimum necessary work that has to be done to increase the area of the crack by unity.

Then, evidently, an increase in the crack area from S_o to some value S_1 should call for an energy expenditure not less than $T_*(S_1 - S_o)$.

Therefore, if we could demonstrate that the external forces and potential energy of deformation do not provide the necessary amount of energy at some stage of crack growth, that would naturally imply that the initial crack is not critical.

Suppose that the external loads are specified. Then it is natural to construct the corresponding energy diagram for the total energy. Let the initial crack contour Γ_o of area S_o be given. If the crack is dangerous, then as the crack grows under invariable loads, the area S will increase and the total energy will take the value $W(S)$; $W(S_o) = W_o$ If we could show that

$$\min_{S_o \le S_{c\infty}} [W(S) - W_o - T_*(S - S_o)] < 0 \qquad (5.22)$$

then it would be a sufficient condition for the crack to be non critical. Of course, a sufficient condition is also given by a more approximate inequality in which $W(S)$ is replaced by some majorant, and W_o by some lesser quantity. This gives a means to make use of the energy bounds. In order to estimate the lower bound of the energy W_o, we may either take a more rigid part of the body, or replace the contour Γ_o by a simpler contour Γ_* enclosed in it, or use both these two approaches.

The concepts of extremal contours can be used in finding the upper bound of W(S). From the already proved inequality W(S)$\leq$ W*(S), where W*(S) is the energy of the body with the crack, whose contour coincides with that of an extremal one of area S. If we modify the body, say by decreasing the rigidity of the material at some of its subdomain, then the corresponding function of energy of extremal contours W_1*(S) will majorize the function W*(S):

$$W_1^*(S) \geq W^*(S) \qquad (5.23)$$

Indeed, let Γ and Γ^1 be the extremal contours of area S in the initial and less rigid bodies respectively. Then on decreasing the rigidity of the initial body without altering the contour Γ, we obtain

$$W(\Gamma) = W^*(S) \leq W_1(\Gamma)$$

on the other hand, by the properties of extremal contours in an "unaltered" body, we have

$$W_1(\Gamma) \leq W_1^*(S_1)$$

Hence, (5.23) follows from this inequality.

6. Bounds for certain integral characteristics of solutions of contact problems of elasticity

Using Theorem 4.1 and arguments similar to those applied in proving it, we may construct double-sided bounds for the displacement of a die (or) the force which acts on the die when the contact area, die shape, and contact conditions vary. Similar theorems are known for contact problems of perfectly plastic bodies.

Theorem 1: Let a rigid die with sharp edges and flat base Ω be pressed into an elastic body by a force acting normal to the die base so that the die moves in the direction of action of the force. (The die is either in sliding contact or in perfect adhesion with the elastic body.)

If the rigidity is decreased (increased) at some part of the body, for the die displacement to be the same it is necessary that the force be diminished (increased).

Indeed, the force is

$$P = \iint_\Omega \vec{\sigma}_n(x) d\sigma \qquad (6.1)$$

Therefore, by virtue of Corollary 2 of Theorem 4.1 (inequality (4.10)) for g(x) = h = const we have

$$hP \geq hP_1 \ , \quad P > P_1 \qquad (6.2)$$

Corollary 1

The force driving the die being constant, a decrease (increase) in the rigidity at some part of the body causes an increase (decrease) in the displacement of a die with flat base. (This is an obvious sequel of Theorem 1).

Corollary 2

In order that two smooth or perfectly adhered dies with flat bases Ω and $\Omega_1 (\Omega > \Omega_1)$ may have the same displacement, it is necessary that the force applied to the die Ω should be greater than the force applied to the die $\Omega_1 (P > P_1)$.

From Theorem 1 it follows that the transition from the die Ω_1 to the die Ω may be regarded as an infintely great increase in the rigidity of the initial body on a part of the boundary Ω/Ω_1.

Corollary 3

For the same driving force, on passing from the die Ω_1 to the die $\Omega > \Omega_1$, the displacement decreases. (This is a trivial consequence of Corollary 2.)

Theorem 1 and its corollaries 1, 2 and 3 give a means for constructing two-sided bounds for the displacement of dies of complicated shapes with flat base Ω from the solution for dies of simpler forms whose bases Ω_1 and Ω_2 are inscribed in or circumscribe $\Omega : \Omega_1 < \Omega < \Omega_2$. Examples of the application of such considerations for a smooth die are given in [12-15]. By way of example we shall estimate the displacement of a square die in perfect adhesion with a semi-space on the basis of the exact solution of the problem of a circular die in adhesion with a semi-space [16].

Let 2α be the side of the square. Consider two circular dies Ω_1 and Ω_2 of radii α and $\alpha\sqrt{2}$ respectively. According to [16] :

$$w_1 = \frac{1}{16\theta(3\mu+\lambda)} \frac{P}{\alpha} \; , \qquad w_1 = \frac{1}{16\theta(3\mu+\lambda)} \frac{P}{\alpha\sqrt{2}} \qquad (6.3)$$

where $\theta = \frac{1}{\pi} \ell n[(3\mu+\lambda)/(\mu+\lambda)], \lambda$ and μ is the Lamé coefficient. Therefore

$$w_2 \leq w \leq w_1$$

If the approximate value of w is taken to be $\frac{1}{2}(w_1+w_2)$ then the error in the determination of w by means of (6.4) and (6.5) will not exceed $[(w_1-w_2)/(w_1+w_2)] \leq 17\%$.

Corollary 4

Consider two dies with nonplanar bases of arbitrary configuration with sharp edges, assuming that their contact areas are the same :

$$x_3 = -\psi_1 (x_1,x_2) \leq 0 \; , \quad x_3 = -\psi_2 (x_1,x_2) \leq 0$$

$$\psi_1 (x_1,x_2) \leq \psi (x_1, x_2) \; , \; \forall \; (x_1, x_2) \epsilon \Omega \qquad (6.5)$$

Then, for the same driving force P, the displacement of the enveloping die ψ_2 is less than that of the enveloped die ψ_1. This is proved by applying Corollary 1 twice : by passing from a die with flat base to the die ψ_2 by successively increasing the rigidity in the region between the flat base and the surface $\psi_1(x_1,x_2)$ and between the surfaces $\psi_2(x_1,x_2)$ and

$\psi_1(x_1,x_2)$.

In particular, for the same driving force $\underline{P}$ and contact area Ω, the displacement of a die with nonplanar base is less than that of the enveloped die with flat base.

In Corollary 4 the contact areas of dies have been taken to be equal. However, as in the proof of Theorem 6.1, we may show that if the initial die with contact region Ω and surface $\psi(x_1,x_2)$ is replaced by some other die of contact region $\Omega_1<\Omega$ and surface $\psi_1(x_1,x_2) \leq \psi(x_1,x_2)$, $\Psi(x_1,x_2)\epsilon\Omega_1$ then the displacement increases for the same driving force.

The same considerations can be used to prove :

Theorem 6.2

Let a die with flat base Ω of arbitrary configuration be pressed by a force P into an elastic body so that one of the following three conditions is satisfied : 1) no friction at the contact region, 2) friction forces act at the contact region, 3) adhesion exists between the die and the elastic space. Then

$$w_1 > w_2 > w_3 \tag{6.6}$$

where w_1, w_2, w_3 are the displacements of the die under contact conditions (1), (2) and (3) respectively.

Proof :

First we shall establish that $w_1 > w_2$. Under contact condition (1), the total energy of the system is

$$W_1(\vec{u}_1) = \int_V A(\vec{u}_1) \ d\tau - w_1P \tag{6.7}$$

where V is the spatial domain occupied by the body.

Choose as the admissible displacement field of a smooth die the field of displacement corresponding to the action of a die under condition (2). Then, by the minimum potential energy principle, we have

$$W_1(\vec{u}_1) \leq W_1(\vec{u}_2) = \int_V A(\vec{u}_2) \ d\tau - w_2P \tag{6.8}$$

On the other hand, when the die acts under friction

$$W_2(\vec{u}_2) = \int_V A(\vec{u}_2) \ d\tau - [w_2P - W_{\psi r}] \tag{6.9}$$

where $W_{\psi r}$ is the work of friction. By virtue of the Clapeyron theorem for equilibrium, for a die under condition (2), we have

$$\int_V A(\vec{u}_2) \ d\tau = \frac{1}{2} w_2P - \frac{1}{2} W_{\psi r} \tag{6.10}$$

and under condition (1)

$$\int_V A(\vec{u}_1) \ d\tau = \frac{1}{2} w_1P \tag{6.11}$$

From (6.7), (6.8), by virtue of (6.10) and (6.11), we obtain

$$- \frac{1}{2} w_2 P - \frac{1}{2} W_{\psi r} \geq - \frac{1}{2} w_1 P \tag{6.12}$$

or

$$w_2 \leq w_1 - \frac{1}{2} \frac{W_{\psi r}}{P} \tag{6.13}$$

Since the work of friction $(- W_{\psi r})$ is negative, Eq. (6.13) shows that the displacement under friction at the contact region is less than the displacement under sliding contact.

The inequality $w_2 > w_3$ is proved in a similar way by taking the field of displacement corresponding to the action of a fully connected die in equilibrium as the admissible displacement field for the action of a die under friction.

APPENDIX A

Consider a three-dimensional elasticity problem for a layer of thickness $H = 2h$ containing an opening-mode crack occupying the domain G in the midplane $x_3 = 0$. Assume that there are no tractions at the layer faces and only normal wedging loads $q(x_1, x_2)$ act on the crack surface. The boundary conditions of the problem are thus of the form :

$$x_3 = 0 \qquad \sigma_{33} = - q(x_1, x_2), \quad (x_1, x_2) \varepsilon G, \quad \sigma_{13} = \sigma_{23} = 0$$

$$x = \pm h \qquad \sigma_{33} = \sigma_{31} = \sigma_{32} = 0 \tag{A.1}$$

We shall reduce this problem to an integral equation for $w(x_1, x_2)$ which is the displacement of points on the crack surface in the direction of x_3. For this purpose, as in the problem of a crack in an infinite space, we shall first calculate the stresses at the layer boundaries $x_3 = \pm h$ assuming that the function $w(x_1, x_2)$, i.e. the shape of the crack, is known. Then we shall write the solution to the problem of a layer (without cracks), whose faces are subjected to the action of tractions equal but opposite to stresses calculated in the first problem. Finally we shall compute the sum of normal stresses which arise in both the problems at the crack and then equate it to the given loads. As a result, we shall obtain the required integral equation.

In expressing the solution of a problem for an infinite space, it is convenient to use the Papkovich-Neuber representations [5,17]. In our case

$$u = - x_3 \frac{\partial B_3}{\partial x_1} - \frac{\partial B_o}{\partial x_1} \quad , \quad v = - x_3 \frac{\partial B_3}{\partial x_2} - \frac{\partial B_o}{\partial x_2}$$

$$w = 2(1 - \nu) B_3 - x_3 \frac{\partial B_3}{\partial x_3} \tag{A.2}$$

$$\frac{1}{2\mu} \sigma_{13} = - x_3 \frac{\partial^2 B_3}{\partial x_1 \partial x_3} \quad , \quad \frac{1}{2\mu} \sigma_{23} = - x_3 \frac{\partial^2 B_3}{\partial x_2 \partial x_3}$$

$$\frac{1}{2} \sigma_{33} = \frac{\partial B_3}{\partial x_3} - x \cdot \frac{\partial^2 B_3}{\partial x_3^2} \tag{A.3}$$

where $\partial B_o/\partial x_3 = (1 - 2\nu)B_3$ (by virtue of symmetry relative to the plane $x_3 = 0$ [17]).

The function B_3 can be represented as

$$B_3 = - \frac{1}{4(1-\nu)} \frac{\partial}{\partial x_3} \iint_G \frac{r(\eta_1,\eta_2) d\eta_1 d\eta_2}{\sqrt{(x_1-\eta_1)^2+(x_2-\eta_2)^2+x_3^2}} \tag{A.4}$$

Then from (A.2) and (A.4) we get

$$w(x_1,x_2,+0) = \begin{cases} r(x_1,x_2) & , \quad (x_1,x_2) \epsilon G \\ 0 & , \quad (x_1,x_2) \bar{\epsilon} G \end{cases} \tag{A.5}$$

i.e. the function $r(x_1,x_2)$ is identical with the displacement of points on the crack surface in the direction of x_3. For a crack in an infinite space, $r(x_1,x_2)$ satisfies the following equation :

$$- \frac{2(1-\nu)}{\mu} q(x_1,x_2) = \Delta_{x_1,x_2} \iint_G \frac{r(\eta_1,\eta_2) d\eta_1 d\eta_2}{\sqrt{(x_1-\eta_1)^2+(x_2-\eta_2)^2}} \tag{A.6}$$

It is more convenient to carry out further calculations with the Fourier transform in (x_1,x_2) with the parameter $\xi = (\xi_1,\xi_2)$. It is not difficult to prove (see, for example, [18]) that

$$\tilde{B}_3 = \frac{\pi}{2(1-\nu)} \tilde{r} e^{-x|\xi|} \tag{A.7}$$

The stresses in the plane $x = h$ are given by the following expressions :

$$\frac{1}{2\mu} (x_1,x_2,h) = \frac{\partial}{\partial x_1} F_\xi^{-1} (\tilde{\tau}) = \tau_1$$

$$\frac{1}{2\mu} (x_1,x_2,h) = \frac{\partial}{\partial x_2} F_\xi^{-1} (\tilde{\tau}) = \tau_2 \tag{A.8}$$

$$\frac{1}{2\mu} \sigma_{33}(x_1,x_2,h) = \frac{\pi}{2(1-\nu)} F_\xi^{-1} \{|\xi|\tilde{r}(1+h|\xi|)\} e^{-h|\xi|} = \sigma \tag{A.9}$$

where

$$\tilde{\tau} = \frac{\pi}{2(1-\nu)} h|\xi|\tilde{r} e^{-h|\xi|} \tag{A.10}$$

Using the formulae given in [17], we shall now write the expressions for Fourier transforms of stresses in a layer under two types of boundary conditions :

$$\text{a)} \quad x_3 = \pm h \quad , \quad \sigma_{33} = - \sigma , \quad \sigma_{31} = \sigma_{32} = 0 \tag{A.11}$$

b) $x_3 = \pm h$, $\sigma_{33} = 0$, $\sigma_{31} = \pm \tau_1$, $\sigma_{32} = \pm \tau_2$ (A.12)

According to [17], under the conditions (a), we have

$$\tilde{\sigma}_{33}^{(a)} = \{h \, ch(h|\xi|)ch(x_3|\xi|) - x_3 sh(h|\xi|)sh(x_3|\xi|) + |\xi|^{-1} sh(h|\xi|) -$$
$$- ch(x_3|\xi|)\} \frac{2\tilde{\sigma}|\xi|}{\Delta} \, , \quad \Delta = 2h|\xi| + sh(2h|\xi|)$$

$$\tilde{\sigma}_{31}^{(a)} = -i\xi_1\{h \, ch(h|\xi| \, sh(x_3|\xi| - x_3 ch(x_3|\xi|)sh(h|\xi|)\} \frac{2\tilde{\sigma}}{\Delta}$$

$$\tilde{\sigma}_{32}^{(a)} = -i\xi_2\{h \, ch(h|\xi|)sh(x_3|\xi|) - x_3 ch(x_3|\xi|)sh(h|\xi|)\} \frac{2\tilde{\sigma}}{\Delta}$$

(A.13)

For tangential tractions of the type (b), according to [17], we have

$$\tilde{\sigma}_{33}^{(b)} = \{x_3 sh(x_3|\xi|) \, ch(h|\xi|) - h \, sh(h|\xi|)ch(x_3|\xi|)\} \frac{2\tilde{\tau}|\xi|^2}{\Delta}$$

$$\tilde{\sigma}_{31}^{(b)} = -i\xi_1 F(x_3, \xi_1, \xi_2, \tilde{\tau})$$

$$\tilde{\sigma}_{32}^{(b)} = -i\xi_2 F(x_3, \xi_1, \xi_2, \tilde{\tau})$$

(A.14)

where

$$F(x_3, \xi_1, \xi_2, \tilde{\tau}) = \{x_3 ch(x_3|\xi|)ch(h|\xi|) - h \, sh(x_3|\xi|)sh(h|\xi|)$$
$$+ |\xi|^{-1} sh(x_3|\xi|)ch(h|\xi|)\} \frac{2\tilde{\tau}|\xi|}{\Delta}$$

Now, using (A.13) and (A.14), we shall calculate the additional stresses $\tilde{\sigma}_{33}^{(ad)} (x_1, x_2, 0)$ which arise in the problem of a layer without any cracks under the joint action of loads (a) and (b) :

$$\frac{2(1-\nu)}{\mu} \tilde{\sigma}_{33}^{(ad)} \Big|_{x_3=0} = \frac{4\pi|\xi|e^{-h|\xi|}\tilde{r}}{\Delta} \{(h|\xi|ch(h|\xi|) + sh(h|\xi|))$$

$$\times (1 + h|\xi|) + h^2|\xi|^2 sh(h|\xi|)\}$$

(A.15)

With the help of Fourier transforms (A.6) and (A.15), finally we obtain the following pseudo-differential equation [18] in $r(x_1, x_2)$:

$$2\pi F_\xi^-\{[-|\xi| + K(h, |\xi|)]\tilde{r}(\xi_1, \xi_2)\} = - \frac{2(1-\nu)}{\mu} q(x_1, x_2)(x_1, x_2)eG$$

(A.16)

where

$$K(h, |\xi|) = \frac{1}{h} K(h|\xi|)$$

$$K(h|\xi|) = \frac{2h|\xi|(1+2h|\xi|+2h^2|\xi|^2 - e^{-2h|\xi|})}{4h|\xi| + e^{2h|\xi|} - e^{-2h|\xi|}}$$

APPENDIX B

To establish the method of convergence of successive approximations applied to Eq. (2.2), we shall first find the magnitude of $(q_{i+1} - q_i)$. From (2.6) we have

$$q_{i+1} - q_i \le \frac{\mu}{2(1-\nu)} \frac{1}{2\pi} \iint_{-\infty}^{\infty} K(h|\xi|) |\tilde{r}_i - \tilde{r}_{i-1}| d\xi \qquad (B.1)$$

From the definition of $\tilde{r}$ it follows that

$$|\tilde{r}_i - \tilde{r}_{i-1}| \le \iint_G w_i - w_{i-1} \, dS = V_i \qquad (B.2)$$

Thus, with the help of (B.1) and (B.2), we obtain

$$w_{i+1} - w_i \le R_\infty |q_{i+1} - q_i| \le \frac{\mu R_\infty}{2(1-\nu)} \frac{1}{2\pi} \iint_{-\infty}^{\infty} K(h|\xi|) \iint_G |w_i - w_{i-1}| dS d\xi \qquad (B.3)$$

Integrating over the domain G we get

$$V_{i+1} \le \left[\iint_G h_\infty \cdot 1 dS \right] \frac{\mu}{2(1-\nu)} \frac{V_i}{2\pi} \int_{-\infty}^{\infty} K(h|\xi|) d\xi \qquad (B.4)$$

After changing to polar coordinates in the plane $\eta = h\xi$ the integral of $K(\eta)$ in (B.4) takes the form

$$T_1 = \frac{1}{2\pi} \iint_{-\infty}^{\infty} K(h|\xi|) d\xi = \frac{2}{h^3} \int_0^{\infty} \frac{r^2(1 + 2r \pm 2r^2 - e^{-2r}) dr}{4r + e^{2r} - e^{-2r}} \qquad (B.5)$$

It is not difficult to calculate the numerical value of the integral which is

$$T = \frac{4.232}{h^3} \qquad (B.6)$$

The integral of $(R_\infty \cdot 1)$ over the domain G is half the volume of a crack occupying the domain G in an infinite space and acted upon by a constant load of intensity 1 on its surface.

If the area of the domain G is S, then as shown in Section 4.2, the volume of crack is not greater than the volume of a penny-shaped crack of the same area, i.e.

$$\iint_G R_\infty \cdot 1 dS \le \frac{4(1-\nu)}{3\mu} \left(\frac{S}{\pi} \right)^{3/2} \qquad (B.7)$$

Then, due to (B.7), from (B.4), we have

$$V_{i+1} - \frac{1}{2} \frac{S^{3/2}}{h^3} V_i = \varepsilon_* V_i \qquad (B.8)$$

115

The convergence of successive approximations method is guaranteed if $\varepsilon_* < 1$, which holds true when

$$S^{\frac{1}{2}}/h < 1.26 \tag{B.8'}$$

If only the crack diameter d is known, then we can derive an estimate less accurate than (B.8). In this case we can write

$$\iint\limits_{G} R_\infty \cdot 1 dS = \frac{2d^3}{\mu} (1 - \nu)\psi(g) \tag{B.9}$$

where $\psi(g)$ is half the volume of a crack occupying the domain g of diameter 1 similar to the domain G. According to the theorem proved in Section 1, we can find the upper bound for $\psi(g)$ by calculating half the volume of a penny-shaped crack of diameter 1 enveloping the crack g :

$$\psi(g) \leq \frac{1}{8} \int_0^{2\pi} \int_0^1 \sqrt{1 - r^2}\, r\, dr\, d\theta = \frac{1}{12} \tag{B.10}$$

From (B.4) we obtain

$$V_{i+1} \leq (d|h)^3\, 0.353\, V_i = \varepsilon V_i \tag{B.11}$$

Successive approximations converge if $\varepsilon < 1$, i.e., if $(d|h) < 1.42$ or $(d|H) < 0.71$.

The bounds (B.12) and (B.8) are coincident only for a penny-shaped crack, a fact which can be easily verified.

APPENDIX C

We shall first estimate the difference $q'-q_1^0$. We have

$$q'-q_1^0 = \frac{\mu}{2(1-\nu)}\, \frac{1}{2\pi h^3} \iint\limits_{-\infty}^{\infty} K(|\eta|)\left[\tilde{r}\left(\frac{\eta}{h}\right)e^{-i}\left(\frac{x_1}{h}\eta_1 + \frac{x_2}{h}\eta_2\right) - r_o(0)\right] d \tag{C.1}$$

Now express the right-hand side as a sum of two integrals by adding and subtracting the expression

$$r(0)e^{-i\left(\frac{x}{h},\eta\right)}$$

in the integrand of (C.1). Thus we get

$$q'-q_1^0 = \frac{\mu}{2(1-\nu)}\, \frac{1}{2\pi h^3} (I_1 + I_2) \tag{C.2}$$

where

$$I_1 = \iint_{-\infty}^{\infty} K(|\eta|) \left[\tilde{r}(\tfrac{\eta}{h}) - \tilde{r}(0) \right] e^{-i\left(\frac{x}{h},\eta\right)} d\eta \qquad (C.3)$$

$$I_2 = \iint_{-\infty}^{\infty} K(|\eta|) \left[\tilde{r}(0) e^{-i\left(\frac{x}{h},\eta\right)} - \tilde{r}_o(0) \right] d\eta \qquad (C.4)$$

Transform the integral I_1 as follows. Since

$$\tilde{r}(\tfrac{\eta}{h}) = \iint_{-\infty}^{\infty} w(u_1,u_2) e^{i\left(\frac{u}{h},\eta\right)} du$$

$$\tilde{r}(0) = \iint_{-\infty}^{\infty} w(u_1,u_2) du$$

and changing over to polar coordinates in the plane η ($\eta_1 = \rho\cos\theta$, $\eta_2 = \rho\sin\theta$) we obtain

$$I_1 = \int_0^{2\pi}\int_0^{\infty} K(\rho) \left\{ \iint_G w(u_1,u_2) \left[e^{i\rho Zu} - 1 \right] du \right\} e^{-i\rho Zx} \rho\, d\rho\, d\theta =$$

$$= \iint_G w(u_1,u_2) \int_0^{2\pi}\int_0^{\infty} K(\rho) \left[\cos\rho(Z_u - Z_x) - \cos\rho Z_x \right] \rho\, d\rho\, d\theta \qquad C.5$$

where

$$Zu = \frac{u_1}{h}\cos\theta + \frac{u_z}{h}\sin\theta \ , \quad Z_x = \frac{x_1}{h}\cos\theta + \frac{x_2}{h}\sin\theta$$

Now we shall estimate I_1 . We have

$$|I_1| = \left| \iint_G w(u_1,u_2) \int_0^{2\pi}\int_0^{\infty} K(\rho) \left[2\sin(\rho\tfrac{Z_u}{2})\sin\rho(Z_x - \tfrac{1}{2}Z_u) \right] \rho\, d\rho\, d\theta\, du \right| \le$$

$$\le \iint_G w(u_1,u_2) \left| \int_0^{2\pi}\int_0^{\infty} K(\rho) Z_u(Z_x - \tfrac{1}{2}Z_u)\rho^3\, d\rho\, d\theta \right| du \qquad (C.6)$$

Since

$$Z_u(Z_x - \tfrac{1}{2}Z_u) = \frac{\sqrt{u_1^2 + u_2^2}}{h} \frac{\sqrt{(2x_1 - u_1)^2 + (2x_2 - u_2)^2}}{h} \sin(\theta - \psi_1)\sin(\theta - \psi)$$

the inequality (C.6) may be extended

$$|I_1| \le \iint_G w(u_1,u_2) 2\pi \cdot \frac{1}{2} \frac{\sqrt{u_1^2 + u_2^2}}{h} \frac{\sqrt{2(x_1 - u_1)^2 + (2x_2 - u_2)^2}}{h} \int_0^{\infty} K(\rho)\rho^3\, d\rho \le$$

$$\le \sqrt{2}\pi d^2 \int_0^{\infty} K(\rho)\rho^3\, d\rho \iint_G w(u_1,u_2) du \qquad (C.7)$$

where due account has been taken of the fact that

$$\sqrt{(2x_1-u_1)}^2 + (2x_2-u_2)^2 \leq 2(3d)^2$$

In (C.7) the integral with respect to ρ is easily calculated and it is equal to

$$\int_0^\infty K(\rho)\rho^3 d\rho = 29.702 \tag{C.8}$$

Further

$$\left|\iint_G w(u_1,u_2)du\right| = \left|\iint_G \left[w_o+(w-w_o)\right]du\right| \leq \iint_G w_o du + \iint_G |w-w_o|du \tag{C.9}$$

By definition

$$V_\infty = \iint_G |w - w_o|\, du \tag{C.10}$$

and according to Appendix B

$$V_\infty \leq V_1 + V_2 + \ldots \leq \frac{V_1}{1-\varepsilon} \tag{C.11}$$

where $\varepsilon = 0.353\,(d|h)^3$.

Thus, from (C.7), by virtue of (C.9) and (C.11), we get the bound of I_1 :

$$|I_1| \leq 3\sqrt{2}\pi\cdot 29.702\left(\frac{d}{h}\right)^2 \left(\iint_G w_o du + \frac{V_1}{1-\varepsilon}\right) \tag{C.12}$$

Using the expressions $V_1 = V_o\varepsilon$ and V_o, we may rewrite (C.12) as follows :

$$|I_1| \leq 3\sqrt{2}\pi 29.702\left(\frac{d}{h}\right)^2 V_o\left(1 + \frac{\varepsilon}{1-\varepsilon}\right) \tag{C.13}$$

Now we shall derive the bounds of the integral I_2. After changing over to polar coordinates in the plane η, we obtain :

$$|I_2| = \left|\int_0^{2\pi}\int_0^\infty K(\rho)\iint_G \left[w(u_1,u_2)e^{-i\rho Z_x} - w_o(u_1,u_2)\right]du\rho d\rho d\theta\right|$$

$$= \left|\iint_G\left\{\int_0^{2\pi}\int_0^\infty K(\rho)\left[w(u_1,u_2)\cos(\rho Z_x)-w_o(u_1,u_2)\right]\rho d\rho d\theta\right\}du\right|$$

$$= \left|\iint_G\left\{\int_0^{2\pi}\int_0^\infty K(\rho)\left[(w(u_1,u_2)-w_o(u_1,u_2))\cos(\rho Z_x)\right.\right.\right.$$

$$\left.\left.\left. - w_o(u_1,u_2)(1-\cos(\rho Z_x))\right]\times \rho d\rho d\theta\right\}du\right| \leq \{\Lambda_1+ \Lambda_2\} \tag{C.14}$$

$$\Lambda_1 = \left\{ \iint_G \left| \int_0^{2\pi} \int_0^\infty K(\rho) [(w(u_1,u_2)-w_0(u_1,u_2))\cos(\rho Z_x)]\rho d\rho d\theta \right| du \right\}$$

$$\Lambda_2 = \left\{ \iint_G \left| \int_0^{2\pi} \int_0^\infty K(\rho) w_0(u_1,u_2)(1-\cos(\rho Z_x))\rho d\rho d\theta \right| du \right\} \qquad (C.15)$$

In the first integral Λ_1 in (C.15), substituting unity for $\cos\rho Z_x$, and then integrating with respect to θ we get

$$\Lambda_1 \leq 2\pi V_\infty \int_0^\infty \rho K(\rho) d\rho \qquad (C.16)$$

where

$$V_\infty \leq \frac{V_1}{1-\varepsilon} \qquad (C.17)$$

The last inequality follows from (C.11). The integral is

$$\int_0^\infty \rho K(\rho) d\rho = 4.232 \qquad (C.18)$$

Therefore

$$\Lambda_1 \leq \frac{V_1}{1-\varepsilon} (2\pi)\cdot 4.232 \qquad (C.19)$$

The second integral, Λ_2, after changing the order of integration, takes the form :

$$\Lambda_2 = \iint_G w_0(u_1,u_2) |\Lambda| du \qquad (C.20)$$

where

$$\Lambda = \int_0^{2\pi} \int_0^\infty K(\rho)(1-\cos\rho Z_x)\rho d\rho d\theta \qquad (C.21)$$

Introduce the function

$$K_1(\rho) = -\int_\rho^\infty r K(r) dr , \qquad K_2(\rho) = -\int_\rho^\infty K_1(r) dr \qquad (C.22)$$

and integrating (C.21) twice in parts, we find that

$$\Lambda = \int_0^{2\pi} Z_x^2 \int_0^\infty K_2(\rho)\cos(\rho Z_x)\,d\rho\,d\theta = \int_0^\infty(\rho)\int_0^{2\pi} Z_x^2\cos(\rho Z_x)\,d\theta\,d\rho \qquad (C.23)$$

Since $\quad Z_x = \dfrac{x_1^2 + x_2^2}{h}\sin^2(\theta-\psi)\ ,\quad \psi = \text{arctg}\ \dfrac{x_1}{x_2}$

(C.23) can be rewritten as

$$\Lambda = \frac{x_1^2+x_2^2}{h^2}\int_0^\infty K_2(\rho)\int_0^{2\pi}\sin^2(\theta-\psi)\cos\frac{\sqrt{x_1^2+x_1^2}}{h}\sin(\theta-\psi)\,d\theta\,d\rho \qquad (C.24)$$

Substituting unity for $\cos\alpha$, we get

$$|\Lambda| \le \frac{x_1^2+x_2^2}{h^2}\,\pi\int_0^\infty K_2(\rho)\,d\rho \le \pi\ (\tfrac{d}{h})^2\int_0^\infty K_2(\rho)\,d\rho \qquad (C.25)$$

Hence, after integrating twice in parts, we obtain

$$|\Lambda| \le (\tfrac{d}{h})^2\,\frac{\pi}{2}\int_0^\infty K(\rho)\rho^3\,d\rho = (\tfrac{d}{h})^2\,\pi\cdot 14.851 \qquad (C.26)$$

Thus,

$$|\Lambda_2| \le \iint_G w_0(u_1,u_2)\,du\ (\tfrac{d}{h})^2\,\pi\cdot 14.851 = V_0(\tfrac{d}{h})^2\,\pi\cdot 14.851 \qquad (C.27)$$

Finally, by virtue of (C.14), (C.19), and (C.27), we obtain a bound for I_2

$$I_2 \le \left\{\frac{\varepsilon V_0}{1-\varepsilon}\ 2\pi\cdot 4.232 + V_0(\tfrac{d}{h})^2\,\pi\cdot 14.851\right\} \qquad (C.28)$$

From (C.2), with due regard for (C.13) and (C.28), we now get

$$|q'-q_1^0|' \le \frac{\mu}{2(1-\nu)^3}\left\{V_0(\tfrac{d}{h})^2\left[3\sqrt{2}\cdot 14.851\ \frac{1}{1-\varepsilon}\right.\right.$$

$$\left.\left. + (7.425 + (d/h)\cdot\frac{0.353}{1-\varepsilon}\ 4.232)\right]\right\} \qquad (C.29)$$

According to (B.11) and (B.12)

$$\frac{1}{h^3}\ \frac{\mu}{2(1-n)}\ V_0 \le \frac{1}{12}\ (\frac{d}{h})^3$$

Therefore

$$|q' - q_1^o| \leq (\frac{d}{h})^5 \; \frac{85}{12} = m \qquad\qquad (C.30)$$

Hence, for $(d/h) < 0.7$ $(m < 1)$ we have

$$q' = q_1^o + (q' - q_1^o) \geq q_1^o - |q' - q_1^o| > 0 \qquad\qquad (C.31)$$

REFERENCES

1. GOLDSTEIN, R.V., ENTOV, V.M., Inter. J. Fracture, 11, 1975.

2. MUSKHELISHVILLI, N.I., Some Fundamental Problems in Mathematical Theory of Elasticity, Moscow, Nauka, 1966.

3. BARENBLATT, G.I., ENTOV, V.M. and SALGANIK, R.L., Kinetics of Crack Growth : Condition for Fracture and Durability, Izv. AN SSSR ser. mekh. tver. tela, N6, 1966.

4. SACH, R.C., KOBAYASHI, A.S., Eng. Fract. Mech., 3, 1971.

5. LUR'E, A.I., Theory of Elasticity, Moscow, Nauka, 1970.

6. POLYA, G. and SEGO, G., Isoperimetric Inequalities in Mathematical Physics, Princeton Univ. Press, Princeton, 1951.

7. SRIVASTAVA, K.N. and DWIVEDI, J.R., Int. J. Eng. Sci., 9, 1971, 399-420.

8. SNEDDON, I.N. and WELCH, J.T., Int. J. Eng. Sci., 1, 1963, 411-419.

9. SHIBUYA, T., NAKAHARA, I., and KOIZUMI, L., ZAMM, B.55, H.718, 1975, S.395-402.

10. TANAKA, T. and ATSUMI, A., Lett. Appl. Eng. Sci., 3, 1975, 155-165.

11. SNEDDON, I.N. and TAIT, R.J., Int. J. Eng. Sci., 1, 1963, 391-409.

12. GALIN, L.A., Contact Problems in Elasticity, Gos. Izd. Tekh. Teor. Lit., Moscow, 1953.

13. ALEKSANDROV, V.M., Some Contact Problems for an Elastic Layer, Prik. Matem. i. Mekh., iss. 4, 24, 1963.

14. BORODACHEV, N.M., On the Determination of Displacement of Rigid Plates and Rocks, Osnov. fundamenty i mekh. gruntov, N4, 1964, 3-5.

15. BORODACHEV, N.M., On One Method of Reducing Certain Contact Problems of Elasticity to Integral Equations of Second Kind. In "Raschet prostranstv. konstrucktsii", 5, Kuibyshev, 1975, 7-13.

16. MOSSAKOVSKII, V.I., Fundamental Mixed-Value Problem in Elasticity for a Semi-Space with Circular Interface of Boundary Conditions, Prik. matem. i mekh., iss. 2, 18, 1954.

17. LUR'E, A.I., Three Dimensional Problems of Elasticity, Gos. Izd. Tekh. Teor. Lit., Moscow, 1955.

18. ESKIN, G.I., Boundary Value Problems for Elliptical Pseudo-Differential Equations, Moscow, Nauka, 1973.

CALCULATION OF STRESS INTENSITY FACTORS
FOR COMBINED MODE BEND SPECIMENS

Wang Ke Jen, Hsu Chi Lin and Kao Hua*

INTRODUCTION

In order to test combined mode fracture criteria experimentally it is necessary to use specimens with a wide range of K_I and K_{II} and to obtain calibrated curves of K_I and K_{II} values for these specimens either by calculation or by experiment. As no such calculated curves of K_I and K_{II} values were readily available for three-point-bend specimens with cracks in an unsymmetrical position (Figure 1), such curves were obtained by use of the boundary collocation method and the finite element method. The boundary collocation method was first used by Gross et al [1,2,3], to calculate K_I values for opening mode specimens. Later the method was used to calculate K_I and K_{II} values for some combined mode specimens [4,5]. As for the finite element method used to determine stress intensity factors, growing interest is now directed to special elements at the crack tip [6,7]. A term in $\sqrt{r}$ is included in the displacement functions of these special elements, where r is the distance from the given point to the crack tip. As this term gives the required singularities of stresses and strains in the vicinity of the crack tip, a higher accuracy can often be obtained with relatively fewer elements. Since the other elements around the special elements are still ordinary ones, whose displacement functions do not contain the terms in $\sqrt{r}$, the results obtained by use of these special elements are not very satisfactory, especially in the case when the size of the elements is decreasing. It should be noted that convergence of the results cannot be insured for elements of diminishing size, if the condition of constant strain is not satisfied by those special elements [8], as is the case with commonly adopted ones. However, the distorted isoparametric elements, proposed by Henshell et al [9] and Barsoum [10], satisfy the constant strain condition and their displacement functions contain the terms in $\sqrt{r}$. The 8-noded isoparametric quadratic and triangular elements with the mid-side nodes near the crack tip at the quarter point have been used in the vicinity of the crack tip and their displacement functions contain the terms in $\sqrt{r}$. Now we have succeeded in including terms in $\sqrt{r}$ in the displacement functions of any isoparametric elements at any arbitrary positions. When these special elements are used in a wider area, not restricted to the vicinity of the crack tip, the accuracy of the calculated results has improved considerably.

Analysis of the energy-momentum tensor, which was proposed by Eshelby [11] and was later used in combined mode fracture criteria by Hellen et al [12], showed that this kind of application is questionable theoretically, and that the results thus obtained are doubtful [13]. In the meantime, an approximate relation between K_I and K_{II} is used to derive the approximate

---

*Institute of Mechanics, Academia Sinica, Peking

formula for K_{II} for the bend specimens mentioned above. The results calculated by this formula are compared with those from the boundary collocation method.

2. BOUNDARY COLLOCATION METHOD

Consider the following expansion of the stress function with the crack tip as the centre:

$$X = \sum_{j=1}^{\infty} r^{\frac{j}{2} + 1} \left\{ C_j \left[- \cos\left(\frac{j}{2} - 1\right)\theta + \frac{\frac{j}{2} + (-1)^j}{\frac{j}{2} + 1} \cos\left(\frac{j}{2} + 1\right)\theta \right] \right.$$

$$\left. + D_j(-1)^{j+1} \left[\sin\left(\frac{j}{2} - 1\right)\theta - \frac{\frac{j}{2} + (-1)^{j+1}}{\frac{j}{2} + 1} \sin\left(\frac{j}{2} + 1\right)\theta \right] \right\}. \quad (1)$$

According to the approach adopted by Gross et al [1,2,3], the expansion is truncated to the first 2N terms and M(M>N) points on the boundary of the specimen are chosen. From the boundary conditions on these M points, 2M equations are obtained and the 2N coefficients in the truncated expansion can be determined. The values of K_I and K_{II} are determined from the first two coefficients:

$$K_I = - C_1\sqrt{2\pi}, \quad K_{II} = D_1 \sqrt{2\pi}. \quad (2)$$

Note that as the term in D_2 in equation (1) is identical with zero, this term should be deleted from the resulting simultaneous equations, otherwise an overflow will take place during the calculation if M is equal to N. The overflow was mentioned in [4] by Wilson et al, and the reason is now explained here.

We take 43 terms and choose 63 collocation points. The calculated results are shown in Table 1. It can be seen from Table 1 that $K_I BW^{3/2}/M$ and $K_{II} BW^{1/2}/Q$ depend on a/W only in a wide range (as long as the crack tip is not very close to the concentrated forces and the support points). It follows that the values of K_I and K_{II} can be determined approximately from the bending moment and the shearing force on the crack section, respectively

3. FINITE ELEMENT METHOD

An 8-noded isoparametric quadratic element is shown in Figure 2. Its shape functions are

$$N_i = \frac{1}{4} (1 + \xi_i\xi)(1 + \eta_i\eta)(\xi_i\xi + \eta_i\eta - 1)$$

for the corner nodes,

$$N_i = \frac{1}{2} (1 - \xi^2)(1 + \eta_i\eta) \quad (3)$$

for the mid-side nodes with $\xi_i = 0$, and

$$N_i = \frac{1}{2} (1 + \xi_i\xi)(1 - \eta^2)$$

for the mid-size nodes with $\eta_i = 0$. Taking the side $\eta = +1$, we have (Figure 3)

$$N_1 = -\frac{\xi(1 - \xi)}{2}$$

$$N_2 = 1 - \xi^2 \qquad\qquad (4)$$

$$N_3 = \frac{\xi(1 + \xi)}{2}$$

By the coordinate transformation used for the isoparametric elements, this side is assumed to be mapped into a segment AB on a line passing through the crack tip O (Figure 3). The lengths of OA and AB are equal to L_0 and L, respectively. The point, $\xi = 0$, is mapped into a point C, which is supposed to divide the segment AB into a ratio of p and (1-p). If the coordinate on the segment AB, after the transformation, is denoted by x, it follows that

$$x = -\frac{\xi(1 - \xi)}{2} L_0 + (1 - \xi^2)(L_0 + pL) + \frac{\xi(1 + \xi)}{2}(L_0 + L). \qquad (5)$$

Let $L_0/L = k$, it can be shown that

$$\xi = -[1 + 2k + \sqrt{4k(k + 1)}] + 2(\sqrt{1 + k} + \sqrt{k})\sqrt{\frac{x}{L}} \qquad (6)$$

when

$$p = \frac{1}{4}[\sqrt{4k(k + 1)} + 1 - 2k]. \qquad (7)$$

By substituting equation (6) into the relevant formulae of the isoparametric element, we obtain expressions for the displacement that include terms in $\sqrt{r}$. Let $k = 0$, then the relations given in [9] and [10] can be obtained. From equation (7), it is easily seen that p approaches 1/2 as k is getting larger. That is to say, that the distorted elements approach the normal (undistorted) ones farther away from the crack tip.

For the 12-noded isoparametric quadratic element (see [8]), if we assume that the mid-side nodes of the distorted elements divide the side into a ratio of p, (q-p) and (1-q) (Figure 4), it can be shown that equation (6) is again established, when

$$p = \frac{1}{9}[1 - 4k + 4\sqrt{k(k + 1)}]$$

$$\qquad\qquad (8)$$

$$q = \frac{1}{9}[4 - 4k + 4\sqrt{k(k + 1)}].$$

The corresponding expressions for the displacement thus obtained contain terms in $\sqrt{r}$ and $r\sqrt{r}$.

To test the method we consider a three-point-bend specimen with a crack at a symmetrical position. The geometry of the specimen and its finite element idealization are shown in Figure 5. In the vicinity of the crack tip we use triangular elements, which were shown to be superior to the quadratic ones [10].

First we use the same procedure as given in [10]. Only those triangular elements in the vicinity of the crack tip are taken to be distorted ones, with the mid-side nodes near the crack tip at the quarter point and all other elements taken to be normal ones. The final results are shown in Figure 6a. By the use of the calculated values of the displacements of the points on the crack edges, the apparent values of the stress intensity factors can be determined from

$$\bar{K}_I = \frac{E}{4(1 - \nu^2)} \sqrt{\frac{2\pi}{r}} \, u \tag{9}$$

$\bar{K}_I$ is plotted against distance, r. By analysing the expansion of the displacement at the crack tip, it can be shown that the apparent value of $\bar{K}_I$ is a linear function of r, if r is sufficiently small. The intersecting point of the straight part of the curve on the vertical axis (r = 0) gives the true K_I value. Some points near the crack tip that deviate from the straight line can be seen in Figure 6a. This indicates that these apparent values of K_I are questionable and should be discarded.

Secondly, we re-calculate the mid-side nodes according to equation (7) for all elements in the shaded area of Figure 5. The final results are shown in Figure 6b in the sense that all points near the crack tip fall on a straight line, as expected from the analysis. The intersecting point gives $K_I BW^{3/2}/M = 7.79$, with a/W = 0.4. This is in good agreement with the result calculated by the boundary collocation method: $K_I BW^{3/2}/M = 7.71$.

4. APPROXIMATE RELATION BETWEEN K_I AND K_{II}

For any plane configuration with a crack as shown in Figure 7 it can be proved that

$$J_1 = -\frac{\partial U}{\partial l} = \int_c Wdy - \underline{T} \cdot \frac{\partial \underline{u}}{\partial x} ds \tag{10}$$

$$J_2 = -\frac{\partial U}{\partial s} = \int_c - Wdx - \underline{T} \cdot \frac{\partial \underline{u}}{\partial y} ds \tag{11}$$

where U is the total potential energy of the system and C is the exterior contour of the configuration. Equation (11) gives the rate of the increase of the total potential energy as the crack translates in the direction perpendicular to the crack. Equation (10) defines the J-integral, whose value is path-independent, as long as the path starts at the lower edge of the crack and ends at the upper edge. As for equation (11), it can be proved that the value of the integral is also path-independent, if the points on the crack edges remain intact [11]. If a contour D sufficiently near the crack tip is taken, it can be shown that

$$J_1' = \int_D Wdy - \underline{T} \cdot \frac{\partial \underline{u}}{\partial x} ds = \frac{(1 + \nu)(1 + \kappa)}{4E} (K_I^2 + K_{II}^2)$$

$$J_2' = \int_D - Wdx - \underline{T} \cdot \frac{\partial \underline{u}}{\partial y} ds = -\frac{(1 + \nu)(1 + \kappa)}{4E} \cdot 2K_I K_{II} \tag{12}$$

where

$$\kappa = \begin{cases} \dfrac{3 - \nu}{1 + \nu} & \text{for plane stress} \\[2mm] 3 - 4\nu & \text{for plane strain.} \end{cases} \tag{13}$$

Due to the properties of the J-integral, J_1 is equal to J_1'. It can be shown that

$$R = J_2' - J_2 = \int_{\Gamma_1 + \Gamma_2} W dx = \frac{(1 + \nu)(1 + \kappa)}{8E} \int_{\Gamma_1 + \Gamma_2} \sigma_x^2 dx \qquad (14)$$

where Γ_1 and Γ_2 are the upper and lower edges of the crack, respectively. For the upper edge the integration proceeds from left to right and for lower edge from right to left. Since σ_x^2 takes the same value on the upper and lower edges in the vicinity of the crack tip and σ_x dwindles when the point moves towards the open end of the crack, it is expected that R will be a small quantity and will not be very sensitive to a small change in the crack length. So it is reasonable to assume that

$$\frac{\partial R}{\partial l} \approx 0 , \qquad (15)$$

Combining equations (15), (10), (11) and (12), we obtain

$$\frac{\partial J_1'}{\partial s} \approx \frac{\partial J_2'}{\partial l} , \qquad (16)$$

from which we obtain the following approximate relation between K_I and K_{II}:

$$K_{II} \frac{\partial K_I}{\partial l} + K_I \frac{\partial K_{II}}{\partial l} + K_I \frac{\partial K_I}{\partial s} + K_{II} \frac{\partial K_{II}}{\partial s} = 0 . \qquad (17)$$

If we further assume that K_I and K_{II} can be determined by the bending moment M and the shearing force Q on the crack section, respectively, we can write

$$K_I = \frac{M}{BW^{3/2}} f_b \left(\frac{a}{W}\right) \qquad (18)$$

$$K_{II} = \frac{Q}{BW^{1/2}} f_s \left(\frac{a}{W}\right) .$$

After substituting equation (18) into equation (17), we get the following equation:

$$f_b \left(\frac{a}{W}\right) \frac{df_s\left(\frac{a}{W}\right)}{d\left(\frac{a}{W}\right)} + \frac{df_b\left(\frac{a}{W}\right)}{d\left(\frac{a}{W}\right)} f_s \left(\frac{a}{W}\right) - \left[f_b\left(\frac{a}{W}\right)\right]^2 = 0 . \qquad (19)$$

The equation is solved to obtain

$$f_s \left(\frac{a}{W}\right) = \frac{1}{f_b\left(\frac{a}{W}\right)} \int_0^{a/W} \left[f_b\left(\frac{a}{W}\right)\right]^2 d\left(\frac{a}{W}\right) . \qquad (20)$$

For $f_b(a/W)$, we make use of the results for pure bending due to Benthem et al [14], the calculated values of $f_s(a/W)$ according to equation (20) are given by Table 2 and are in reasonably good agreement with the results calculated by the boundary collocation method.

5. CONCLUDING REMARKS

The paper has outlined the results of three methods used in the calculation of K_I and K_{II} for combined mode bend specimens. If an estimate is to be made at the design stage of an experiment, the results (Table 2) calculated from the approximate relation of Section 4 can be used. K_I and K_{II} can be determined from the crack length a/W, the bending moment and the shearing force on the crack section. The final calculation for a specimen may be made by the boundary collocation method or the finite element method.

REFERENCES

1. GROSS, B., SRAWLEY, J. E and BROWN, W. F., Jr., "Stress factors for a single-edge-notch tension specimen by boundary collocation of a stress function" NASA TN D-2395, 1964.
2. GROSS, B. and SRAWLEY, J. E., "Stress-intensity factors for single-edge-notch specimens in bending or combined bending and tension by boundary collocation of a stress function: NASA TN D-2603, 1965.
3. GROSS, B. and SRAWLEY, "Stress-intensity factors for three-point-bend specimens by boundary collocation" NASA TN D-3092, 1965.
4. WILSON, W. K., CLARK, W. G. and WESSEL, E. T., "Fracture mechanics technology for combined loading of low to intermediate strength metals" AD-682754, 1968.
5. WILSON, W. K., "Numerical method for determining stress intensity factors of an interior crack in a finite plate" Trans. ASME Series D, 93, 1971, 685.
6. HILTON, P.D. and SIH, G. C., "Application of the finite element method to the calculation of stress intensity factors" Methods of Analysis and Solutions of Crack Problems, G. C. Sih, Ed., Noordhoff International Publishing, 1972.
7. WILSON, W. K., "Finite element methods for elastic bodies containing cracks" Methods of Analysis and Solutions of Crack Problems, G. C. Sih, Ed., Noordhoff International Publishing, 1972.
8. ZIENKIEWICZ, O.C., The Finite Element Method in Engineering Science, McGraw-Hill Publishing Company Limited, 1971.
9. HENSHELL, R. D. and SHAW, K. G., "Crack tip finite elements are unnecessary" Int. J. Numerical Method in Engineering, 9, 1975,495.
10. BARSOUM, R. S., "On the use of isoparametric finite elements in linear fracture mechanics" Int. J. Numerical Methods in Engineering, 10, 1976, 25.
11. ESHELBY, J. D., "The continuum theory of lattice defects" Solid State Physics, 3, 1956.
12. HELLEN, T. K., et al., "The calculation of stress intensity factors for combined tensile and shear loading", Int. J. Fracture, 11, 1975, 605.
13. "Energy-momentum tensor and its application in fracture mechanics", (in Chinese) Research Paper of Institute of Mechanics, Academia Sinica 1976.
14. BENTHEM, J.P. and KOITER, W. T., "Asymptotic approximations to crack problems", Methods of Analysis and Solutions of Crack Problems, G.C. Sih, Ed., Noordhoff International Publishing, 1972.

Table 1 Calculated results for K_I^* and K_{II}^* for three-point-bend specimens with s/W=4 by the boundary collocation method

a/W	2s₁/s	0	1/6	2/6	3/6	4/6	5/6	11/12
0.4	K_I^*	7.71	8.50	8.55	8.36	8.33	8.50	8.50
	K_{II}^*	0	1.032	1.400	1.350	1.298	1.376	1.644
0.45	K_I^*	8.86	9.67	9.72	9.38	9.48	9.55	
	K_{II}^*	0	1.142	1.562	1.488	1.466	1.464	
0.50	K_I^*	10.27	11.48	11.50	11.60	11.15	11.59	11.53
	K_{II}^*	0	1.410	1.864	1.840	1.664	1.660	1.760
0.55	K_I^*	12.11	13.30	13.60	13.03	12.90	13.46	
	K_{II}^*	0	1.588	1.980	2.050	1.976	2.100	
0.60	K_I^*	14.47	14.25	14.65	14.91	14.88	14.74	14.50
	K_{II}^*	0	2.348	2.248	2.276	2.320	2.294	2.090

$$K_I^* = K_I BW^{3/2}/M \qquad K_{II}^* = K_{II} BW^{1/2}/Q$$

Table 2   Calculated results due to Equation (20)

a/W	$f_b(a/W)$	$f_s(a/W)$	$f_s'(a/W)$	Difference in percentage
0.05	2.54	0.0636		
0.10	3.51	0.180		
0.15	4.26	0.327		
0.20	4.97	0.496		
0.25	5.67	0.667		
0.30	6.45	0.857		
0.35	7.32	1.080		
0.40	8.35	1.317	1.350	-2.5
0.45	9.60	1.557	1.488	4.4
0.50	11.12	1.838	1.840	-0.1
0.55	13.09	2.125	2.050	3.5
0.60	15.66	2.441	2.276	6.8
0.65	19.17	2.794		
0.70	24.15	3.077		

Note: $f_b(a/W)$ and $f_s(a/W)$ are identical with K_I^* and K_{II}^* in Table 1. $f_s'(a/W)$ is calculated by the boundary collocation method for the case $2s_1/s = 3/6$.

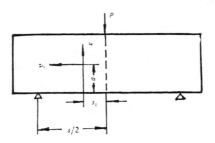

Figure 1   Three-point-bend specimen with crack in unsymmetrical position

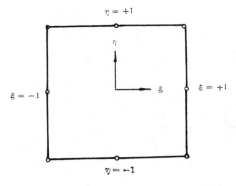

Figure 2   8-noded isoparametric quadratic element

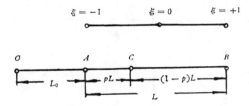

Figure 3   Distorted side of 8-noded isoparametric element

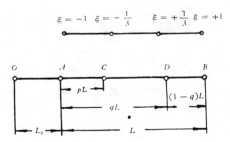

Figure 4   Distorted side of 12-noded isoparametric element

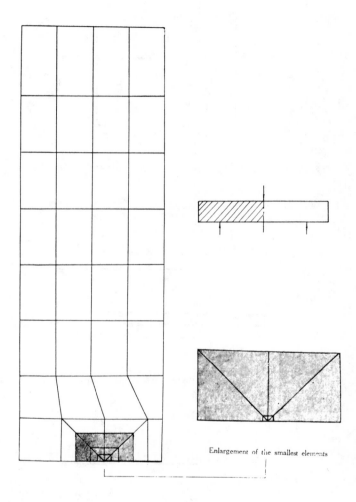

Figure 5   Finite element idealization of a three-point-bend
specimen with a crack in a symmetrical position,
a/W = 0.4

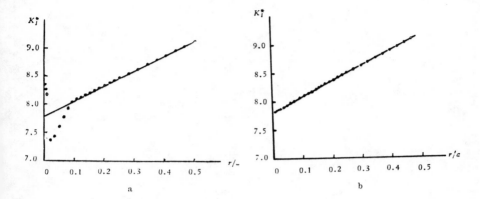

Figure 6 Apparent values of stress intensity factors for
specimen shown in Figure 5

a. Special elements are restricted in vicinity of crack tip
b. Special elements are not restricted in vicinity of crack tip.

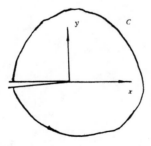

Figure 7 Plane configuration with crack

FRACTURE CRITERIA FOR COMBINED MODE CRACKS

Wang Tzu Chiang*

1. INTRODUCTION

Linear elastic fracture mechanics (LEFM) has been successfully employed in solving the problem of the unstable growth of opening mode cracks, but in engineering practice cracks are usually under a combined mode state of deformation in which K_I, K_{II} and K_{III} are all present. Crack branching will take place in cases where the loading is unsymmetrical, the crack is in an unsymmetrical position, the material is anisotropic, or the crack is propagating with a high velocity. Therefore, investigation of the fracture criteria for combined mode cracks is important theoretically and has wide practical relevance.

There are two kinds of criteria for combined mode fracture, i.e., energy release rate criteria [1-3] and stress parameter criteria [4,5]. The problem of crack branching was analysed by Anderson [1], who was among the first to make an attempt to solve the problem by a complex variable method. Hussain et al [3] gave a detailed analysis of the energy release rate criterion, but it appears to the author that there are some points in this derivation which are questionable.

The complex variable method is employed in this paper to analyse the energy release rate for combined mode cracks. A functional integral equation, which contains no singularity, is derived for a branched crack problem by a functional transformation. The integrand $\phi_1'(z)$ is expanded in eigenfunctions. The energy of fracture criterion for the combined mode (K_I and K_{II}) cracks is then derived when the propagation branch is made to approach zero. An energy of fracture criterion is also presented for the case when a K_{III} is present. In addition, a new fracture criterion for combined mode cracks based on the stress parameters is proposed.

2. FUNDAMENTAL EQUATION AND ITS TRANSFORMATION

Consider a crack branch, which makes an angle γ with the main crack, as shown in Figure 1. According to [3], we have the following formulae for the mapping function $\omega(\zeta)$:

$$\omega(\zeta) = \frac{A}{\zeta} (\zeta - e^{i\alpha_1})^{\lambda_1}(\zeta - e^{i\alpha_2})^{\lambda_2} \tag{1}$$

*Institute of Mechanics, Academia Sinica, Peking

$$\lambda_1 = (1 - \gamma/\pi) \qquad \lambda_2 = (1 + \gamma/\pi) \tag{2}$$

$$\left.\begin{array}{l}
\lambda_1 \ \mathrm{ctg}\left(\dfrac{\alpha_1 - \beta_1}{2}\right) + \lambda_2 \ \mathrm{ctg}\left(\dfrac{\alpha_2 - \beta_2}{2}\right) = 0 \ , \\[12pt]
\lambda_1 \ \mathrm{ctg}\left(\dfrac{\alpha_1 - \beta_2}{2}\right) + \lambda_2 \ \mathrm{ctg}\left(\dfrac{\alpha_2 - \beta_2}{2}\right) = 0 \ . \\[12pt]
r_1 = 4A \ \sin\left(\dfrac{\alpha_1 - \beta_1}{2}\right)^{\lambda_2} \sin\left(\dfrac{\alpha_2 - \beta_1}{2}\right)^{\lambda_2} \\[12pt]
r_2 = 4A \ \sin\left(\dfrac{\beta_2 - \alpha_1}{2}\right)^{\lambda_1} \sin\left(\dfrac{\alpha_2 - \beta_2}{2}\right)^{\lambda_2}
\end{array}\right\} \tag{3}$$

Denoting

$$\varepsilon = \left(\frac{\alpha_2 - \beta_2}{2}\right) \ , \qquad \delta = \left(\frac{\beta_2 - \alpha_1}{2}\right) \tag{4}$$

we have,

$$\left.\begin{array}{l}
\delta = \mathrm{tg}^{-1}\left(\dfrac{\lambda_1}{\lambda_2} \ \mathrm{tg} \ \varepsilon\right), \\[10pt]
\beta_1 = (\varepsilon - \delta) - (\varepsilon + \delta)\gamma/\pi \ , \\[10pt]
\beta_2 = (\delta - \varepsilon) - (\varepsilon + \delta)\gamma/\pi + \pi, \\[10pt]
r_1 = 4A(\cos \varepsilon)^{\lambda_1}(\cos \delta)^{\lambda_2} \\[10pt]
r_2 = 4A(\sin \delta)^{\lambda_1}(\sin \varepsilon)^{\lambda_2}
\end{array}\right\} \tag{5}$$

In the limit as ε approaches zero, r_2, δ and β_1 approach zero, α_1, α_2 and β_2 approach π, and r_1 approaches $4A$. The boundary value problem of elasticity can be reduced to the problem of finding $\phi(\zeta)$ and $\psi(\zeta)$,

$$\sigma_x + \sigma_y = 4\mathrm{Re}\{\phi'(\zeta)/\omega'(\zeta)\} \tag{6}$$

$$\sigma_y - \sigma_x + 2i\tau_{xy} = 2\{[\overline{\omega(\zeta)}/\omega'(\zeta)][\phi'(\zeta)/\omega'(\zeta)]' + \psi'(\zeta)/\omega'(\zeta)\}. \tag{7}$$

$\phi(\zeta)$ and $\psi(\zeta)$ are holomorphic in the exterior of a unit circle and satisfy the following boundary conditions:

$$\phi^-(\sigma) + \frac{\omega(\sigma)}{\overline{\omega'(\sigma)}} \overline{\phi'^-(\sigma)} + \overline{\psi^-(\sigma)} = 0 \ , \qquad \sigma \in L \tag{8}$$

Denoting

$$\phi_*(\zeta) = (\zeta - e^{i\beta_1})(\zeta - e^{i\beta_2})\phi(\zeta) \tag{9}$$

we obtain

$$\phi_*(\zeta) = G_\infty(\zeta) - M_0(\zeta) + G_0'(\zeta) + \frac{(1 - e^{-2\gamma i})}{2\pi i} \int_{L_2} \frac{\overline{\phi'^-(\sigma)} \ g_*(\sigma)}{(\sigma - \zeta) \ \sigma} d\sigma, \quad \zeta \in D^- \tag{10}$$

after some manipulation (Appendix 1), where

$$
\begin{cases}
G_\infty(\zeta) = (\zeta - e^{i\beta_1})(\zeta - e^{i\beta_2})(\Gamma A\zeta + A_0) \\
\quad + A_1(\zeta - \gamma_1 - \gamma_2) + A_2 , \\
M_0(\zeta) = \overline{\Gamma}' A e^{i(\beta_1 + \beta_2)}/\zeta, \\
G_0'(\zeta) = \overline{\Gamma} A e^{i(\alpha_1 + \alpha_2)}/\zeta.
\end{cases}
\tag{11}
$$

Equation (10) is the fundamental equation after the transformation. The coefficients Γ, Γ', A_0, A_1 and A_2 are determined by the behaviour of functions $\phi(\zeta)$ and $\psi(\zeta)$ at the infinity.

A further manipulation gives that

$$
\phi'^-(\gamma_2) = \phi_0'(\gamma_2) + \frac{1}{(\gamma_2 - \gamma_1)}
$$

$$
\cdot \left\{ \frac{1}{2} f_0''^-(\gamma_2) - \frac{f_0'^-(\gamma_2)(\gamma_2 - \gamma_1) - f_0^-(\gamma_2) + f_0^-(\gamma_1)}{(\gamma_2 - \gamma_1)} \right\}
\tag{12}
$$

where

$$
\gamma_1 = e^i \quad , \quad \gamma_2 = e^i
$$

$$
\phi_0(\zeta) = \Gamma A\zeta + A_0 - \frac{A(\overline{\Gamma} + \overline{\Gamma}')}{\zeta}
\tag{13}
$$

$$
f_0(\zeta) = \frac{(1 - e^{-2\gamma i})}{2\pi i} \int_{L_2} \frac{\overline{\phi'^-}(\sigma) g_*(\sigma)}{\sigma(\sigma - \zeta)}
\tag{14}
$$

$$
g_*(\sigma) = (\sigma - e^{i\alpha})(\sigma - e^{i\alpha_2}) .
\tag{15}
$$

In the limit as the length of the branch goes to zero, it can be shown (Appendix 2) that

$$
\phi'(\gamma_2) = \phi_0'(\gamma_2) - \frac{1}{4}(1 - e^{-2\gamma i}) \cdot C^* \cdot \overline{\phi'^-(\gamma_2)} ,
\tag{16}
$$

where

$$
C^* = C_1^* + iC_2^*
\tag{17}
$$

$$
C_1^* = \left(\frac{\lambda_1}{\lambda_2}\right)^{\gamma/2x} \cdot \left\{ P(t_2) + \frac{1}{2} \Omega(t_2) P_2(t_2) \right\}
$$

$$
C_2^* = \frac{\Omega(t_2)}{\pi} \left(\frac{\lambda_1}{\lambda_2}\right)^{\gamma/2x} \cdot \left\{ \frac{1}{\Omega(t_2)} \int_0^1 \frac{P(t) - P(t_2)}{(t - t_2)} dt - \left[\int_0^{t_2-\xi} + \int_{t_2+\xi}^1 \right] \frac{dt}{(t - t_2)^3 P(t)} \right.
$$

$$
+ \frac{1}{2} \int_{t_2-\xi}^{t_2+\xi} \frac{P_2(t) - P_2(t_2)}{t - t_2} dt - \frac{1}{2\xi}[P_1(t_2 + \xi) + P_1(t_2 - \xi)]
$$

$$
\left. + \frac{1}{2\xi^2}\left[\frac{1}{P(t_2 + \xi)} - \frac{1}{P(t_2 - \xi)}\right] + \frac{P(t_2)}{\Omega(t_2)} \ln \frac{(1 - t_2)}{t_2} \right\} .
\tag{18}
$$

Functions $\Omega(t)$, $P(t)$, $P_1(t)$ and $P_2(t)$ are given in Appendix 2. The result given in reference [3] is equivalent to the case $C^* = 1$. The calculated values of C_1^* and C_2^* are listed in Table 1. As the length of the branch approaches zero, the stress intensity factors at the branch tip approach the following limiting values:

$$K_I - iK_{II} = \frac{(\alpha - \bar{\alpha}\beta)}{1 - \beta\bar{\beta}} \tag{19}$$

where

$$\alpha = (\overset{\circ}{K}_I - i\overset{\circ}{K}_{II})e^{\gamma i} \left(\frac{\lambda_1}{\lambda_2}\right)^{\gamma/2\pi} \tag{20}$$

$$\beta = \frac{1}{4}(e^{2\gamma i} - 1) \cdot C^* \tag{21}$$

and $\overset{\circ}{K}_I$ and $\overset{\circ}{K}_{II}$ are the stress intensity factors of a crack which does not have a branch.

3. ENERGY RELEASE RATE AND ENERGY OF FRACTURE CRITERION

In the vicinity of any crack tip, the stresses and the strains are determined by

$$\sigma_r = \frac{1}{2\sqrt{2\pi r}}\{K_I(3 - \cos\theta)\cos\frac{\theta}{2} + K_{II}(3\cos\theta - 1)\sin\frac{\theta}{2}\}$$

$$\sigma_\theta = \frac{1}{2\sqrt{2\pi r}}\{K_I(1 + \cos\theta) - K_{II} \cdot 3\sin\theta\}\cos\frac{\theta}{2} \tag{22}$$

$$\tau_{r\theta} = \frac{1}{2\sqrt{2\pi r}}\{K_I\sin\theta + K_{II}(3\cos\theta - 1)\}\cos\frac{\theta}{2}$$

$$u_r = \frac{1}{4\mu}\sqrt{\frac{r}{2\pi}}\{K_I[(2\kappa - 1)\cos\frac{\theta}{2} - \cos\frac{3\theta}{2}] - K_{II}[(2\kappa - 1)\sin\frac{\theta}{2} - 3\sin\frac{3\theta}{2}]\}$$

$$u_\theta = \frac{1}{4}\sqrt{\frac{r}{2\pi}}\{K_I[-(2\kappa + 1)\sin\frac{\theta}{2} + \sin\frac{3\theta}{2}] - K_{II}[(2\kappa + 1)\cos\frac{\theta}{2} - 3\cos\frac{3\theta}{2}]\}$$

$$\tag{23}$$

from which it can be seen that the displacements on the upper and the lower edges are equal in magnitude and opposite in sign (apart from a uniform displacement of the crack tip). When a branch of length r_2 at an angle θ to the main crack is developed from the main crack, the energy released from the elastic system is equal to

$$G \cdot r_2 = \frac{1}{2}\int_0^{r_2}\{\overset{\circ}{\sigma}_\theta u_\theta^{(1)} + \overset{\circ}{\tau}_{r\theta}u_r^{(1)}\}dr - \frac{1}{2}\int_0^{r_2}\{\overset{\circ}{\sigma}_\theta u_\theta^{(2)} + \overset{\circ}{\tau}_{r\theta}u_r^{(2)}\}dr$$

$$= \int_0^{r_2}\{\overset{\circ}{\sigma}_\theta u_\theta^{(1)} + \overset{\circ}{\tau}_{r\theta}u_r^{(1)}dr\} = \frac{(\kappa + 1)}{16\mu}r_2\{K_I\overset{\circ}{f}_1 + K_{II}\overset{\circ}{f}_2\}.$$

Therefore, the energy release rate is

$$G = \frac{\kappa + 1}{16\mu} \{K_I \overset{\circ}{f_1} + K_{II} \overset{\circ}{f_2}\} \tag{24}$$

$$\overset{\circ}{f_1} = \{\overset{\circ}{K_I}(1 + \cos\theta) - \overset{\circ}{K_{II}} \cdot 3 \sin\theta\} \cos\frac{\theta}{2}$$

$$\tag{25}$$

$$\overset{\circ}{f_2} = \{\overset{\circ}{K_I} \sin\theta + \overset{\circ}{K_{II}}(3 \cos\theta - 1)\} \cos\frac{\theta}{2}$$

where the superscript $^\circ$ is used to denote the functions and the physical quantities of the crack which does not have a branch. The case of Figure 1 is equivalent to the case $\theta = -\gamma$.

According to the energy of fracture criterion, the crack will propagate in the direction in which the energy release rate is maximum and it will start to propagate when this maximum energy release rate G_{max} reaches a critical value. The calculation of equation (24) leads to the following results: for a crack in the sliding mode, the fracture angle is $\gamma = 76.2°$, and $K_{IIc} = 0.724 K_{Ic}$, while according to the maximum σ_θ criterion, $K_{IIc} = 0.87 K_{Ic}$ and the fracture angle is $\gamma = 70.5°$, and the criterion of the minimum strain energy density gives $K_{IIc} = 0.96 K_{Ic}$ and $\gamma = 82.3°$ (with $\nu = 0.3$).

For the case of uniaxial tension with an inclined crack, the fracture angles are shown in Figure 3, and the correlation curve of $\overset{\circ}{K_I}$ and $\overset{\circ}{K_{II}}$ in the critical state is shown in Figure 4. Also shown in these figures are the available experimental results, which have a rather wide scatter band.

4. ENERGY OF FRACTURE CRITERION INCORPORATING K_{III}

As shown in Figure 5, due to the combined action of the axial stress σ and the antiplane shear stress τ at infinity, all K_I, $\overset{\circ}{K_{II}}$ and $\overset{\circ}{K_{III}}$ are present and they are

$$\left.\begin{aligned} K_I &= \sigma\sqrt{\pi a}\ \sin^2\beta \\[4pt] K_{II} &= \sigma\sqrt{\pi a}\ \sin\beta\cos\beta \\[4pt] K_{III} &= \tau\sqrt{\pi a}\ \sin\beta \end{aligned}\right\} \cdot \tag{26}$$

Since the antiplane shear produces only the displacement w, in the direction perpendicular to the plane, the in-plane displacements u and v are both equal to zero in the case where K_{III} alone is present. As the axial traction is responsible only for the strains in the plane, it has no contribution to the strains in the direction perpendicular to the plane. When an infinitesimal branch is developed, the stress intensity factors at the branch tip are

$$K_I - iK_{II} = \frac{\alpha - \alpha\bar{\beta}}{1 - \beta\bar{\beta}} \tag{27}$$

$$K_{III} = \overset{\circ}{K_{III}}\left(\frac{1 - m}{1 + m}\right)^{m/2} \tag{28}$$

where $m = \gamma/\pi$. Equation (28) was derived in reference [9].

According to reference [7], the total potential energy released during the formation of the new crack surfaces C_1' and C_2' can be calculated by

$$-\Delta\Pi = -\frac{1}{2}\int_{C_1'+C_2'} T_i \Delta u_i \, dS \tag{29}$$

where T_i are the tractions acted on the surfaces C_1' and C_2' before the crack has extended, and Δu_i are the additional displacements produced after the crack has extended. With the action of the antiplane shear, the stresses and the strains at the crack tip are

$$\tau_{zr} - i\tau_{z\theta} = \frac{K_{III}}{\sqrt{2\pi r}} \sin\frac{\theta}{2} - i\cos\frac{\theta}{2} \tag{30}$$

$$w = \frac{1}{\mu}\sqrt{\frac{2r}{\pi}} \cdot K_{III} \sin\frac{\theta}{2} \ . \tag{31}$$

Consider the propagation branch shown in Figure 1. The stresses along OB before crack extension are

$$\overset{\circ}{\tau}_{zr} - i\overset{\circ}{\tau}_{z\theta} = -\frac{\overset{\circ}{K}_{III}}{\sqrt{2\pi r}} \sin\frac{\gamma}{2} + i\cos\frac{\gamma}{2} \tag{32}$$

and the additional displacements after crack extension are

$$w = \pm\frac{1}{\mu}\sqrt{\frac{2(r_2 - r)}{\pi}} \ K_{III} \tag{33}$$

where K_{III} is the stress intensity factor at the tip of the propagation branch B after the extension. Substituting equations (32) and (33) into (29), the energy release rate is obtained:

$$G = \underset{r_2 \to 0}{\text{Lim}} -\left(\frac{\Delta\Pi}{r_2}\right) = \frac{1}{2\mu} K_{III}\overset{\circ}{K}_{III}\cos\frac{\gamma}{2} \tag{34}$$

Combining equations (34) and (24), we obtain the energy release rate under the combined action of K_I, K_{II} and K_{III}:

$$G = \frac{(1-\nu^2)}{E}\left\{\frac{1}{2}[K_I\overset{\circ}{f}_1 + K_{II}\overset{\circ}{f}_2] + \frac{1}{(1-\nu)} K_{III}\overset{\circ}{K}_{III} \cos\frac{\gamma}{2}\right\} \ . \tag{35}$$

According to the energy of fracture criterion and equation (35), it follows that

$$K_{IIIc} = \sqrt{(1-\nu)} \ K_{Ic} \ . \tag{36}$$

The correlation curves of $\overset{\circ}{K}_I$, $\overset{\circ}{K}_{II}$ and $\overset{\circ}{K}_{III_o}$ in the critical state are shown in Figure 6. The correlation curve of $\overset{\circ}{K}_I$ and $\overset{\circ}{K}_{III}$ with $\overset{\circ}{K}_{II}$ equal to zero is shown in Figure 7. The curve can be represented by the following equation

$$\left(\frac{\overset{\circ}{K}_I}{K_{Ic}}\right)^2 + \left(\frac{\overset{\circ}{K}_{III}}{K_{IIIc}}\right)^3 = 1 \quad . \tag{37}$$

It can be seen from Figure 7 that the theory is in fairly good agreement with with the experimental data.

5. STRESS PARAMETER CRITERION FOR COMBINED MODE FRACTURE

Among the stress parameter criteria for combined mode fracture, the maximum σ_θ criterion and the minimum strain-energy-density criterion are those commonly used [4]. Both are based on a comparison of the mechanical quantities on circles with the crack tip as their centre. This kind of comparison has a clear geometrical significance, but it can be argued that the different points on the circle are not under the same mechanical state (Figure 8).

Consider the strain energy density in the front of the crack

$$W = \frac{1}{\pi r} (a_{11}K_I^2 + 2a_{12}K_I K_{III} + a_{22}K_{II}^2) \tag{38}$$

where

$$a_{11} = \frac{1}{16\mu} \{(1 + \cos\theta)(\kappa - \cos\theta)\}$$

$$a_{12} = \frac{1}{16\mu} \{2\cos\theta - (\kappa - 1)\}\sin\theta \tag{39}$$

$$a_{22} = \frac{1}{16\mu} \{(\kappa + 1)(1 - \cos\theta) + (1 + \cos\theta)(3\cos\theta - 1)\}$$

and

$$\kappa = \begin{cases} 3 - 4\nu & \text{for plane strain} \\ \dfrac{3 - \nu}{1 + \nu} & \text{for plane stress} \end{cases} \tag{40}$$

We choose the strain energy density W as a mechanical measure to characterize brittle fracture and consider the lines with equal strain-energy-densities (the iso-W lines) (Figure 9). For example, if $W = a_0$ on an iso-W line Γ_0 the points A_0, B_0 and C_0 on the line will have the same strain-energy-density. Since the elements, with the points A_0, B_0, C_0 etc., as their centres, contain the same quantity of strain energy, these points can be compared with one another and in the direction of the point where the circumferential stress σ_θ is a maximum fracture is most apt to occur. Thereby a new criterion is obtained to determine the direction along which the crack will start to propagate, that is, the crack will start to grow in the direction where the circumferential stress σ_θ is maximum on an iso-W line. Let the fracture angle be θ_0, then

$$(\sigma_\theta)_{\substack{\theta=\theta_0 \\ W=a_0}} = \max(\sigma_\theta) \quad .$$

The load at which the crack will start to grow can be determined by

$$\lim_{r \to 0} \sqrt{2\pi r} \; (\sigma_\theta)_{\theta=\theta_0} = K_{Ic} \; . \tag{42}$$

On the iso-W lines we have

$$W = \frac{S}{r} = a_0, \tag{43}$$

where S is the strain-energy-density factor given by

$$S = \frac{1}{\pi} (a_{11}K_I^2 + 2a_{12}K_I K_{II} + a_{22}K_{II}^2) \; . \tag{44}$$

In the front of the crack we have

$$\sigma_\theta = \frac{1}{2\sqrt{2\pi r}} \{K_I(1 + \cos\theta) - 3 \sin\theta K_{II}\} \cos\frac{\theta}{2} \; . \tag{45}$$

From equation (43) we have

$$r = \frac{S}{a_0} \; , \tag{46}$$

Substituting equation (46) into equation (45), we obtain

$$\sigma_\theta = \frac{a_0}{2\sqrt{2\pi S}} \{K_I(1 + \cos\theta) - 3 \sin\theta K_{II}\}\cos\frac{\theta}{2} \; , \tag{47}$$

Equation (47) gives the relationship between the circumferential stress σ_θ and θ on the iso-W lines. Since a_0 is a positive constant, the fracture angle θ_0 can be determined by the point where the following function f is maximum:

$$f(\theta) = \frac{1}{\sqrt{\pi S}} \{K_I(1 + \cos\theta) - 3 \sin\theta K_{II} \cos\frac{\theta}{2} \tag{48}$$

Calculated results for the in-plane shear of a plate with a central crack are given in Table 2. A fracture test is proposed in reference [3] on a 152 mm wide by 406 mm long panel of 0.05 mm thick steel foil containing a circular crack, where a pure shear state at the crack tip can be realised. The measured fracture angles have an average value of -75.4°, which is in good agreement with the theory just described. The fracture angles for the case of uniaxial tension with an inclined crack are shown in Table 3 and they are in good agreement with the experimental data.

APPENDIX 1

$\phi(\zeta)$ and $\psi(\zeta)$ are holomorphic functions in the exterior of a unit circle in the image plane and satisfy the following boundary condition:

$$\phi^-(\sigma) + \frac{w(\sigma)}{\overline{w'(\sigma)}} \; \overline{\phi'^-(\sigma)} + \overline{\psi^-(\sigma)} = 0 \; , \qquad \sigma \in L \tag{1}$$

According to reference [1], we have

$$\frac{w'(\zeta)}{w(\zeta)} = \frac{(\zeta - e^{i\alpha_1})(\zeta - e^{i\alpha_2})}{\zeta(\zeta - e^{i\beta_1})(\zeta - e^{i\beta_2})} = \frac{1}{\zeta g(\zeta)} \quad , \tag{2}$$

By the mapping function $w(\zeta)$ a deflected crack in the physical plane is mapped onto a unit circle in the ζ-plane, as shown in Figure 2, where the arcs L_1 and L_2 are the images of the main crack and the propagation branch in the physical plane, respectively. Hence we have

$$\frac{w(\sigma)}{\overline{w'(\sigma)}} = \begin{cases} \overline{\sigma g(\sigma)} \,, & \sigma \in L_1 \\ \overline{\sigma g(\sigma)} e^{-2\gamma i} \,, & \sigma \in L_2 \end{cases} \tag{3}$$

We locate the branch cut along a secant $\bar{L}_2$ for the mapping function $w(\zeta)$, so $w(\zeta)$ and $w'(\zeta)$ are continuous across the unit circle (apart from two points $e^{i\alpha_1}$ and $e^{i\alpha_2}$).

Introducing a jump function $h(\sigma)$, as

$$h(\sigma) = \begin{cases} 1 & \sigma \in L_1 \\ e^{-2\gamma i} & \sigma \in L_2 \end{cases} \tag{4}$$

and noting that $\overline{g(\sigma)} = -g(\sigma)$, equation (1) can be written as

$$\phi^-(\sigma) - \frac{g(\sigma)}{\sigma} h(\sigma) \overline{\psi'^-(\sigma)} + \overline{\psi'(\sigma)} = 0 \,. \tag{5}$$

Let

$$f_*(\sigma) = (\sigma - e^{i\beta_1})(\sigma - e^{i\beta_2}) \tag{6}$$

$$g_*(\sigma) = (\sigma - e^{i\alpha_1})(\sigma - e^{i\alpha_2}) \tag{7}$$

$$\phi_*(\sigma) = f_*(\sigma)\phi(\sigma) \tag{8}$$

and multiplying equation (5) by the function $f_*(\sigma)$, we have

$$\phi_*^-(\sigma) - \frac{g_*(\sigma)}{\sigma} h(\sigma) \overline{\phi'^-(\sigma)} + f_*(\sigma) \overline{\phi^-(\sigma)} = 0 \,, \qquad \sigma \in L \tag{9}$$

Assuming that the function $\psi(\zeta)$ has poles of order one at the points $\zeta = e^{i\beta_1}$ and $\zeta = e^{i\beta_2}$, it can be shown that the function $f_*(\zeta) \psi(1/\zeta)$ is holomorphic in the interior of the unit circle, except for the origin. From equation (9), using the extended Cauchy's integral formula, we obtain

$$-\phi_*(\zeta) + G_\infty(\zeta) - \frac{1}{2\pi i} \oint_L \frac{h(\sigma)g_*(\sigma)}{\sigma(\sigma - \zeta)} \overline{\phi'^-(\sigma)} \, d\sigma - M_0(\zeta) = 0, \quad \zeta \in D^- \tag{10}$$

where $G_\infty(\zeta)$ is the main part of the function $\phi_*(\zeta)$ in the neighbourhood of $\zeta = \infty$ and $M_0(\zeta)$ is the main part of the function $f_*(\zeta) \psi(1/\zeta)$ in the neighbourhood of $\zeta = 0$.

Assume that at infinity we have

$$\phi(\zeta) = \Gamma A\zeta + A_0 + \frac{A_1}{\zeta} + \frac{A_2}{\zeta^2} + \ldots \tag{11}$$

$$\psi(\zeta) = \Gamma'A\zeta + B_0 + \frac{B_1}{\zeta} + \frac{B_2}{\zeta^2} + \ldots \tag{12}$$

From equation (10) we can obtain

$$\phi_*(\zeta) = G_\infty(\zeta) - M_0(\zeta) + G_0'(\zeta) + \frac{(1-e^{-2\gamma i})}{2\pi i} \int_{L_2} \frac{\overline{\phi'^-(\sigma)}}{(\sigma-\zeta)} \, g_*(\sigma) \, d\sigma \tag{13}$$

where $G_0'(\zeta)$ is the main part of the function

$$\frac{g_*(\zeta)}{\zeta} \, \bar{\phi}' \left(\frac{1}{\zeta}\right),$$

holomorphic in the interior of the unit circle, in the neighbourhood of $\zeta = 0$. From equations (11) and (12), we have

$$G_\infty(\zeta) = (\zeta-\gamma_1)(\zeta-\gamma_2)(\Gamma A\zeta+A_0) + A_1(\zeta-\gamma_1-\gamma_2) + A_2 \tag{14}$$

$$M_0(\zeta) = \bar{\Gamma}'A\gamma_1\gamma_2/\zeta \tag{15}$$

$$G_0'(\zeta) = \bar{\Gamma}A\sigma_1\sigma_2/\zeta \tag{16}$$

where

$$\gamma_1 = e^{i\beta_1}, \quad \gamma_2 = e^{i\beta_2}, \quad \sigma_1 = e^{i\alpha_1}, \quad \sigma_2 = e^{i\alpha_2}.$$

Let

$$f_0(\zeta) = \frac{(1 - e^{-2\gamma i})}{2\pi i} \int_{L_2} \frac{\overline{\phi'^-(\sigma)}}{(\sigma-\zeta)} \, \frac{g_*(\sigma)}{\sigma} \, d\sigma \tag{17}$$

Equation (13) becomes

$$\phi_*(\zeta) = (\zeta-\gamma_1)(\zeta-\gamma_2)(\Gamma A\zeta+A_0) + (\zeta-\gamma_1-\gamma_2)A_1 + A_2$$
$$+ \frac{A}{\zeta}(\sigma_1\sigma_2\bar{\Gamma} - \gamma_1\gamma_2\bar{\Gamma}') + f_0(\zeta). \tag{18}$$

In the limit as ζ approaches γ_1 and γ_2 from outside of the unit circle, we have

$$\left. \begin{array}{l} -A_1\gamma_2 + A_2 + \dfrac{A}{\gamma_1}(\sigma_1\sigma_2\bar{\Gamma} - \gamma_1\gamma_2\bar{\Gamma}') + f_0^-(\gamma_1) = 0 \\[2mm] -A_1\gamma_1 + A_2 + \dfrac{A}{\gamma_2}(\sigma_1\sigma_2\bar{\Gamma} - \gamma_1\gamma_2\bar{\Gamma}') + f_0^-(\gamma_2) = 0 \end{array} \right\}, \tag{19}$$

from which we obtain

$$A_1 = - A(\bar{\Gamma} + \bar{\Gamma}') - \frac{f_0^-(\gamma_2) - f_0^-(\gamma_1)}{(\gamma_2 - \gamma_1)} \tag{20}$$

$$A_2 = \frac{\gamma_1 f_0^-(\gamma_1) - \gamma_2 f_0^-(\gamma_2)}{(\gamma_2 - \gamma_1)}$$

Substituting equation (20) into equation (18), after re-arrangement, we have

$$\phi(\zeta) = \phi_0(\zeta) + \frac{1}{(\gamma_2 - \gamma_1)} \left\{ \frac{f_0^-(\zeta) - f_0^-(\gamma_2)}{(\zeta - \gamma_2)} - \frac{f_0(\zeta) - f_0^-(\gamma_1)}{(\zeta - \gamma_1)} \right\} \tag{21}$$

where

$$\phi_0(\zeta) = \Gamma A \zeta + A_0 - \frac{A}{\zeta}(\overline{\Gamma} + \overline{\Gamma}') \tag{22}$$

and

$$\phi'(\zeta) - \phi_0'(\zeta) + \frac{1}{(\gamma_2 - \gamma_1)} \left\{ \frac{f_0'(\zeta)(\zeta - \gamma_2) - f_0(\zeta) + f_0^-(\gamma_2)}{(\zeta - \gamma_2)^2} \right.$$

$$\left. - \frac{f_0'(\zeta)(\zeta - \gamma_1) - f_0(\zeta) + f_0^-(\gamma_1)}{(\zeta - \gamma_1)^2} \right\} \tag{23}$$

Using Taylor's formula with remainder, we have

$$f_0(\zeta_1) = f_0(\zeta) + f_0'(\zeta)(\zeta_1 - \zeta) + \frac{1}{2} f_0''(\zeta + \theta(\zeta_1 - \zeta))(\zeta_1 - \zeta)^2 \quad \zeta, \zeta_1 \in D^- \tag{24}$$

As ζ_1 goes to γ_2, we obtain

$$f_0^-(\gamma_2) = f_0(\zeta) + f_0'(\zeta)(\gamma_2 - \zeta) + \frac{1}{2} f_0''(\zeta + \theta(\gamma_2 - \zeta))(\gamma_2 - \zeta)^2 \tag{25}$$

Substituting equations (24) and (25) into equation (23) and let ζ go to γ_2, we obtain

$$\phi'(\gamma_2) = \phi_0'(\gamma_2) + \frac{1}{(\gamma_2 - \gamma_1)}$$

$$\cdot \left\{ \frac{1}{2} f_0''^-(\gamma_2) - \frac{f_0^-(\gamma_2)(\gamma_2 - \gamma_1) - f_0^-(\gamma_2) + f_0''^-(\gamma_1)}{(\gamma_2 - \gamma_1)^2} \right\} \tag{26}$$

APPENDIX 2

Let the function $f_0(\zeta)$ be

$$f_0(\zeta) = \frac{(1 - e^{-2\gamma i})}{2\pi i} \int_{L_2} \frac{\phi'^-(\sigma) g_*(\sigma)}{\sigma(\sigma - \zeta)} d\sigma, \qquad \zeta \in D^- \tag{27}$$

If the Goursat functions in the physical plane (z-plane) are $\phi_1(z)$ and $\psi_1(z)$, we have

$$\phi(\zeta) = \phi_1(w(\zeta)), \ \phi'(\zeta) = \phi_1'(w(\zeta))w'(\zeta)$$

Let the region between the arc L_2 and the secant $\overline{L}_2$ be denoted by T_2, as shown in Figure 10, then the function $\overline{w}(1/\zeta)$ is holomorphic in T_2 and takes the same values on L_2 as the function $w(\zeta)e^{2\gamma i}$, sectionally holomorphic with the cut $\overline{L}_2$. Therefore in T_2 we have this identity:

$$\overline{w}\left(\frac{1}{\zeta}\right) = w(\zeta)e^{2\gamma i}, \quad \zeta \in T_2 \tag{28}$$

Function $\phi'(1/\zeta)$ is also holomorphic in T_2, therefore,

$$f_0(\zeta) = \frac{(1 - e^{-2\gamma i})}{2\pi i} \int_{\overline{L}_2} \frac{\overline{\phi}'(1/\sigma) g_*(\sigma) d\sigma}{\sigma(\sigma - \zeta)}, \qquad \zeta \in D^- \tag{29}$$

where the integration path is already shifted to the secant $\overline{L}_2$. We have the following relations:

$$\overline{\phi}' \; \frac{1}{\zeta} \;=\; \overline{\phi' \; \frac{1}{\overline{\zeta}}} \;=\; \overline{\phi_1' \; w \; \frac{1}{\zeta} \; w' \; \frac{1}{\zeta}}$$

$$=\; -\overline{\phi_1'}(w(\zeta)\,e^{-2\gamma i})\; \frac{\zeta(\zeta-\gamma_1)(\zeta-\gamma_2)}{(\zeta-\sigma_1)(\zeta-\sigma_2)}\; w(\zeta)\,e^{2\gamma i}, \qquad \zeta \in T_2 \qquad (30)$$

As is well known, the function $\phi_1(z)$ can be expanded into the following series:

$$\phi_1(z) \;=\; \sum_{n=1}^{\infty} A_n (z-z)^{\nu_n}, \tag{31}$$

$$\phi_1'(z) \;=\; \sum_{n=1}^{\infty} \nu_n A_n (z-z_2)^{\nu_n-1}, \qquad \nu_n = \frac{n}{2} \tag{32}$$

in the neighbourhood of the propagation branch tip $z = z_2$. Hence

$$\overline{\phi_1'\left(w\left(\frac{1}{\zeta}\right)\right)} \;=\; \sum_{n=1}^{\infty} \nu_n \overline{A}_n \left(\overline{w}\left(\frac{1}{\zeta}\right) - \overline{z}_2\right)^{\nu_n-1} \;=\; \sum_{n=1}^{\infty} \nu_n \overline{A}_n (z-z_2)^{\nu_n-1} e^{2\gamma i(\nu_n-1)},$$

when $\zeta \epsilon T_2$. Substituting this expression into equation (29), we have

$$f_0(\zeta) \;=\; -\frac{(1-e^{-2\gamma i})}{2\pi i} \int_{\overline{L}_2} \frac{w^-(\sigma)(\sigma-\gamma_1)(\sigma-\gamma_2)}{(\sigma-\zeta)} \sum_{n=1}^{\infty} \nu_n \overline{A}_n (w^-(\sigma)-w(\gamma_2))^{\nu_n-1} e^{2\gamma i \nu_n} d\sigma$$

$$=\; -(1-e^{-2\gamma i}) \sum_{n=1}^{\infty} \nu_n \overline{A}_n f_n(\zeta)\, e^{2\gamma i \nu_n} \tag{33}$$

where

$$f_n(\zeta) \;=\; \frac{1}{2\pi i} \int_{\overline{L}_2} \frac{w^-(\sigma)(\sigma-\gamma_1)(\sigma-\gamma_2)}{(\sigma-\zeta)} [w^-(\sigma) - w(\gamma_2)]^{\nu_n-1} d\sigma \tag{34}$$

and $w^-(\sigma)$ refers to the values of $w(\zeta)$ on the secant $\overline{L}_2$ as ζ goes to $\overline{L}_2$ from inside of the region T_2. Introduce the following linear transformation

$$\zeta \;=\; \sigma_1 + s(\sigma_2-\sigma_1) \tag{35}$$

by which the exterior of the unit circle in the ζ-plane is mapped onto the exterior of the circle L* in the s-plane, and the secant $\sigma_1\sigma_2$ on to the segment $(0,1)$ on the real axis. Then

$$w(\zeta) \;=\; Ae^{-\pi\lambda_2 i}\; \frac{(\sigma_2-\sigma_1)^2}{\sigma_1}\; \Omega(s) \tag{36}$$

$$\Omega(s) \;=\; \frac{s^{\lambda_1}(1-s)^{\gamma_2}}{(1+es)} \tag{37}$$

$$e \;=\; \frac{\sigma_2-\sigma_1}{\sigma_1} \tag{38}$$

Substituting equations (36), (37) and (38) into equation (34), we obtain

$$f_n(\zeta) = \left(\frac{Ae^{-\pi\lambda_2 i}}{\sigma_1}\right)^{\nu_n} \frac{(\sigma_2-\sigma_1)^{2\nu_n+1}}{2\pi i}$$

$$\cdot \int_0^1 \frac{\Omega(t)(t-s_2)}{(t-s)} [\sigma_1-\gamma_1 + t(\sigma_2-\sigma_1)][\ (t)\ -\ (s\ )]^{n-1} dt. \tag{39}$$

$f_n(\zeta)$, $f_n'(\zeta)$ and $f_n''(\zeta)$ exist everywhere in the exterior of the unit circle and on the unit circle, including the point $\zeta = \gamma_2$, except for the points σ_1 and σ_2. Using the above expressions, it can easily be shown that for the case that $n \geq 2$, $f_n(\zeta)$, $f_n'(\zeta)$ and $f_n''(\zeta)$ all approach zero in the limit as the length of the propagation branch goes to zero. Hence in order to find $f_0(\zeta)$ in the limiting case it is only necessary to calculate $f_1(\zeta)$, which is

$$f_1(\zeta) = (\sigma_2-\sigma_1)^2 \sqrt{\frac{Ae^{-\pi i\lambda_2}}{\sigma_1}} \int_0^1 \frac{\Omega^-(t)(t-s_2)\{\sigma_1-\gamma_1 + t(\sigma_2-\sigma_1)\}}{(t-s)\ \sqrt{\Omega^-(t) - \Omega(s_2)}} dt$$

and

$$f_1''(\zeta) = 2 \sqrt{\frac{Ae^{-\pi i\lambda_2}}{\sigma_1}} \int_0^1 \frac{\Omega^-(t)(t-s_2)\{\sigma_1-\gamma_1 + t(\sigma_2-\sigma_1)\}}{(t-s)^3\ \sqrt{\Omega^-(t) - \Omega(s_2)}} dt \tag{40}$$

where s_2 is the image of γ_2:

$$s_2 = \frac{(\gamma_2-\sigma_1)}{(\sigma_2-\sigma_1)}.$$

On the other hand, since the numerator of the function $\Omega^-(t)$ takes real values when t varies on the interval $[0,1]$ of the real axis, $^-(t)$ can be extended analytically from the lower half plane to the upper half plane through the interval $[0.1]$. Therefore, the function $\Omega(s)$ can be expanded into a Taylor's series in the neighbourhood of $s = s_2$ with a circle of convergence including some part of the interval $[0,1]$. Since $w'(\gamma_2) = 0$ and $\Omega'(s_2) = 0$, we have

$$P_0(t) = \frac{\Omega^-(t)-\Omega(s_2)}{(t-s_2)^2} = \sum_{n=2}^{\infty} \frac{\Omega^{(n)}(s_2)}{n!} (t-s_2)^{n-2} \tag{41}$$

in the interval. Denoting

$$P(t) = - i \sqrt{P_0(t)} \tag{42}$$

$$Q(s_2) = \frac{1}{2\pi} \int_0^1 \frac{-\Omega^-(t)dt}{(t-s_2)^3 P(t)} \tag{43}$$

and integrating by parts, we obtain

$$Q(s_2) = \left\{\int_0^1 \frac{P(t)-P(s_2)}{(t-s_2)} dt + P(s_2)\ln\left(\frac{s_2-1}{s_2}\right)\right\}\frac{1}{2\pi}$$

$$- \Omega(s_2)\left\{\left[\int_0^a + \int_b^1\right]\frac{dt}{(t-s_2)^3 P(t)} + \frac{1}{2}\int_a^b \frac{\left[\frac{1}{P(t)}\right]'' - \left[\frac{1}{P(t)}\right]''_{t=s_2}}{(t - s_2)} dt\right.$$

$$\left. + \frac{1}{2}\left[P^{-1}(t)\right]''_{t=s_2} \ln \frac{(b-s_2)}{(a-s_2)} - \frac{1}{2}\frac{\left[\frac{1}{P(t)}\right]'}{(t-s_2)}\Big|_a^b - \frac{1}{2}\frac{\left[\frac{1}{P(t)}\right]}{(t-s_2)^2}\Big|_a^b\right\}\frac{1}{2\pi} \tag{44}$$

where all integrals are Riemman integrals in the ordinary sense. In the limit, as the length of the propagation branch approaches zero, we have

$$s_2 \to t_2 = \frac{\lambda_1}{2} \ , \quad \Omega(t) \to t^{\lambda_1}(1-t)^{\lambda_2} \tag{45}$$

$$D = \underset{r_2 \to 0}{\mathrm{Lim}} Q(s_2) = \frac{\Omega(t_2)}{2\pi} \left\{ \frac{1}{\Omega(t_2)} \int_0^1 \frac{P(t)-P(t_2)}{t-t_2} \, dt \ - \left[\int_0^{t_2-\xi_0} + \int_{t_2+\xi}^1 \right] \frac{dt}{(t-t_2)^3 P(t)} \right.$$

$$+ \frac{1}{2} \int_{t_2-\xi}^{t_2+\xi} \frac{P_2(t)-P_2(t_2)}{t-t_2} \, dt - \frac{1}{2\xi} \left[P_1(t_2+\xi) + P_1(t_2-\xi) \right]$$

$$+ \frac{1}{2\xi^2} \left[\frac{1}{P(t_2+\xi)} - \frac{1}{P(t_2-\xi)} \right] + \frac{P(t_2)}{\Omega(t_1)} \ln \frac{(1-t_2)}{t_2} - \pi i \left(\frac{P(t_2)}{\Omega(t_2)} + \frac{1}{2} P_2(t_2) \right) \right\} \tag{46}$$

where

$$P_1(t) = \frac{1}{2} \frac{P_0'(t)}{P_0(t)} \frac{1}{P(t)} \tag{47}$$

$$P_2(t) = \frac{1}{2} \left\{ P_0(t)P_0''(t) - \frac{3}{2} P_0'(t)P_0'(t) \right\} \frac{1}{P_0^2(t)P(t)} \tag{48}$$

and ξ is an arbitrary positive number that satisfies the following condition:

$$2\xi \le \lambda_0 = \min \{\lambda_1, \lambda_2\} \tag{49}$$

It is easily shown that [13]

$$\left| \Omega^{(n)}(t_2) \right| \le \Omega(t_2) \left| \cdot (n+1)! \left(\frac{2}{\lambda_0} \right)^n , \quad n \ge 3 \tag{50}$$

Hence the following Taylor's series exists:

$$\Omega(t) = \Omega(t_2) + \Omega'(t_2)(t-t_2) + \cdots + \frac{\Omega^{(n)}(t_2)}{n!} (t-t_2)^n + \cdots \tag{51}$$

when $|t-t_2| \le \xi$. From equations (40) and (46), we have

$$\underset{\varepsilon \to 0}{\mathrm{Lim}} \ f_1''^-(\gamma_2) = -4\sqrt{\frac{Ae^{-\pi\lambda_2 i}}{\sigma_1}} \ D \tag{52}$$

$$\underset{\varepsilon \to 0}{\mathrm{Lim}} \ f_1'^-(\gamma_2) = 0 \tag{53}$$

Due to equation (26), we have

$$\phi'^-(\gamma_2) = \phi_0'(\gamma_2) - \frac{1}{4} f_0''^-(\gamma_2) \tag{54}$$

as $\varepsilon \to 0$. Using equation (33), we can obtain

$$\frac{1}{4} f_0''^-(\gamma_2) = -\frac{1}{4}(1-e^{-2\gamma i}) \frac{1}{2} \bar{A}_1 e^{\gamma i} f_1''^-(\gamma_2) = \frac{1}{2}(1-e^{-2\gamma i})\bar{A}_1\sqrt{Ae}^{\gamma i} e^{-\pi(\lambda_2+1)i/2} D \tag{55}$$

It was shown in reference [1] that

$$\frac{1}{\sqrt{e^{\pi\lambda_1 i}}} \; A_1 = \frac{1}{\sqrt{2\pi}} \; (K_I - iK_{II}) = \frac{\sqrt{2} \; \phi'^{-}(\gamma_2)}{\sqrt{e^{\pi\lambda_1 i}} \, w''(\gamma_2)} \tag{56}$$

Substituting equation (56) into equations (54) and (55), we have

$$\phi'^{-}(\gamma_2) = \phi_0'(\gamma_2) - \frac{1}{4} \, (1 - e^{-2\gamma i}) C^* \, \overline{\phi'^{-}(\gamma_2)} \tag{57}$$

where

$$C^* = 2 \left(\frac{\lambda_1}{\lambda_2}\right)^{\gamma/2\pi} \qquad Di = C_1^* + iC_2^* \tag{58}$$

$$C^* = \left(\frac{\lambda_1}{\lambda_2}\right)^{\gamma/2\pi} \left\{ P(t_2) + \frac{1}{2} \, \Omega(t_2) P_2(t_2) \right\} \tag{59}$$

$$C_2^* = \frac{\Omega(t_2)}{\pi} \left(\frac{\lambda_1}{\lambda_2}\right)^{\gamma/2\pi} \left\{ \frac{1}{\Omega(t_2)} \int_0^1 \frac{P(t)-P(t_2)}{(t-t_2)} \, dt - \left[\int_0^{t_2-\xi} + \int_{t_2+\xi}^0 \right] \frac{dt}{(t-t_2)^3 P(t)} \right.$$

$$+ \frac{1}{2} \int_{t_2-\xi}^{t_2+\xi} \frac{P_2(t)-P_2(t_2)}{(t-t_2)} \, dt - \frac{1}{2\xi} \, [P_1(t_2+\xi)+P_1(t_2-\xi)]$$

$$\left. + \frac{1}{2\xi^2} \left[\frac{1}{P(t_2+\xi)} - \frac{1}{P_2(t_2-\xi)} \right] + \frac{P(t_2)}{\Omega(t_2)} \, \ln \frac{(1-t_2)}{t_2} \right\} \tag{60}$$

REFERENCES

1. ANDERSON, H., J. Mech. Phys. Solids, 17, 1969, 405.

2. PALANISWAMY, K., and KNAUSS, W.G., Int. J. Fracture Mech., 8, 1972, 114.

3. HUSSAIN, M.A., PU, S.L., and UNDERWOOD, J., ASTM STP 560, 1974.

4. SIH, G.C., Methods of Analysis and Solutions of Crack Problems, G.C. Sih, Ed., Noordhoff International Publishing, 1972.

5. ERDOGAN, F., and SIH, G.C., J. Basic Engineering, 85, 1963.

6. ANDERSON, H., J. Mech. Phys. Solids, 18, 1970, 437.

7. BUECKNER, H.F., Trans. ASME, 80, 1958, 1225.

8. SHAH, R.T., ASTM STP 560, 1974.

9. SIH, G.C., J. Appl. Mech., 32, 1965.

10. BILBY, B.A., and CARDEW, G.E., Int. J. Fracture, 11, 1975, 708.

11. POOK, L.P., Engineering Fracture Mech., 3, 1971, 205.

12. "A new criterion for combined mode fracture", (in Chinese), Research Paper of Institute of Mechanics, Academia Sinica, 1976.

13. "Energy release rate for combined mode cracks", (in Chinese), Research Paper of Institute of Mechanics, Academia Sinica, 1975.

Table 1   Values of C_1^* and C_2^*

γ	0°	5°	10°	15°	20°
C_1^*	1.00	1.0003	1.0010	1.0023	1.0042
$-C_2^*$	0	4.137×10^{-3}	8.297×10^{-3}	1.250×10^{-2}	1.678×10^{-2}

γ	25°	30°	35°	40°	45°
C_1^*	1.0066	1.0095	1.0131	1.0173	1.0222
$-C_2^*$	2.116×10^{-2}	2.566×10^{-2}	3.031×10^{-2}	3.515×10^{-2}	4.022×10^{-2}

γ	50°	55°	60°	65°	70°
C_1^*	1.0279	1.0343	1.0417	1.0500	1.0594
$-C_2^*$	4.555×10^{-2}	5.118×10^{-2}	5.178×10^{-2}	6.361×10^{-2}	7054×10^{-2}

γ	75°	80°	85°	90°	
C_1^*	1.0700	1.0821	1.0957	1.1110	
$-C_2^*$	7.804×10^{-2}	8.624×10^{-2}	9.524×10^{-2}	0.1052	

Table 2 $-\theta_0$ for in-plane shear (in degrees)

ν	0	0.1	0.2	0.3	0.4
S-criterion	70.5	74.5	78.5	82.3	86.2
Present criterion	70.5	72.3	74.5	76.5	79.5

Table 3   Fracture angles of inclined crack under uniaxial
tension (in degrees)

β	30	40	50	60	70	80
Max. σ_θ criterion	60.2	55.7	50.2	43.2	33.2	19.3
S-criterion	63.5	56.7	49.5	41.5	31.8	18.3
Present criterion	62.4	56.2	49.9	42.4	32.6	18.7
Test results [4]	62.4	55.6	51.1	43.1	30.7	17.3

Figure 1   Crack with branch

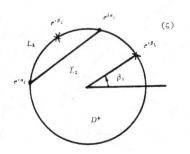

Figure 2 ζ-plane

151

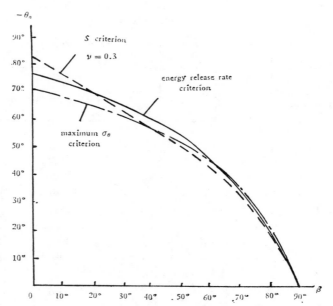

Figure 3   Fracture angles for inclined crack under uniaxial tension

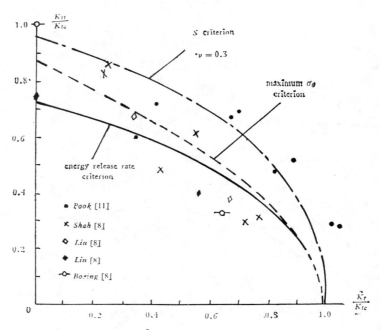

Figure 4   Critical $\overset{o}{K}_I$ and $\overset{o}{K}_{II}$ for inclined crack under uniaxial tension

σ
β
2a

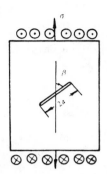

Figure 5 Combined action of axial stress σ and antiplane shear stress τ

KII/KIc
0.8
0.6
0.4
0.2
0
ν = 0.3
KIII/KI = 2.0 1.4 1.2 1.0 0.8 0.6 0.4 0.2 0
0.2
0.4
0.6
0.8
1.0
KI/KIc

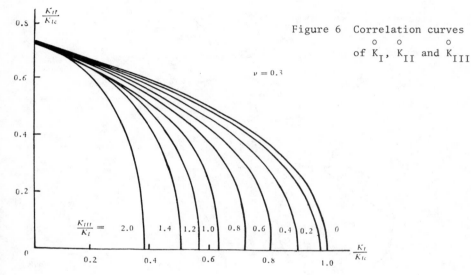

Figure 6 Correlation curves of $\overset{\circ}{K}_I$, $\overset{\circ}{K}_{II}$ and $\overset{\circ}{K}_{III}$

KI/KIc
1.0
0.8
0.6
0.4
0.2
0
0.2
0.4
0.6
0.8
1.0
KIII/KIIIc

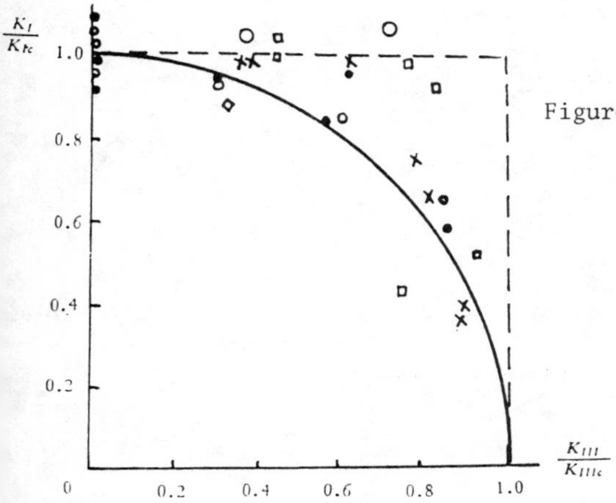

Figure 7 Theoretical and experimental results of crack under action of $\overset{\circ}{K}_I$ and $\overset{\circ}{K}_{III}$

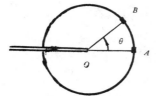

Figure 8   Comparison on circle

Figure 9   Iso-W line

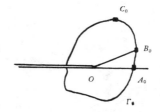

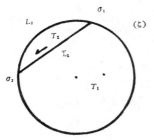

Figure 10   Circle on ζ-plane

Figure 11   s-plane

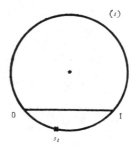

A STUDY OF THE PHYSICAL NATURE OF FRACTURE IN
COMPOSITE MATERIALS

V. R. Regel*

INTRODUCTION

This paper represents an attempt to pose problems arising in research
into the physical nature of fracture of composite materials in terms of
modern ideas encompassed in the kinetic concept of the strength of solids.

According to this concept, fracture should be considered as resulting from
a progress of accumulation of defects developing in time from the moment
of load application. This process should be characterized not by some
critical stress alone, but rather by the rate of defect accumulation at a
given temperature regime of loading, or by the reciprocal lifetime of the
specimen under load from the moment of loading until failure (τ). An
investigation of the temperature - stress dependence of the lifetime
leads to the conclusion that for homogeneous solids of various structure
the dependence of τ on σ and T over a sufficiently broad range of stress σ
and temperature T can be described by a formula which is the same for all
solids (single and polycrystals, glasses, polymers) [1].

$$\tau = \tau_0 \exp \frac{U_0 - \gamma\sigma}{KT} \qquad (1)$$

This expression is frequently called Zhurkov's formula.

The properties of the coefficients τ_0, U_0, and γ, entering this formula
were studied experimentally on various materials. Experiments showed the
prefactor τ_0 to be practically constant for all the solids studied and
equal to 10^{-13} s within one or two orders of magnitude, i.e. close to the
period of atomic vibrations in solids.

The quantity U_0 which may be treated as an activation energy of fracture
varies from one material to another while turning out to be structurally
insensitive. The value of U_0 for metals and crystals is found to be
closer to the energy of sublimation than self diffusion, and to coincide
with the activation energy of the initial stage of the thermal destruction
process in the case of polymers. The alloying of metals, their mechanical
and thermal treatment, the plastification and orientation of polymers,
which change their structure substantially, do not noticeably affect the
magnitude of U_0.

The coefficient γ turns out to be the only structurally sensitive coefficient in Eq. (1) and varies markedly with variation of material structure.
The value of γ, which has the dimensions of volume, ordinarily exceeds the
volume occupied by atoms in a solid by a factor of several hundred.

*Moscow University, U.S.S.R.

The form of Zhurkov's formula (1) and the properties of the coefficients τ_0, U_0 and γ give grounds for assuming fracture to be a thermally activated process. This process is based on elementary events in thermofluctuation and rupture of atomic bonds in a solid. When subjected to mechanical stress, the energy U_0 required for atomic bond rupture decreases by $\gamma\sigma$, the probability of their recombination being reduced.

Such a concept of the nature of fracture of solids was put forward earlier only on the basis of phenomenological studies of the temperature - stress dependence of lifetime found for a large number of materials. The validity of these ideas has been confirmed thereafter by an investigation of the fracture process using direct methods of physical experiment capable of following the development of fracture at the atomic and molecular level. The application of IR spectroscopy, EPR and mass spectrometry make it possible to follow the extension under load of individual macromolecules in oriented polymers, and the rupture of drawn atomic bonds in molecules. It was shown that it is essentially the kinetics of atomic bond rupture accumulation that determines the kinetics of fracture of oriented polymers. This work was described in detail in the monograph [2].

We start from the assumption that application of the kinetic approach to the study of composite fracture will turn out to be fruitful in understanding the physics of fracture of this new and important class of materials. In our opinion it would be wrong to wait for the final solution of nuclear problems in the physics of strength of solids by investigating homogeneous materials with a subsequent transition to the physics of composites.

It is reasonable to start studying the kinetics of fracture of composite materials and to apply to them the kinetic concept of strength from the simplest kind of this material, i.e. unidirectional fibrous composites with continuous filaments. Just as in the case of homogeneous solids, studying the fracture of composites within the framework of the kinetic concept of strength should be started with systematic experiments on the kinetics of fracture and, primarily, with an investigation of the temperature and stress dependence of the lifetime of these materials. Such an investigation may yield valuable information on the nature of fracture in components. By analogy with studies performed on homogeneous solids [2], one should thereafter employ direct experimental techniques which permit one to follow the process of fracture at the atomic and molecular level. Considering the specific nature of the composite materials, particular attention should be paid to the process of fracture at the interface between the constituents in order to reveal the role of adhesion-bonding between the constitutents in determining the strength of composites.

We are able at present to analyse some results obtained in studies of the fracture of composites reinforced with continuous unidrectional filaments, and to formulate problems which should be solved in the future.

An Investigation of the Temperature - Stress Dependence of the Lifetime of Fibrous Composites

The first problem that arises when studying the temperature - stress dependence of the lifetime of composites consists in checking whether the functional dependence (1) of lifetime τ on acting mean stress σ and test temperature T for composites is the same as for homogeneous solids.

There is experimental evidence showing Zhurkov's formula (1) to be valid
for various fibrous composites subjected to tensile stress with specimens
cut along the filament direction. As an illustration, Figure 1 presents
data on the temperature - stress dependence of lifetime of a tungsten fiber-
copper matrix composite [3]. Similar data are available for polymeric
matrix composite [4]. Experimental data similar to those presented in
Figure 1 show equation (1) to hold for composites within a certain stress
and temperature range. This means that the fracture of composites, just as
that of the other solids, should be considered as a thermally activated
process of gradual accumulation of defects.

The next problem which can be posed in a systematic investigation of the
temperature - stress dependence of lifetime of composites is to study the
properties of parameters τ_0, U_0, γ in equation (1) and revealing their
physical meaning. The final purpose of such studies may be a development
of "kinetic rules of mixture", i.e. rules permitting determination of parameters τ_0, U_0, γ in equation (1) for composites having data on the corresponding parameters of the composite constituents and taking into account
the effect of the mechanical properties of the interface between them. One
should develop three rules for the determination of each of three parameters
τ_0, U_0 and γ in equation (1), accordingly, for composites with different
volume ratio of the fibres.

On the basis of the available experimental data on the properties of τ_0,
U_0 and γ in equation (1) for composites, and on the kinetic rules of mixture, the following can be said at present.

(a) *First rule of mixture*

The magnitude of τ_0 for all the composites studied up to now (their number
is small) turned out to be constant within experimental errors and equal
to $\tau_{0comp}=10^{-13\pm2}$ sec, i.e., about equal to the prefactors in equation (1)
for the composite constituents. Therefore the first rule of mixture may
be formulated in the following way: the prefactors in equation (1) for
the composite τ_{0c}, for fibres $\tau_{0\phi}$ and for matrix τ_{0m} are approximately
constant and equal to 10^{-13} sec:

$$\tau_{0c} = \tau_{0m} = \tau_{0\phi} = 10^{-13}\,\text{sec} \qquad (2)$$

In accordance with this rule, the physical meaning of τ_{0comp} does not
differ from that attributed to the prefactor τ_0 in Zhurkov's formula for
homogeneous solids.

(b) *The second rule of mixture* relates to the determination of the activation
energy for the fracture of composites, U_{0comp}. Since any composite contains at least two constituents, each of them being characterized by its
own activation energy for fracture, U_{0f} and U_{0m}, a question arises as to
the magnitude of U_{0comp} at different volume contents of the constituents.

We will limit ourselves here to a simple case where U_{0f} and U_{0m} for the
original specimens of the constituents outside the composite are numerically
equal to U_{0f} and U_{0m} of these constituents in the composite. However, even
in this case, a question arises as to how the activation energy of fracture
of this composite changes with the changing fiber volume ratio.

The answer to this question can be obtained in experiments on specimens of
tungsten fiber-copper matrix composite with different tungsten fiber volume ratio [3,4]. One first studied the temperature - stress dependences

of the lifetime of the constituents of this composite, and obtained the activation energies of fracture of copper and tungsten. Experiments show the activation energy of copper fracture to be $U_{0m} = U_{0Cu} = 80$ kcal and that of the fracture of tungsten to be $U_{0f} = U_{0w} = 230$ kcal.

Figure 2 represents a graph of the activation energy of fracture of the copper-tungsten composite vs. volume content of the tungsten fibers, $U_{0comp}(V_f)$. At low tungsten fiber content the activation energy of composite fracture is seen to coincide numerically with that of the copper matrix fracture, $U_{0comp} = U_{0m} = 80$ kcal/mole, whereas at a sufficiently high tungsten fiber content U_{0comp} coincides numerically with the activation energy of fracture of the tungsten reinforcement $U_{0comp} = U_{0f} = 230$ kcal/mole.

Figure 2 shows also that the change of activation energy from $U_{0m} = 80$ kcal/mole up to $U_{0f} = 230$ kcal/mole occures practically in a jump at a certain value of the fiber volume ratio, $V_f = V_f^*$. Assuming that at this fiber volume ratio V_f^* the strengths of the matrix and the reinforcement in the composite become equal, i.e., $\sigma_f V_f^* = \sigma_m(1-V_f^*)$, then for an estimate of V_f^* we may write

$$V_f = \frac{\sigma_m}{\sigma_f + \sigma_m} = \frac{\gamma_f}{\gamma_f + \gamma_m \left(\dfrac{U_{0f} - C}{U_{0m} - C} \right)} \tag{3}$$

where

$$C = RT \ln \frac{\tau}{\tau_0}$$

such an evaluation yields for our case $V_f^* = 7.3\%$ which is not in disagreement with experiment.

Thus the second rule of mixture for the simplest case of U_{0f} and U_{0m} remaining in the composite the same as outside it may be expressed in the following way:

$$U_{0comp} = U_{0m} \quad \text{for } V_f < V_f^*$$
$$U_{0comp} = U_{0f} \quad \text{for } V_f > V_f^* \tag{4}$$

As already pointed out, in more complex cases U_{0comp} may differ from U_{0f}, and sometimes from U_{0m}, at $V_f > V_f^*$ and $V_f < V_f^*$ respectively.

(c) *The third rule of mixture* should permit us to estimate the magnitude of γ_{comp} from data on γ_f and γ_m and from information on the mechanical properties of the interface between the constituents. The experimental data available are, however, insufficient to be able to formulate this rule for a general case. One may expect the development of this rule to turn out to be a particularly difficult problem because of the structural sensitivity of the coefficient γ in equation (1).

One may consider here the simplest case to the well-known "static rule of mixture" [5].

$$\sigma_{comp} = \sigma_f \cdot V_f + \sigma_m \cdot V_m \tag{5}$$

where σ_f, σ_m are the strengths, and V_f, V_m the volume contents of the fibers and matrix, respectively in the composite.

Then for γ_{comp} the following relation should hold:

$$\frac{1}{\gamma_{comp}} = V_f \cdot \frac{1}{\gamma_f} \frac{U_{0f}-C}{U_{0m}-C} + V_m \frac{1}{\gamma_m} \qquad \text{for } V_f < V_f^*$$

$$\frac{1}{\gamma_{comp}} = V_f \cdot \frac{1}{\gamma_f} + V_m \cdot \frac{1}{\gamma_m} \frac{U_{0m}-C}{U_{0f}-C} \qquad \text{for } V_f > V_f^*$$

(6)

where

$$C = RT \cdot \ln \frac{\tau}{\tau_0} .$$

One may expect, however, that an estimate of γ_{comp} by equation (6) will frequently be wrong. This formula, just as equation (5), does not take into account the effect of defects at the interface between the constituents on the composite strength. Equations (5) and (6) do not also include the fact that the structure and strength properties of the matrix, and sometimes of the fibers as well, may change in the process composite manufacture. This means that the magnitudes of γ_f and γ_m in the composite are different from those for the original constituents. Further work on improvement of the third rule of mixture (as well as in the determination of corrections to the static rule of mixture) should take into account these complicating points.

An analysis of the parameters τ_0, U_0 and γ in equation (1) for composites and of problems in the development of kinetic rules of mixture shows that the systematic investigation of the temperature - stress dependences of the lifetime of model composites may yield valuable information on the process of composite fracture. Therefore further studies of the kinetics of composite fracture both on composites of other type and on specimens cut at an angle to the filament orientation, as well as at other regimes of loading and different kinds of stresses should be considered as promising. The studies of the temperature - stress dependence of lifetime should be accompanied by direct observations on the appearence and growth of cracks at different stages of the process, and with a recording of the distribution of stress in the fibers and matrix in time. This will permit to develop in more detail the kinetic rules of mixture and to reveal the physical meaning of the parameters τ_0, U_0 and γ in the formula for lifetime.

The mechanism of fracture in polymer matrix composites can be explored by the same experimental techniques as the ones used in the study of polymer fracture. Among them are IR spectrometry, radiospectroscopy (EPR and NMR), mass-spectrometry and small-angle x-ray scattering. The possibilities offered by these methods in studies of fracture of composites can be illustrated by results of the research performed in Leningrad Physical-Technical Institute, Academy of Sciences of the USSR [6-8]. These results can be summarised as follows.

A model polymer - polymer composite made up of strong polyvynil alcohol filaments embedded in an epoxy matrix has been studied by IR spectroscopy [6]. Bifunctional diisocyanate was introduced between the epoxy and the filaments as a coupling agent. Diisocyanate was shown to form chemical bonds with both the matrix and the filaments. In the course of heat treatment of this composite, new absorption bands naturally appear in IR spectrum.

These chemical bonds at the interface between the constituents turned out to be carrying considerable stress thus strengthening substantially the composite.

This is evidenced by the frequency shift of the corresponding absorption band maxima under load.

IR spectroscopic measurements carried out on composites made up of polypropylene filaments in a polyethylene matrix showed also that the matrix which was originally nonoriented becomes partially oriented in a layer adjoining the oriented reinforcement. Such effects should be taken into account when calculating the strength of a composite.

Mass spectrometry was employed to investigate thermal desorption of very thin polymer layers from metal surfaces [7]. The interaction of polymers with the substrate, as was shown, can reduce substantially the activation energy of thermal degradation of polymers. The same authors studied the kinetics of water liberation from the surface of glass filament heated in vacuum. The presence of a mono-molecular water layer on the surface of glass filaments was shown to reduce considerably the activation energy of their fracture and, accordingly, their strength. This phenomenon is explained by dissociation of stressed chemical bonds between the silicon and oxygen atoms in the surface layer of glass facilitated by the reaction of hydrolysis in a presence of water. This effect accounts for the experimentally observed changes in the activation energy of fracture of glass-fiber reinforced composites depending on the presence or absence of even traces of water at the interface between the glass filaments and the polyme matrix.

Small-angle x-ray scattering permitted to detect the formation of submicrocracks of two types in model polymer-polymer composites under load [8]. One of them is ascribed to the reinforcement, and the other, to matrix or the interface between the constituents. When combined with the acoustic technique, this method will yield valuable information on the formation and kinetics of growth of micro- and submicrocracks in composites.

The examples presented indicate that the methods employed earlier in the studies of the nature of fracture of homogeneous solids provide important data also on the fracture of composites at the microlevel. Combination of these methods with phenomenological studies of the kinetics of fracture of composites should be used as a basis for research into the nature of the fracture of composites.

REFERENCES

1. ZHURKOV, S. N., Vestnik AN SSSR 3, 1968, 46.
2. REGEL, V. R., SLUTSKER, A. I. and TOMASHEVSKII, E. E., Kinetic Nature of Strength of Solids, Izv. Nauka, Moscow, 1974.
3. REGEL, V. R., BOBONASAROV, H. B., BETECHTIN, V. I. and KISSELOV, E. A Proc. VIII Conf. on Plasticity and Strength of Metals and Alloys, Kuibishev, 1976, 71.
4. REGEL, V. R., Izv. AN SSSR, Phys. Section 40, n7, 1976, 1376.
5. Composite Materials (L.V. Broutman and R.H. Krock, Eds.), New York, London, Academic Press, V.I. - VIII, 1974.

6.   GABARAEVA, A.D., REGEL, V.R., PHILLIPOV, N.N. and LEKSOVSKII, A.M., Proc. III Int. Conf. on Polymer Composites, Berlin, 1977.
7.   POSDNJAKOV, O.F., AMELIN, A.V., REGEL, V.R., and REDKOV, B.P., Shalimov, Proc. of VI Int. Symp. on Mechanoemission and Mechanochemistry, Berlin, 1977.
8.   LEKSOVSKII, A.M., ORLOV, L.G. and REGEL, V.R., Proc. of   Int. Conf. on Mechanics and Technology of Composites, Varna, 1976.

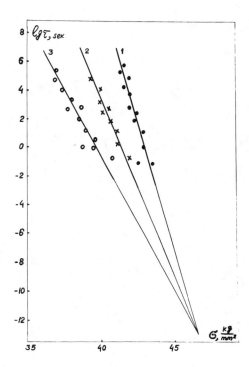

Figure 1   The temperature - stress dependence of the lifetime of tungsten-fiber copper-matrix composite. The volume ratio of tungsten fibers V_f = 10%. Temperatures, °C : 1   +18; 2   +150; 3   +300.

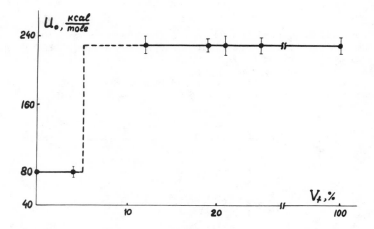

Figure 2   The activation energy of fracture (U_c) of Cu - W composite
versus the volume ratio of tungsten fibers.

OVERVIEW OF WORKSHOPS

R.F. Smith

A total of 332 papers was included in the Workshop Sessions, which took place on alternate evenings and afternoons during the first four days of the Conference. All the papers underwent a rigorous refereeing procedure. On receipt of a manuscript at Waterloo, copies were despatched to the appropriate member of the Editorial board, who would handle its review. He would, in turn, pass it to two referees for their opinion, on which he would base his recommendation to the Editor-in-Chief. The final decision on acceptance would then rest with him; and would be based on both the evaluation of the paper's technical merit and a weighing of factors such as the overall balance of the Conference and the origin of the paper. This process extended over the period from April 1976, when the first manuscript was received, to March 1977, when the last full manuscript was received and later accepted. Subsequent to this, several more papers were received in abstract form, and were accepted on this basis for presentation, where space in the programme permitted.

The workshop papers were classified into 7 parts:

	Part	No. of Papers
I	Physical Metallurgy	43
II	Voids, Cavities, Forming	44
III	Fatigue: Micromechanisms	40
IV	Fatigue: Mechanics	45
V	Analysis and Mechanics	67
VI	Applications	43
VII	Non-Metals	50

The exact scheduling of the Workshop Programme evolved from repeated iterations involving members of the Canadian Fracture Committee, the Local Arrangements Committee and the Editorial Board. Every attempt was made to classify papers correctly and to minimize clashes and overlaps. The task was in essence the organization of seven medium-sized conferences, in addition to the Plenary Conference!

The structure of the sessions was designed with the aim of producing active discussion and of developing continuity of themes. As all the papers were pre-published the authors were instructed to make only short (8 minute) presentations, stressing their major points and highlighting controversial and innovative areas. After the consecutive presentation of roughly five papers half the session time was devoted to discussion. This was led by a well briefed Workshop Foreman adopting the "rapporteur" approach. Integration of the sessions in each Part was fostered by Part Chairmen, who monitored the progress of their respective Parts through the week and attempted to develop and emphasise the connecting links between papers and in the discussion. This format was generally felt to be very successful and was instrumental in producing much meaningful debate. It was noticeable throughout the week that many of the lecture rooms held a capacity, or over capacity, audience : ample evidence that the transactions there had greater appeal than discussion in the foyer, which is frequently *not* the case. This stimulating *full house* atmosphere was particularly apparent in many of the sessions on Microstructures; Voids and Cavities; and Non-Metals.

No detailed record of discussion during the Workshop Sessions was taken. Instead the Part Chairmen were asked to compile what must necessarily be a rather subjective *Overview* of their sessions, and these follow below. These are not only useful now in directing attention to matters of greatest current concern, but will prove of considerable interest when set against any comparable documents which emerge from ICF5. It is indeed hoped that a similar approach to Workshop Sessions is adopted at ICF5 to develop further this *ICF style*.

I PHYSICAL METALLURGY

Introduction

The aim of this Part of the Conference was to examine the ways in which the various features of metallurgical microstructures affect the onset of fast fracture. The fracture modes considered were primarily those of transgranular and intergranular cleavage, together with some discussion of void-growth or fast shear mechanisms, which inevitably overlapped some of the content of Part II. Consideration of intergranular fracture encompassed embrittlement by trace impurity elements, by hydrogen and by liquid metals. 43 papers were presented, and points were raised in what was mostly very active discussion on virtually all of these. Thus it is not possible to do other than provide a general forward-looking summary of the events in these four vigorous sessions.

The main features of a microstructure that might be expected to be of importance in fracture processes are listed and discussed below.

Grain Size

Effects may arise, either because large grains can accommodate long dislocation arrays, which facilitate crack initiation or because (long) microcracks can propagate more easily in coarse-grained material. If "grain size" is then to be interpreted as the feature that provides a microstructural barrier to slip or to microcracks, it may turn out that the important parameter is the simple, as-annealed, grain diameter, the cell size in pre-strained materials, or the martensitic lath or packet width in a quenched steel. Additionally, if any texture has been introduced into the materials during working, barriers between some pairs of grains may be weaker than usual, leading, for example, to cleavage facet diameters larger than the average grain diameter. In material containing a large volume fraction of hard matrix particles, it is not clear that grain size *per se* should be of great significance, unless the heat-treatment used to obtain the grain size has also produced grain-boundary embrittlement or hardening.

Second-Phase Particles

These may be divided into two sets : one composed of those which are deliberately produced to give matrix strength, such as carbides in steels or
intermetallics in aluminium alloys : the other, comprising those present
as non-metallic inclusions, such as sulphides, de-oxidation products, or
coarse (> 0.5 μm diameter) intermetallics. Additionally, in steels, there
may be particles, such as NbC, added in the first instance for grain re-
finement, which fit somewhat into both categories : larger particles
pinning grain boundaries and acting rather more as inclusions; fine dis-
persions inducing matrix strengthening.

Dislocations interact strongly with the first set, producing high local
stresses, such that, as the plastic strain is increased, the particle may
shear, crack or cavitate. A particle which shears at low plastic strains
induces planar slip, which gives sharp slip-bands, high stresses at the
end of pile-ups, and easy penetration of protective surface films. Grain
size might be expected to be of significance in such an alloy. Stronger
particles may induce cross-slip and dislocation tangles/cell structures if
the stacking fault energy of the matrix is not too low : the grain size
should then be of less importance. Even in this second case, flow may
localise to give intense shear bands at higher applied strains, perhaps
because the particles crack or cavitate, by decohering from the matrix.
In the former case, the strength of the particle itself is of importance :
in the latter case, the strength of the particle/matrix interface is critical. Here, the segregation of impurities to the interface may strongly
affect the work of decohesion. The strengths of carbides and carbide/
matrix interfaces are of critical importance in the initiation of transgranular and intergranular cleavage fracture, and the cleavage fracture
stress of the materials may often be related to carbide sizes, using
Griffith-type arguments.

Large non-metallic inclusions are not expected to be strongly bonded to
the matrix in a chemical sense, but tensile or compressive residual,
"tessellated", stresses may exist near the particle, due to differential
thermal contraction on cooling. Residual tension in the matrix may aid
crack initiation (but the effect of the stress is very short-ranged) :
residual tension in the particle may enhance particle fracture. As the
size of the inclusion is reduced (fine-scale deoxidation products in weld-
metals, 0.1 - 0.2 μm diameter intermetallics in aluminium alloys, fine
sulphides in ESR steels), there may be evidence of particle/matrix bonding so that dislocation arrays are necessary to initiate voids : the more
usual situation is, however, that in which voids are nucleated at very
low plastic strains. Not only the volume fraction, but also the distribution of inclusions, may be important in controlling a material's
ductility or toughness : for example, a redistribution of sulphide particles at austenite grain boundaries during "overheating", or fine-
scale inter-dendritic MnS particles in a cast or incompletely worked
steel, can give low toughnesses.

Dislocation Structures

Dislocations produced by transformation are usually associated with martensitic transformations. In low carbon steels, there is some evidence
that yield strength is governed by lath width, but that cleavage resistance is controlled by packet diameter, although few pictures of microcracks arrested at packet boundaries have been shown. In higher carbon
steels, plate martensite and retained austenite may also be present, but

evidence that cleavage fracture remains trans-lath has been presented.
Other possibilities are that the fracture runs along lath or packet boundaries, but these have not been proved unambiguously.

The effect of the distribution of dislocations produced by deformation is
also of importance : planar slip being induced by low stacking-fault energy
or easily-sheared particles (particularly in ordered alloys). Sharp slip-
bands seem generally to be deleterious, (although they are associated with
high cyclic hardening properties), but the conditions under which shear
localisation can occur in a plastically deforming matrix, which has until
then been deforming in a uniform manner, remain unexplained.

Further work also needs to be done to explain the effects of dislocation
structures in controlled-rolled steels or thermo-mechanically treated
aluminium alloys.

Trace Impurity Elements/Hydrogen

It is now fairly well understood that the segregation of certain impurity
elements to grain boundaries (Bi in Cu; P, Sb, Sn, As, etc., in alloy
steels) can in some way lower grain boundary cohesion, so that a crack may
run at low stress along the boundary, and the toughness of the material is
reduced. Often, hydrogen, supplied externally from a hydrogen atmosphere
or an acidic corroding solution, seems to act in a similar fashion, giving
intergranular fracture in an alloy with impure grain boundaries which
would otherwise cleave in a transgranular manner. In many cases, interactions between impurity elements and major alloying elements can affect
the details of segregation. It is conceivable that hydrogen may interact
with matrix elements to form hydrides; however, there is no experimental
support for this idea. The precise mechanism of embrittlement is as yet
unclear, although a simple lowering of cohesive stress, by modification of
the normal iron-iron interionic potential, seems preferable to any pinning
mechanism.

Other Features

These include the degree of texture present in a material, which also
affects the anisotropy of elastic constants and plastic flow. Additionally, the grains may have elongated and inclusions may be aligned pre-
ferentially in some directions. Segregation of alloying elements and
carbides may also give rise to non-uniformity of properties in cast and
wrought alloys.

Discussion

In the light of the many features of microstructure outlined above, it is
hardly surprising that many of the workshop papers were not as clear-cut
with respect to the operative micro-mechanisms as might have been hoped.
The ensuing discussion often tended to concentrate on pointing out some of
the other variables, which had not been taken fully into account : for
example, that the heat-treatments used to produce variations in grain size
in steel might, at the same time, produce differences in carbide sizes and
distributions in dislocation sub-structures, and in the distribution of
minor impurity elements. Appropriate plenary papers were assigned to
particular sessions, but the authors of such papers were unfortunately
not always present, so that the discussion could not always include comments on the models proposed in the plenaries. A glaring example of disagreement between workshop and plenary papers was shown in the very last

paper in the session on liquid metal embrittlement : figures presented in the workshop, and endorsed by most of the audience, were almost an order of magnitude different from those quoted in the appropriate plenary paper. Discussion in the final session, on embrittlement, was particularly interesting and constructive. Even here, as generally with previous sessions, the only overall conclusion to be drawn is that most microstructures are very complicated and that the details of the ways in which they fracture, under aggressive or non-aggressive environments, are idiosyncratic in the extreme.

J. F. Knott
C. J. McMahon, Jr.

II VOIDS, CAVITIES, FORMING

This part of the Conference was concerned with a variety of aspects of plasticity which occurred with concomitant void nucleation and growth including both low temperature and high temperature deformation modes. There was considerable discussion of the problem of void nucleation which emphasised the theoretical problems of void production at a variety of heterogeneities such as second phase particles, grain boundaries and substructural features. This is clearly an area in which more detailed criteria and mechanisms are needed. However, as a cautionary note, it should also be recognised that experimentally it is difficult clearly to rule out nucleation at second phase particles because this can occur at particles less than 10^{-2} μm in size. These can readily arise at very low overall impurity levels when, in addition, their spacing may be large.

In regard to the growth of holes some definitive experiments were reported which clearly related the displacement rate and cavity shape to the stress system around the growing void. The results are at variance with current theoretical models and it is clear that more precise constitutive relations are required for solids in which plasticity occurs without volume conservation. Further, it was clear from a number of papers that the criterion for strain localisation and its influence in prescribing the allowable strain to failure is of major theoretical and practical interest.

Clearly, these problems are important in forming operations. A number of important problems emerged during discussions including the need to relate R values, texture and the extent to which plane strain deformation is enforced on a microscopic scale. In addition, the majority of ductility models is related simply to the volume fracture of inclusions. However, in many forming operations, where fracture is initiated at the surface, the failure criteria may well reflect the distribution of inclusion spacing and be related to the minimum inclusion spacing. This may be of importance in such technological problems as the forming of plates which possess a sheared edge.

Turning to the question of the applicability of void nucleation and growth to high temperature deformation, it is apparent that both the theoretical models and experimental evidence are less clear cut. The evidence regarding cavitation during superplastic flow suggests that there is some disagreement regarding the occurrence of cavitation in all types of material at large plastic strains. This aspect is clearly important for defining the range of applicability of various constitutive equations for superplastic flow and the mechanisms of mass transport involved in achieving the overall strain.

In the field of high temperature deformation two salient features emerged from the discussion. First, it is clear that in many materials micro-structural instabilities occur which have a profound effect both on the range of utilisation of constitutive equations and the definition of the mode of failure in terms of the operative stress and temperature required. Secondly, it is clear that more effort is required in formulating cumulative damage rules. For example, the ability to link microstructural models of damage, such as hole growth by diffusive flux or dislocation motion, to phenomena which involve path dependent deformation would be of immense value in predicting creep life under variable loading conditions.

Finally, it is germane to indicate that currently much effort is being expended on the production of fracture or failure maps for both low temperature and high temperature conditions. Although these concepts permit the rationalisation of a variety of existing data, it is of importance to extend their utilisation to path dependent phenomena and to recognise the limitations imposed on their use by microstructural changes and environmental effects.

In conclusion, it can be stated that the area of ductile failure and hole growth both at high and low temperatures is a fruitful one for future work based on the development of new constitutive equations and carefully controlled experimental studies. In particular, the areas of defining the total volume of material involved in hole growth, the criteria for strain localisation and the path dependence of cumulative damage models should provide fruitful areas for research in the next decade.

J. D. Embury

B. F. Dyson

III and IV   FATIGUE : MICROMECHANISMS AND MECHANICS

The eighty-five papers contained in these two sessions attest to the current high level of interest in the fatigue process. Scheduling of these papers into the two parts had, of necessity, to be somewhat arbitrary, resulting in many delegates having to choose between two sessions of equally interesting sets of papers. However, almost all sessions in both parts were well attended and generated enthusiastic and often heated discussion.

Part III, under the theme of "micromechanisms", considered, in order : crack initiation mechanisms and their dependence on microstructure; the interaction of creep and fatigue at elevated temperatures; the effect of environment on fatigue; and finally the strain concentrating effect of discontinuities. Clearly this range of topics precludes giving, in a short review, more than a cursory view of the trend of the discussions. Characterizing the sessions in brief, it appears that the first three sessions had in common a pragmatic approach of detailed observations of the importance of individual microstructures in fatigue process rather than a preoccupation with general models. It seems reasonable at present to conclude that for some time practical advances in predicting the effects of creep-fatigue interactions will be aimed at particular temperature and deformation regimes of existing alloy systems. There is further evidence that for many applications the effect of environment must be included in determining creep-fatigue resistance.

Papers in the environmental session particularly exemplified a material by material study of processes. Here again it appears that we have much to learn before generalized models can be put forward. In contrast to the first three sessions the last session on the effect of discontinuities on fatigue dealt with a topic which is reasonably well understood. Indeed, this session includes an interesting set of applications of notch analyses and gives a state of the art review of fatigue analysis techniques for notched components.

Part IV, in its first and last sessions, covered current observations and models of fatigue crack propagation. Discussion pointed out both the relatively good predictive abilities of current models and the still not completely resolved nature of crack closure and the threshold stress intensity. A particularly interesting observation relating to the former was reported by McEvily, who pointed out that recent observations indicated that the crack was, to a large degree, propped open by the surface layers in a plate, and that removal of these layers could strongly affect closure and related phenomena such as crack retardation. The consequent influence of specimen thickness on delay in variable amplitude loading appears to be a high priority problem. The two middle sessions dealt with a variety of special topics and applications. They, as a group, pointed out both that approaches outlined in the other two sessions could be applied to crack propagation and that the crack initiation phase, as noted in Part III, still requires special treatment. In addition, both papers and discussions considered features of metal deformation, biaxial straining, metallurgy and failure modes in non metallics so far neglected in simplified fatigue analyses. Each of these topics will no doubt receive further attention. The influence of metallurgical features on crack initiation and threshold stress intensity, as well as specimen thickness on crack closure, are, as yet, not satisfactorily resolved. However, it appears that at least the outline of a consistent approach to crack propagation has emerged. The approach at present relies strongly on empirical, but broadly based, observations. This suggests that, while advances in the mechanics of crack tip behavior and in metallurgical models will probably refine, amplify and explain present predictive methods, the approaches are essentially sound.

T. H. Topper
L. Coffin

V ANALYSIS AND MECHANICS

Part V of ICF4 itself comprised virtually a full conference. Sixty-four workshop papers were scheduled, and three were added at the last minute. The authors represented at least fifteen countries, and discussers and audience provided an even wider resource of interest, commentary, and evaluation. Papers were scheduled in six sessions (meeting in just four three-hour periods!) entitled Crack Growth, Dynamic Processes and Stress Intensity Factors, Fracture Analysis, J-Integral and COD, Non-Planar and 3-Dimensional, and Non-Linear. For the most part, papers were grouped well and thus stimulated a good deal of discussion.

It would oversimplify matters to lump nearly seventy papers into just a few categories, for of course each author or group of authors tends to work independently. The alternative of commenting here on each paper individually is repetitive, for the thrust of the work appears concisely in these Proceedings. It is more to the point, however, to make limited observations on certain *problems* which have or have not attracted the attention of the researchers at ICF4.

In broad terms, the community seems now less preoccupied with determining stress intensity factors by any of several methods; interest has shifted to what may be termed characterizing parameters, and to simulating crack growth processes. The previously dominant exercises of stress analysis (almost apparently for their own sake) are now to be found in efforts of greater scope, i.e. identifying and quantifying these characterizing parameters, including crack opening and energy related quantities. In this connection, it is to be noted that a main function of analysis and mechanics is to model an event or a process so that such parameters may be inferred, or perhaps a theory constructed. Observe, however, that, analysis by itself leads not to criteria - these require experimental confirmation.

What was heard at ICF4, therefore, was a variety of presentation whose objective appeared to be determination of the nature of one characterizing parameter or another, for a range of materials, geometries, and load configurations. Some papers focussed on mixed-mode excitation of a crack in an elastic medium, a particular issue of recent interest. The two key problems here, of course, are the stress modelling and the condition of criticality. Recent literature has been more concerned over the second, while the first - for compressive loading - remains not altogether resolved. Relatively little attention was given at ICF4 to finite element methods *per se*, although there is still viable activity to be found on this topic, and modelling of crack problems done in conjunction with numerical analyses was discussed but little. It was almost as though a consensus had been achieved to the effect that such questions are resolved and attention could pass to crack growth and characterizing parameters.

Accordingly, we were treated to discussions of J and other integrals, measures of crack opening, and other stress or energy based parameters. Without going over these papers one by one, we can make some observations. Owing to the fact that such work usually involves non-linear behaviour (i.e., plasticity and, sometimes, finite deformations), the requisite analyses typically proceed on the computer, employ finite elements as a solution tool, and are expensive. Some work, of course, is more analytical and relies on a simple model, e.g., the Dugdale zone. In all cases, however, it is prohibitive to investigate such aspects as the effect of modelling itself, numerics, variations in material representation and/or load configuration, and other implications of the assumptions which underlie a given study. Moreover, we know of no published work which gives a comparative evaluation of various analytical methods, computational or otherwise, which could serve to confirm that any one analysis is concerned. In a strict sense, therefore, these fundamental data cannot be regarded as wholly unique.

It follows that the behaviour of various characterising parameters may be peculiar to the data sets from which they were extracted. While we acknowledge that this difficulty may prove eventually to be negligible, the fact is that we do not presently know whether such is the case. Hence the characterising parameters, as described by different investigators, may be expected to relate only remotely to laboratory behaviour and, eventually to service performance. Moreover, the <u>definitions</u> of at least some of these parameters may be different when <u>applied, first</u>, to numerical results, and then to experimental observations.

We have then a rather loosely tied set of results. Crack opening, for example, is defined differently in different (analytical) models ranging from a built-in blunting, to a separation at various positions along a crack's flanks - including an extrapolated position, to a stretch or strain at the crack's end; at the same time, experimental data relate to three-dimensional configurations of actual materials (ranging from polycarbonate to carbon steel). Contour integral quantities are well defined mathematically, have some appealing properties for certain idealisations of material behaviour, and are easily evaluated on the computer; at the same time, it is not they but their presumed analogues which exist experimentally. Crack growth in fact occurs, and it is modelled by approaches ranging from constant velocity behaviour in an elastic continuum to quasi-static node by node release of points in a finite element array, a technique requiring decision as to node spacing, conditions for release, simulation of material unloading and reloading.

Such are typical of problems attempted by the analysts reporting at ICF4 and such are the difficulties encountered in their work. We can applaud their efforts and dedication, and indeed do so. It should be recognised, however, that the final words have yet to be written on these problems, for the models and methods will be refined in the years ahead. In this connection, we note that analysts and mechanicists must increasingly turn to laboratory observations as a critical resource for the formulation of their work. Further, the laboratory will eventually become a proving ground so that results of the type presented in Part V of ICF4 will be both rigorously based in terms of mechanics and firmly corroborated, physically. Finally, the information we all seek ultimately to develop will provide bases for the design procedures which are needed to ensure new levels of integrity and reliability in critical structures.

J. L. Swedlow

VI APPLICATIONS

Attendance at the *Applications* sessions was generally quite good, and the discussions following the technical presentations always enthusiastic. The visual, aural and technical quality of these was universally excellent. The authors clearly laboured diligently in their preparation and all attending profited as a result. Authors whose first language is not English are especially congratulated for the fine presentations they made using a foreign language.

General observations gleaned from the technical presentations and discussions are as follows:

1. A number of investigators reported developing specimen configurations which closely resembled structural components. The resulting data were quite useful in predicting the behaviour of the actual components and have, at times, correlated quite well with data generated using standard fracture toughness specimens.

2. Other investigators reported developing specimen configurations designed to be more economical, or to have more complex stress states, than the standard fracture toughness specimens. The results presented indicate generally good correlation between test data generated using the standard configuration.

3. Several researchers developed stress intensity solutions for cracks in

various bodies, e.g. in stiffened panels, in holes, and in cracked lugs. These solutions will greatly aid analysts in predicting fatigue-crack propagation and fracture in structural components containing cracks.

4. The effects of various metallurgical parameters on fracture toughness were reviewed. These parameters, including hydrogen, oxygen and nickel contents, pore density and heat treatment all influence fracture toughness. The information on these parameters will be quite valuable to designers and fabricators as they select materials for structural use.

5. Several experimenters reported on the effects of temperature on fatigue crack propagation. Generally, fatigue-crack propagation is not affected by reductions in temperature as is fracture toughness, and the experiments reported helped to clarify the reasons for this.

6. Studies of both static and dynamic fracture toughness in metals were described. In some instances, unique light techniques were used to measure stress intensity factors during the tests. Correlations between the static and dynamic fracture toughness and the static and dynamic yield strength were reported. These techniques and correlations will prove quite helpful in predicting failure in rapidly-loaded structures.

7. Some researchers studied the effectiveness of various nondestructive test techniques. In some instances, acoustic emission techniques were very useful for detecting stress corrosion cracking. However, ultrasonic techniques can seriously underestimate flaw sizes (by factors of 4½ in some instances) in components. Such underestimations can lead to gross overestimations of the remaining life of the component. It is hoped that future *Applications* sections will deal with the reliability of nondestructive test techniques in greater detail. The reliability of any life prediction made using linear elastic fracture mechanics hinges critically on knowing the actual size of the initial flaw.

8. Virtually all authors added valuable test data to the technical literature. These data are invaluable to the analysts who must perform fracture mechanics analyses but who lack their own laboratory facilities.

In conclusion, *Applications* covered a wide range of interests. It provided the technical literature with many new and exciting concepts for future research, and provided the tools for solving some old problems. This section of ICF4 must be classified an unqualified success.

C. Michael Hudson

VII NON METALS

In this part, the fracture of a great diversity of materials was considered : ceramics, glasses, composite materials, polymers, biomaterials and concrete. Even metals appeared (in the form of prosthetic devices) under the guise of biomaterials. The workshop sessions were generally very well attended and the discussions vigorous, sometimes to the point of becoming heated. This led to many useful conclusions, which will be dealt with separately according to the type of material, in what follows.

Ceramics and Glasses

These two sessions will be considered together, since the topics discussed in each were closely allied, and fell into three general areas :
1. Factors affecting fracture toughness (also the topic of the Plenary paper), 2. Subcritical crack growth, and 3. Use of fracture mechanics data in engineering design. In the area of fracture toughness, the paper by Hubner sparked a lively discussion. His measurements on alumina demonstrated R-curve behaviour, in that the toughness increased as the crack extended. This mechanism was also discussed by Evans in his Plenary paper, and by Buresch in his workshop paper, and the subsequent debate over R-curve behaviour centred around the validity of the techniques used for fracture toughness measurement. The paper by Wiederhorn on the most reliable crack velocity -- stress intensity relation for subcritical crack growth in glass provoked some controversy. Wiederhorn felt that the data was best represented by relating crack velocity to a power function of stress intensity factor, although this did not have any theoretical basis. J.S. Williams pointed out that polymers also obey this type of relation, and that he had been able to derive the power law equation using the relationships for the time-dependent behaviour of yield stress and modulus which are established for many polymeric materials. (They show an inverse power law dependence on time.) Wiederhorn's response was that these power law dependencies were also empirical.

Composites

The Plenary paper on this subject showed that much progress has been made in understanding the factors that contribute to toughness in fibre reinforced composites. There are difficulties in applying fracture mechanics to these materials, however, and these are in large part due to the size of the "process zone" at the crack tip. They can therefore be avoided to some extent by testing samples with sufficiently large initial cracks, as is done with ductile metals. The workshop papers covered a wide diversity of composite types, including joints, laminated sheets, and planar composites with two components, as well as fibre reinforced materials. The scope for mathematical analysis was shown to be very great, but many participants felt that the paper on strength criteria was a solution in search of a problem.

Polymers

The Plenary paper on impact testing of polymers showed that, when suitably instrumented and interpreted, impact tests can give reliable data on the fracture properties of polymers. In contrast to this, the Plenary papers which attempted to relate the fracture of polymers to the behaviour of the individual polymer molecules indicated a complete divergence of opinion as to what was occurring. The evidence from recent experiments seemed to support Kausch rather than Peterlin. It was agreed that one of the most important phenomena occurring during the fracturing of polymers is crazing, i.e. the nucleation, growth, and coalescence of very large numbers of very small cracks. Annealing has a very great effect on crazing, and during the workshop sessions the importance of annealing craze-prone polymers before testing was stressed. The effect of environment on crazing can also be serious, and the silicone fluid used for applying pressure to polymers for multiaxial stressing was found to promote crazing. Crazing is also important in fatigue, where failure occurs by enhancement of craze growth and coalescence in polymers which craze.

Biomaterials

The micromechanical processes occurring during the fracture of bone are of great complexity, owing to the highly inhomogenous and anisotropic nature of this material. These generated food for thought when it was pointed out, in the Plenary paper, that in humans, well-used bones have very superior mechanical properties to those not much used. Prosthetic devices have to be mechanically compatible with the relevant part of the body, as well as unaffected by body fluids. The demands made on these devices are often extremely severe.

Concrete

Attempts to apply fracture mechanics to concrete aroused lively discussion, and participants questioned the validity of tests in which the artifically induced cracks were of the same order of magnitude as the dimensions of natural discontinuities in the material (e.g. interfaces and coarse aggregates). This problem is common to all composites. Concrete presents many problems of its own as well. There is the paradox of dry concrete, which is extensively cracked, being demonstrably stronger than wet concrete, which is uncracked. The active role of water in dimensional changes and strength reduction is nothing to do with lubrication.

Many other lively discussions, too numerous to detail here, took place during the generous time allotted for the purpose. It was noteworthy that the bringing together of fracture experts from several disciplines was a significant factor in promoting discussion, as often workers from different fields view a particular phenomenon in different ways.

                                                M. R. Piggott

RESULTS OF AN OPINION POLL OF DELEGATES TO ICF4

At the end of ICF4, each delegate was invited to complete and return a questionnaire to provide information for the help of future organisers of ICF Conferences. In all 143 such forms were returned, sufficient for the opinions so collected to be regarded as coming from a reasonable fraction of the total audience. The results of this survey are summarised in this brief report.

The first 15 questions required a straight forward check-list answer, such that the results can be summarised quantitatively as shown in Table 1. This makes it clear that the organisers had made good choices on many of the most important aspects of the Conference. The total length of the Conference, the timetabling of free afternoons with late evening sessions, the number of social functions, the number of topics covered, the number of panel discussions, the length of each plenary presentation and panel discussion period were all points which the majority of those delegates who returned questionnaires believed had been selected just right. There was also considerable support for the use of author-presentation rather than rapporteurs, and the use of workshop foremen to guide the discussion. The convenience of holding the Conference on the very pleasant Waterloo University campus and so lodging almost all the delegates in the one centralised accommodation was also widely appreciated because it helped them to make contact with each other so that they could arrange informal meetings and discussions.

On points on which there was some wish for change, about half of those replying thought that the permitted length of each workshop paper was too short, as was their presentation time. There was considerable support for a reduction in the number of workshop sessions, associated with a significant majority who felt that there were too many workshop papers at ICF4. There was also considerable support for a reduction in numbers of plenary papers.

The final three questions of the questionnaire called for comments on things that should not be repeated, on aspects which should be introduced and for general comments. Any review of the replies to these must of course be somewhat subjective since the wording of each reply was individual and included many fine shades of opinion. Moreover, it appeared that several of those completing the forms duplicated the same comments in various of their replies. In the main, as was to be expected, the written comments reflected the check list answers to the questions summarised in Table 1. The one overwhelming comment was praise for the Waterloo Conference, for the site, for the organisation, the domestic arrangements, the friendly approach, the excellent social arrangements, the efficient office staff and for the publishing arrangements. These points were mentioned in almost every one of the replies, many being expressions of gratitude as well as of compliments. The main source of complaint was in relation to the quality and large number of papers - and several suggestions were made for dealing with this by pre-announcement and selection of favoured topics, by careful and strict refereeing, and by encouragement of other methods of publication. Associated with these comments was the frequently expressed wish for fewer parallel sessions and also for increased time for individual workshop presentations. Another point made in the several answers was that a list of delegates showing residential location issued early in the Conference would be a major help in making contacts. There was some criticism of the choice of plenary speakers, perhaps associated with suggestions that there should be fewer such contributions, each of which should provide an overview of the present world status on a particular topic and indicate areas where discussion would be of value and appropriate for future work. The need for question time and discussion at the plenary sessions was emphasised. There were of course very individual comments, some of which provide valuable ideas which will be communicated to the organisers of future conferences.

The overall impression gained by reading the questionnaires was one of pleasure and satisfaction with the arrangements provided at Waterloo, and a wish to express the thanks of all those that attended to the organisers.

R. W. Nichols
President (1977 - 1981)
International Congress on Fracture
September 6, 1977.

Table 1. Summary of Replies from 143 Questionnaires

Aspect	Percentage			
	Too many	Right	Too few	No answer
Number of days of Conference	15	83	1	1
Number of plenary lectures	52	46	1	1
Number of workshop sessions	43	45	6	6
Number of topics covered	29	62	4	4
Number of panel discussions	14	55	11	20
Number of social functions	0	80	5	15
Number of workshop papers	57	32	2	9
	Too long	Right	Too short	No answer
Length of each plenary presentation	3	67	27	3
Length of each workshop paper	0	46	49	5
Length of each workshop presentation	1	41	47	11
Length of panel discussions	20	44	7	29
	Valuable	Not Valuable	No answer	
Workshop Foreman system	66	25	9	
Rapporteur Presentation	34	50	16	
Free afternoons with late evening sessions	67	27	6	
"Centralized" accommodation	53	42	5	

FRACTURE AND SOCIETY

An ICF4 Interview with Sir Alan Cottrell FRS

As Cottrell pointed out in his opening address at ICF2, everyone is concerned, from a very early age, with why things break. Children's toys
break, we break our bones, the engines of our motor cars and washing machines
fracture - but more importantly, advanced large-scale structures can fracture - pipelines, bridges, skyscrapers, nuclear reactors, ships and aircraft - even the very earth itself fractures in earthquakes. The understanding and alleviation of all such fractures are the special concern of
the scientists and engineers who gather together every four years at each
International Conference on Fracture. The inspiration of these conferences
has been Professor Takeo Yokobori, Founder-President of ICF. It is the
principal purpose of ICF to regularly bring together, from every corner of
the world, the major workers in all aspects of fracture for a re-assessment
of the advances made and to provide a basis for sound and relevant scientific
and engineering work in the future. This purpose will surely be achieved
at ICF4. But, in Canada, in preparing for ICF4, we were especially conscious of the larger purpose of placing all this research in the full context of society as a whole. As the complexity of our technological systems
increases, so do the possible catastrophic consequences of failure. By way
of emphasis, one may cite the Presidential Campaign of 1976 in the United
States where the consequences of fractures in nuclear reactors, and hence
their safety, played a significant role. The safety of many of our energy
systems including reactors, offshore structures, super-tankers, LNG ships,
pipelines, is now of very wide social concern and is discussed regularly
and thoroughly in the ordinary press. Accordingly, it is both an obligation
and extremely prudent that we, at this conference, address ourselves to our
responsibilities to the safety of the technological world at large.

Thus, the dominant themes of ICF4 are the *applied* aspects of fracture and
especially the application to large-scale engineering structures. At the
same time, the broad purpose of bringing together workers in every aspect
of fracture has not been forgotten. But, to ensure that the social implications of our work can be fully appreciated and discussed, two Plenary
Panel Discussions have been organized under the general heading *Fracture
and Society*. The first of these focusses upon Educational issues whilst
the second is concerned with the broad Socio-Political context itself. Very
early in the organizing of ICF4 (1973), Sir Alan Cottrell was approached
for his views on the most useful orientation for ICF4. It will be remembered that Sir Alan has very special qualifications for being so consulted.
He is amongst the handful of major contributors to our understanding of
fracture processes in solids. His background includes experience in Chairs
at two major English Universities and in the nuclear industry and the full
accolade of the world scientific community for his own research and his many
books. But, for many years, in more recent times, he has been privy to
the Corridors of Socio-Political Power through his appointment as Chief
Scientific Advisor to the British Government. In this role, he became a
household name in Britain and led a team to Canada to investigate the CANDU
reactor system. He became well known in several parts of the world for
lectures on *Science and Society* and for his broad consciousness and intellectual grasp of the issues confronting the development of technologically

advanced nations. More recently still, he has begun another career as Master of Jesus College at Cambridge and as a rather youthful Elder Statesman. Thus, there could surely be no one better qualified to address ICF4 on the topic chosen for these panel discussions - *Fracture and Society* - in 1976, Sir Alan was approached accordingly. Unfortunately for ICF4 (but fortunately for his University) he had just been elected Vice-Chancellor and convocation precluded his absence from Cambridge. After various discussions about possible alternates, it was realized that the purpose would be best served through a published interview with Sir Alan - somewhat in the style of some American and British magazines. Such an interview would be structured around prepared questions (taken on notice) on the topic *Fracture and Society* and would be probing and wide-ranging so as to provide the best possible foundation for discussion in the ICF4 Panels - and elsewhere.

Through the good offices of Dr B. Ralph, a Fellow of Jesus, and Dr. J. F. Knott, I was able to arrange, at short notice in December 1976, a meeting in which the proposition could be put forward in detail. I was advised to approach the proposition rather slowly and not until, perhaps, the third course of our dinner in Hall. Somewhat untowardly, I discarded this advice and before we had finished our soup, Sir Alan had gladly accepted the proposition and asked me to forward to him the questions, so that he could look them over prior to the recording session. I also arranged to send him copies of a selection of the Plenary and Workshop papers for his scrutiny and we were able to sketch over some of the questions and the overall scope of the interview.

The details of the actual interview were as follows. Through discussions with various scientists, amongst whom figured Tetelman, Eshelby, Embury, Knott, Bilby, Smith, Averbach, Ralph, a list of questions was compiled and forwarded to Sir Alan during January 1977. The interview took place over a period of two hours on February 16, 1977 in the Master's Lodge at Jesus. Dr Brian Ralph and Dr John Knott, both former research students of Sir Alan, conducted the actual recording session on behalf of ICF4. The interview was tape-recorded and the printed transcript that follows has been subjected only to minor editorial changes.

On behalf of all the participants of ICF4, I wish to record here our gratitude and appreciation to Sir Alan for the task he undertook. I am sure it will be judged to be of very considerable substance, value and interest. It now only requires that in the Panel Discussions we apply ourselves equally to the task which we address. Fracture researchers are a relatively cohesive and harmonious group. It should be possible, thereby, to arrive at some sound and relevant conclusions in what has become, increasingly (Newsweek, May 9, 1977) not only a model and structure sensitive subject but also politically and socially sensitive.

D.M.R. Taplin
May, 1977

THE INTERVIEW

ICF4:   *Concentrating initially on the science of fracture, do you think that it's fair to describe Griffith as the father of the science of fracture?*

Sir Alan: Yes, I think he is the father of the science of fracture. Griffith took the key step of treating the growth of a crack as a thermodynamic process; and his formula, because it balances the volume and the surface energies incrementally, has a universal application to all forms of true fracture.

ICF4:   *If we now follow on with the second question: looking back over the development of our understanding of fracture since Griffith, what are the four or five major steps in that progress?*

Sir Alan: One can think of lots of steps. I would list a few of them as follows: One major step was the reaching of ideal fracture strengths in fibres of brittle substances and in rods of glass from which the natural Griffith cracks have been removed.

Another very important achievement was the extension of Griffith's theory to include effects of plasticity at the tips of cracks. This was done first by Orowan, who formally introduced a plastic work term into Griffith's surface energy. Later it was dealt with more comprehensively by a treatment of the crack in terms of dislocation theory, so that fracture theory could be linked up with plasticity theory.

A third important step was the realisation that in practice, more or less all fractures, even those in the most brittle substances, are preceded by some plastic deformation. I think that's very significant. A more practical step of great value was the application of crack theory in the form known as fracture mechanics to practical engineering problems (Irwin).

Another important step, was the realisation that plastic ruptures are not true fractures, but are forms of plastic necking and sliding-off. Yet another was the realisation that brittleness is not the same thing as fragility. The thing that the engineer wants to avoid is fragility and it is possible to have brittle materials which nevertheless are very tough; that, of course, is the principle which underlies the design of strong fibrous composites.

ICF4:   *On that last point, do you think that such design was realised in a theoretical sense or was it something that came out of people experimenting with fibre composites? In other words, when did the realisation of the toughness, of what was supposed to be brittle material, arise as an experimental result? Was it really a theoretical prediction?*

Sir Alan: I think it arose as a theoretical result. Tracing back, you'll find that people were thinking theoretically along the lines of using weak interfaces as crack stoppers even before they had done much work on strong fibrous materials. Of course, staring you in the face all the time was wood and for some reason people never though about wood. It was too familiar.

ICF4:   *This must have occurred also with wrought iron. To some extent
the brittle fracture problem as such only came about when
engineers stopped using wrought iron and started using steel.*

Sir Alan: Yes, but for some reason, people working in the theory of the
subject didn't think about those things at all. They only began
thinking seriously when they began to ask themselves what they
could do with whiskers or strong fibres and it was such thoughts
that led people to the principle that brittle materials can be
tough.

ICF4:   *How has it come about that so many good scientists have ended
up working in the field of fracture, for at least some limited
part in their scientific lifetime? It seems to me that a great
many of the more notable scientists in the fields of Physics,
Engineering and Materials Science and Technology have worked
in some aspect of fracture.*

Sir Alan: Well, it's a very attractive subject. It obviously has a very
practical interest because of big engineering failures and that
sort of thing, but also it's a straight scientific challenge.
What are the laws that determine the formation and growth of a
crack? It's obviously something to do with atomic structures,
so that, after the heroic phase of the 1920's when the theories
of atomic structures were being built up, crack theory became
one of the areas where one could apply these ideas, so that one
virtually had both scientific and practical challenges in a
field very rich in phenomena as well.

ICF4:   *How much further do you think we ought to go in pursuing the
theory of fracture?*

Sir Alan: It depends what you want to do. I think there are some
areas of the subject which are not understood at all: a lot
of fatigue, and fractures of that kind. We understand some
aspects of fatigue fracture now but many of the others I think
are still not understood at all, in terms of the the atomic
processes going on. I think that's also true of some of the
things to do with stress corrosion and corrosion fatigue fractures and some forms of the high temperature fractures, so there's
a lot of atomic mechanisms which are still not understood, there,
and if one is trying to build up a full atomic science of all
these things there are some challenges. On the other hand, I
think that for engineering design purposes, a lot of that may
prove of limited value. It may well be that from the point of
view of the design engineer we know what is useful to know
about fatigue already: e.g., simply knowing that it is a plastic process and therefore you've got to make the stuff hard and
particularly hard at the surface where the cracks begin.

ICF4:   *Yes, that obviously is of value for initiation controlled fatigue
Clearly, a lot of structures have defects and cracks there already and one has to apply something which cannot prevent
fatigue but can control its rate of growth.*

*I think there's a number of small points I'd like to take up;
if we just start with monotonic loading of cracks. Work that
has been done has separated out the face-centred cubic ductile*

> *materials and strongly covalently or ionically bonded brittle*
> *materials and we're still left with perhaps the most useful*
> *and interesting material of all: iron. Iron seems to come*
> *out as a borderline case in all calculations that have been done.*
> *Do you think that there is still an intellectual challenge here:*
> *to find out why iron behaves in the way that it does? Should*
> *scientific effort rather than engineering effort be put into*
> *this particular study?*

Sir Alan: I think that the reason why it behaves as a borderline material
is really a problem for quantum mechanics of the solid state
rather than fracture theory. It's a matter of how the forces
between atoms work out and some people find that a very challenging field to be in. I think it is an area where you can
put in a lot of hard effort with not a great deal to show for it
at the end of the day - that's a personal view! On the other
hand, because iron is a borderline case, it shows an extra-
ordinary richness of phenomena, there's no end to the variety
of things that it will do! For that reason, it's a very rich
substance to work on experimentally: just discovering in the
laboratory all the various phenomena and getting them sorted
and classified and understood.

ICF4: *But you think that, in terms of understanding of fundamental*
mechanisms, there's not that much further to go in monotonic
loading; although fatigue is almost completely not-understood
in terms of how atoms come apart?

Sir Alan: Yes, that's a bit over-simplified but that is broadly the way I
would see it.

ICF4: *Obviously where corrosive environments are involved our under-*
standing is even more limited than in fatigue. One of the
things we were talking about earlier is really what the
aim of the game in fracture research is all about: whether
we are trying to probe the fundamentals of the subject or
whether our knowledge of fundamentals is sufficient for our
purpose, and the attention of all these people working in the
field should be taken towards application to engineering design.

Sir Alan: Well, I suppose different people have different motives in the
field: some would be in it as pure scientists, and I would
have thought that, for them, parts of the subject are now complete. If they want to go on pursuing pure science then that
should be done in the fatigue field and in some of the more
complicated conditions of fracture where you've got to get down
to corrosive effects as well. But for those who are interested
in the practical side of the subject, I think that the great
challenge is really turning what we know about the science of
fracture into engineering design procedures. I still think
there's a great deal to be done to work out a theory of how
materials behave in service.

ICF4: *I think that, indeed, is where a gap is still in existence.*
You have outlined steps in the progress of our understanding
of fracture and I think that those steps come over very well.
But, suppose we were to rephrase the question in terms of
the application of that knowledge to engineering design? Even

> *if we take the fracture mechanics approach, the number of*
> *situations in which <u>that</u> can be precisely applied is really*
> *rather few.*

Sir Alan: That is true. The direct links between fracture theory and
engineering design in operation are rather small. There is the
fracture mechanics approach, which is but one link, but nevertheless it is a very good one, and fracture toughness is now used
as a standard design parameter in lots of engineering designs.
That is directly attributable to all the fracture research work.
I think that, at present, all the rest of fracture theory does
not get through to engineering design, but that does not mean
to say that it is not useful, because it gets through to metallurgical design to provide better materials for the engineer to
use. The whole business of getting fine-grain materials, fine
microstructures, so that you can get great hardness without
losing ductility, is again directly attributable to the theory
because in the end that comes out of Griffith's equation. That
has been another great area of application, although all the
design engineer sees of that is that he's suddenly got some
better materials than he had before.

ICF4: *Yes, and one can even go a bit further than just that. By a*
knowledge of the behaviour of things like inclusions and impurity
elements, you can reduce the <u>scatter</u> in material quality.

Sir Alan: Yes, that again is offering the engineer something of better
quality.

ICF4: *I think that this is something of importance. I may be jumping*
ahead of some of the questions but there is certainly a tendency
now: I think it comes out in Tetelman's paper and in the con-
tribution that Wells made to the Rosenhain conference towards
a probability approach to design, so that one ends up saying
'this has got a life expectancy', of whatever it may be and you
get to the answer by multiplying a number of probabilities,
like the probability of finding a crack and the probability of
obtaining a given value of toughness. It seems to me that this
is a useful diagnostic process because one can investigate what
the effect of narrowing the standard deviation is on any of the
parameters.

Sir Alan: Of course, that principle has been used in the analysis of things
such as nuclear pressure vessels. You go into the probability
of detecting cracks of various sizes, the probability of the
fracture toughness deviating and all that sort of thing and out
of that you get an estimate of the chance of a failure during
a lifetime.

ICF4: *In the long term you save materials because you're tightening*
up the engineering specification and you can get away with
lighter sections. Although I don't want to jump into the topic
of future research at the moment, I think that there is a point
which ought to come through when one is looking at this. That
is, where the effort should really be put? If one thinks of
the parameters that are collected together in a fracture analysis
stress, toughness, defect length; then there are three fields
of interest and we concentrate as metallurgists on material

*toughness and quality. However, look at the variability in
that and compare it with the variability of say stress analysis
in a structure where one doesn't know perhaps what residual
stresses are acting, or the variability of non-destructive testing
in terms of the accuracy with which you can measure crack lengths!
It seems to me that we might be pouring more and more research
effort into a field in which the benefits are smaller as time
goes on and that one could pour the same efforts into other
fields which would aid the engineer rather more.*

Sir Alan: I think that, as regards more work on the fundamentals, if it's
going to produce anything useful it will produce better materials.
It may be that the more important work would be in applying what
we already know to the production of better materials. You mentioned non-metallic inclusions; it may be that we could do more
scientific work on how you get rid of the inclusions, or how you
refine them. I think this will be physical/chemical research
rather then fracture mechanics research. That is probably one
example where there is useful work to be done. Coming back to
engineering design, I would make again a point I made a little
earlier, that the application of all this science to the development of a new science of materials in service is still a problem
for the future. There are not many places where a respectable
science in service exists and mostly it's just a compendium of
empirical experiences and bits of know-how. Let me give you one
example, it's not a fracture one but it's a failure one, where
a proper understanding of materials behaviour has helped this
significantly. This is in the theory of the plastic design of
engineering structures. You mentioned a little while ago the
business about internal stresses. A stress analysis of a complicated structure will not tell you what internal stresses are
in it. The great thing about the theory of plastic design,
however, is that it shows that in most cases it doesn't matter.
You can work out the strength and stability of your structure by
ignoring all the fine details and that rests on a sophisticated
understanding of the stress/strain curve and the onset of plasticity. One feels that there must be other opportunities like
this of applying what we now know about strengths, fracture and
cracks and all that sort of thing. That's really an example of
what I would call a proper science of materials in service. But
I think that fracture mechanics and the use of fracture toughness
is also an excellent example.

ICF4: *Yes, I think that comes in, even if not into the original design,
which is on some plastic collapse basis. It certainly comes in
to the assessment of the danger of defects which are subsequently
found to be there.*

Sir Alan: I think that the things that will be useful to the engineer will
not be sophisticated atomic theory. I don't see you handing the
engineer something like a Schrödinger equation, which he will
then plug in and work out as design. It's not going to be like
that at all. I think the principles will be rather simple ones.
I'll give you an example of a valuable principle that's come out
of this; this is the principle of "leak before-break" in pressure systems. This again basically is another application of
Griffith's crack principle. If the wall thickness is small compared with the critical crack length, then your system will leak

long before it breaks - that's a valuable safety principle and
I think structural engineers will become more and more convinced
that they have to design that feature into structures where
there are big consequences of a breakdown.

ICF4:   *I wonder on that point if I could go back into history a little
and probably quite unfairly ask for comments on the tests that
were done at Risley many years ago with pressure vessels which
are pressurized. They contained initially very long cracks
which were then sealed, and at that stage pneumatic pressuriza-
tion caused completely catastrophic failure although the material
was well above its ductile transition temperature. Do you think
that was a misleading set of experiments in terms of the dangers
that really occur in pressure vessels and pipelines?*

Sir Alan: Well, I don't really know. It goes back a long time and I've
forgotten the details.

ICF4:   *The details are something like a pressure vessel of one inch wall
thickness, 5 ft. diameter or something of that order, containing
an artificial crack sawn in to the order of about a foot in
length. This was then sealed with some aluminum foil or neoprene
seal and what it enabled the material to do was start a fast
running shear fracture at a fairly low hoop stress because there
was a long initial defect. Even above the transition temperature
this gave rise to catastrophic failure. On a leak before break
concept, presumably the initial defects would be less than wall
thickness, in other words less than an inch to start with, and
I think there's been a lot of recent discussion that the sort
of results that were obtained with the artificially long cracks
could not be obtained in normal practice with the sorts of
defects that could be found in pressure vessels.*

Sir Alan: Yes, I think you're right. I think to that extent the cutting
of the very long cracks and then sealing them up so they didn't
leak is unfair to the material in that sense. You're suppressing
the leak-before-break in an artificial way. On the other hand
I think those tests do show in a valuable way that even well
above the transition temperature a big crack running away is
unstoppable.

ICF4:   *Yes, that point is accepted, it's just that I wanted to bring
it out at the moment because it is apparently something that
doesn't mix with leak-before-break unless you think about it
rather carefully. It was also the factor that I think was
largely responsible for attention being paid to the prevention
of fracture initiation rather than propagation. If we look at
the list, may we take up the point concerning disappointments?*

Sir Alan: Well, again, I would say that I think the disappointments have
been that we still don't understand a great deal about fatigue
or some of the corrosion forms of failure even though there
has been a vast amount of work done on the subjects. I think
this is because these are processes at the atomic level where
you've really got to see what is going on and we still haven't
got instruments that will quite allow us to do that fully.

ICF4:   *When you say we don't understand fatigue do you mean that we*
cannot simply rationalise it in terms of firstly hardening up
microstructure by repeated to and fro plastic flow, and then
being able to localise deformation along a specific slip plane;
together with some gas adsorption or something of that nature
which prevents complete rehealing?

Sir Alan: I think there's a whole lot of things we don't understand. For
example, there are problems in the work hardening stage, because
we don't really understand why there is the localisation in
specific active slip planes. Also, the oscillating mechanical
system is geometrically reversible (I use the work "geometrically"
because it's obviously not thermodynamically reversible) and
ideally the atoms should all go back again. I'd say that there
is some second order effect which is opening up a crack in the
surface of these active slip bands and that is something else
that we don't understand. Maybe we know the answer already
but haven't proved it. It may be true that gaseous adsorption
makes it structurally irreversible but we don't have a hard
proof of that. It is just a good surmise.

ICF4:   *And you think if that research is done we would stand a better*
chance of being able to advise the engineer on how to design
against fatigue failure?

Sir Alan: We might do. I don't hold out strong hopes on that point. But
I think there are still other things that we don't understand
about fatigue. The crack starts off down the active slip plane
and then at a certain depth it turns away sharply, for some
unexplained reason, and becomes the much more conventional plastic crack which I think we understand now. At least, we partly
understand it, although there isn't yet an elastic/plastic crack
growth theory that works in a rigorous mathematical way in the
fully plastic range.

ICF4:   *Yes, there's certainly been a number of attempts to carry over*
the simple models.

Sir Alan: A sort of full post-yield fracture mechanics: you can go a
little way beyond but you can't get into the full plastic range.
I think there's a whole bunch of problems there. But whether
the solutions of those would help in the question that was
asked, I think is quite doubtful. I think we know enough of
the qualitative principles to be able to see what you need to
do to improve the fatigue strength of materials, but we can't
get much beyond that point.

ICF4:   *Yes, and certainly the local modes for the way in which a*
fatigue crack propagates in an elastic continuum with local
plastic flow must be very similar to the way in which cracks
initiate: there will be local sliding on slip planes inclined
to the line of the crack front and the same sort of processes
must go on there, and some experiments done in vacuum have shown
that the cracks go a lot slower.

Sir Alan: I think it's no criticism of those that have worked in the field
that we still don't understand it. It's just a hideously tough
problem where I think the vital events are so extremely localised

and unless you get down there with some sort of super-microscope
to see what's happening there's always doubt whether you've got
the right picture when you form a theory.

ICF4:   *But how do you direct that research effort to answer some of
these questions? This raises the question of how you activate
people to concentrate on those aspects where solutions are
needed, or where the solutions you've already found need to be
applied?*

Sir Alan: I think you have to distinguish between a pure science and its
applications. If you know the science you can run large teams
on the applications. I think that one could probably run a
large team now and make good progress in developing what we call
the science of materials in service. I think that if you're
wanting to understand what's happening at the tip of the fatigue
crack it's a bit like the cancer problem, in a way. If you run
a big co-ordinated programme you may at the end of the day make
no advance on it. I think the only way you can get at that sort
of problem is to have a healthy university research environment
and let people go round pretty freely to exchange ideas.

ICF4:   *Yes, those obviously are the two extremes, I think if we move
from the research to the politics behind the research, it's a
question of the relative effort that one ought to put into
letting people go their own way and in design.*

Sir Alan: This is an aspect of the much more general question of how
much university research a country ought to do. There's nothing
unique about fracture research in this. I think I might say
that my own personal view is that too much "university" research is done these days. Not too much in money terms that is.
I don't think the money expenditure is such a crucial factor.
But university research is producing too many young men who have
expectations of research careers that are not available to them
afterwards. That's a great disappointment to them and it's why
I think that we're overdoing the amount of "university" research
at present. I don't worry about the money, because it is still
relatively small compared with the cash flows through the other
sectors of the economy.

ICF4:   *It certainly seems to me, from a number of comments, that the
thing that is lacking is the application of what is already
known. We've touched on this already with regard to the
fracture field and I'm sure it's true in a much broader sense
than that.*

Sir Alan: That's right. A lot of university researchers are bred from
the textbooks and the textbooks talk about atoms and Schrödinger
equation and that sort of thing. It's all so exciting; and

inevitably lots of research groups set out to do work of that same kind. There is a great lack of contact with industry here, and of course industry's not very helpful, because its so pre-occupied with its immediate problems that it can spare little for building bridges to the universities.

ICF4: *Do you think the situation between university/industry is worse in the United Kingdom that it is elsewhere in the rest of the world or do you think we're typical?*

Sir Alan: I suspect that it is much about the same in the USA. I think in Europe the situation is rather different, I think there's a different European tradition: for subjects such as this to be done in the applied institutes of technology, or technical universities where there is a much closer linkage with industry and a more practical approach to matters.

ICF4: *It's interesting that you think the situation is about the same in the States as it is here, because it is often held that the American academic has closer contacts with industry because he's forced to do consulting work and things of that nature. I have a question now which is, "how to teach a comprehensive knowledge of fracture to engineers in a modern university?"*

Sir Alan: It is a difficult problem, and I think the mistake is to teach it in the way that would be most interesting to people of one's own kind: to teach it as a pure science; to repeat all the steps of the argument by which you've reached full conviction that this is what happens: to go through all the evidence and analyse it critically; all the things whereby you become a good research man in materials science. That is not what the engineer wants at all. To the engineer a material is just a black box. He has to pay more money for it than he would like and he hopes it will accept the right inputs at one end and deliver the right outputs at the other end. That's all he asks of it: that there should be a satisfactory relation between its inputs and outputs and it shouldn't cost too much money. How you teach engineers in that case is a difficult problem, and the only way to teach it, I think, is as a form of intelligent engineering design rather than as advanced pure science. That means that they're not going to get a rigorous proof of some of the basic propositions of the subject but nevertheless I think that a lot of it does lend itself to a rather qualitative, even pictorial, approach. You can make very nice moving pictures of dislocations moving about and that sort of think, so that people can see in a qualitative way what's going on and immediately realise that it's reasonable, without having to go through all the mathematics of elasticity theory and checking to see if everything is going right. So I would say that the basic science should be done in a qualitative, illustrative way, and all the emphasis should be on teaching as an aspect of engineering design.

ICF4: *In that are you incorporating a description of the sort of poly-phase, polycrystalline structure that one normally has for a material. My criticism in teaching engineers or in supervising engineers who are lectured to on the subject of fracture, is that they have a good knowledge of the basic equations: they know what K_{IC} is; but they have very little physical interpretation*

> *of it, so it's just a number they pull out of the hat and plug into the right equation. You calculate that you must not have cracks more than 11 mm in diameter, without actually realising that there are material variables that go in which mean that 11 mm isn't quite the safest number to have pulled out of an equation.*

Sir Alan: Yes, well this is another difficult problem. I think what bores the engineer stiff is to have a long exposition of the theory of microstructures and phase diagrams. We're talking about mechanical properties here; and polyphase material from this point of view is really a little engineering system; each grain is a linkage in the system: a nut and bolt in the system. Perhaps a useful intermediate step is an analysis of the strength and fracture behaviour of some of the composite materials because there the engineer actually puts them together: he puts the wires into the plastic and pulls it all apart. Its not so very different from working out what happens to a nut and bolt in a structure, so I think you can take that argument along and you can tell him, from that point onwards, when you have materials with natural structures in them, they're on a finer scale but you've got the same principles working.

ICF4: *But you do agree that he has to have some feeling for the microstructure even if its not from the thermodynamic/kinetic/ phase diagram point of view?*

Sir Alan: That's right, and of course he needs it in his engineering: he needs it when he works out the strength of a bridge. He's got to work out the strength of the girders and the brackets and the nuts and bolts, the rivets and all that, so he's already got a "microstructure" to his bridge and metallurgical microstructure from his point of view is only doing the same thing on a finer scale.

ICF4: *One of the disappointments is that, for some reason, engineers are quite happy to do sets of calculations on fairly macroscopic structures and yet throw up their hands in horror when those things are reduced in size by a factor of about a hundred to talk about the parts of microstructure that really matter. I think it's because there is too much leading up to the microstructure in the way of phase diagrams and things of that nature which does bore them.*

Sir Alan: Yes, I think the thing to avoid is to take the engineer through the long story of how you get those microstructures, that's somebody else's job, but, given the microstructures as a *fait accompli,* then I think he needs to know something about how they work as engineering systems.

ICF4: *Yes, that's true and the importance of directionality and inclusions I think is about the limit in this area. I see that as being very safe when one is designing for the classical ductile/ brittle situation, but if we took the cases you were talking about a moment ago - fatigue and creep - where there's the possible interaction with the plastic part, which can in fact disrupt that microstructure, then how do you get over the fact that you've got to teach him something about the thermodynamic stability of material: reversion and all that sort of thing?*

Sir Alan: I think in this case you've simply got to put down warnings to him and say "if you're getting into the high temperature range, or its a corrosive environment, you're getting into a dangerous game, you've got to bring other specialists in who will help you". In this sense he's a general practitioner: he's got to know, if he finds an acute case of kidney failure, that he's got to go to the renal specialist.

ICF4: *But surely that has been the problem in the past, he's gone to the specialist when an acute case has happened rather than before it happens. Very rarely have people of our type been called in at the design stage of a project. We're called in when Flixborough happens or something like that, to explain it from a materials standpoint afterwards.*

Sir Alan: That's right, you need some sort of information centre so that the designer can link up at an early stage, that he can say "I designed this to work under these conditions; it looks alright to me but I am not an expert on the metallurgy or corrosion or whatever, what do the experts say about it?"

ICF4: *How do you generate an infra-structure like that? How do you make him aware in the educational sense of how limited his knowledge is because students don't like being told 'you only know so much at this point, you've got to go and ask somebody else' and then, if you tell them that, where is this source of information that they go to?*

Sir Alan: Well, it all exists embryonically: a good liaison between the engineering professional institutions and the metallurgical ones should take care of that.

ICF4: *But I think bringing in materials people at the design stage is still a very hit and miss procedure.*

Sir Alan: It's done in the biggest projects, but in the more day-to-day matters there aren't the staff to do this.

ICF4: *There must I think be a degree of overlap between a materials man's training and an engineer's training so that, at least on their particular interface, they are speaking the same language and know what concerns the other person. It's a question of the degree to which this is done because I think in some of the older courses, we have seen metallurgy and engineering go along rather different routes, and the end products haven't been able to talk to each other at all. I don't know if you have any specific views?*

Sir Alan: I think this is right. The very old metallurgical training was really a kind of mining training rather than an engineering design training, so I don't think that there was anything for the materials man to say to the engineer then. I think a lot of the more modern courses have been training the materials scientists, and again I think he hasn't had much to say to the engineer because I think his outlook has been different. The engineer has got the practical problem; he's got to produce a design that will go into the drawing office in a couple of months time, whereas the materials scientist will say "We don't know anything about the behaviour of this material under that environment, we've got to stop and do research, and it'll be three years before I can say anything at all." This is hopeless for the engineer. The engineer has got to take his best chance with the thing and I think that this is where you need a new outlook in the materials line, an outlook that is much more sympathetic to the engineer's problem. The outlook has got to be that you find your satisfaction in helping to produce a successful design, rather than helping to understand some fact of nature.

ICF4: *Putting it very broadly, has it not really been that metallurgists and materials people have been concerned primarily with the making, production, and fabrication of their materials; the end point being the microstructures that are obtained, and then that material satisfying a number of, to them, fairly arbitrary criteria, like yield strength, U. T. S. These were fed in from outside and they could develop their material to meet those requirements, but they never asked the question why these requirements were needed. Is that fair or not?*

Sir Alan: Yes, I think that is a quite fair possibility. I think there is another quite different factor you have to bear in mind in this. The materials man has not found it really attractive to make a career in helping the engineer to design things. I think the materials man has known he would always be only an assistant in that kind of work. He would never become the chief designer and never become the head of the firm. It does not prove such an attractive avenue for materials people as some of the other careers.

ICF4: *Yes, I think that's an extremely valuable point. Nobody likes being in a position that can only end up as subservient.*

Sir Alan: So I think that a materials man who goes into the design side should really make himself, as well, a fully-fledged engineer so that he can have access to all parts of an engineer's career.

ICF4: *There are now some broad social questions and one of them is concerned with the view that society in general takes about failures and fractures: public enquiries and how much impact these have on public thinking, perhaps on the amount of research that is sponsored on fracture? There is also the question of how the senior political person, without any engineering background, responds to both real and perceived problems of fracture, and, generally, how the non-technically trained person should assess the risk of fracture and the needs of doing work or research or engineering design work to overcome it.*

Sir Alan: Well, the senior political person doesn't ever think about fracture from one years' end to another, and quite rightly so! He's got other problems on his plate and if he's going to think about things like fracture he won't be good at the job he has taken on. I think that the job of the senior political person, in this respect, is to make sure that the country has a good Health and Safety Inspectorate, or whatever the Inspectorate is called, and that it's working actively on all these things. I think also he must give it strong moral support when it has to make unpopular decisions, because a health and safety inspectorate is under a very difficult set of working conditions. In a sense, the better he does his job the more unpopular he becomes. To be able to stop people in their tracks and say 'no you must not do that' he must have the backing of his ministers or whoever the authority is, otherwise his life becomes impossible. That is really the senior political person's responsibility - to see that there is a good safety inspection system that is active and, secondly, to be prepared to stand by them when they have to make unpopular decisions.

ICF4: *That's at the moment, but how does a political structure antici-*
pate a problem of the Comet or Flixborough type and put the
resources of the country to work? From the inspectorate point
of view I can understand what you're saying: a sort of quality
control along the line, and we need more of it; but how does
a political structure, which, after all, in the long run, defines
where the science and technology policy of the country goes,
make sure that it's putting enough resources into the right sort
of places so that situations like the Comet or Flixborough occur
less frequently?

Sir Alan: Well it doesn't try. It leaves all this to the professional institutions and, if a new type of aircraft crashes then ulti- mately it's because the designers in the professional institu- tions have not been fully up to the job: they've missed a few tricks.

ICF4: *Or is it because they haven't 'lobbied', for enough money to*
investigate that particular aspect?

Sir Alan: On the whole, I don't think that is so. I think that it is impossible for the politicians to intervene in these things beforehand. I think that if they find that a certain part of the engineering activity of the country is causing a lot of trouble then they can intervene. With the box girder bridges for example - they set up a professional enquiry into that to see what's going on, raised questions about the standard of the profession and all that sort of thing, and at the moment I think the government is about to set up another enquiry into the educa- tion of engineers. This is the sort of thing that the politicians can do, but I don't think they themselves can get so far down into the technical "guts" of the country to anticipate these sorts of things going wrong. It's just impossible for them because fracture is only one of the million ways in which a country can go wrong.

ICF4: *And I think Flixborough and the Comet are really very different*
examples. Flixborough, despite what actually happened and the
details of it, is basically due to a patched up repair job by
a non-qualified person. The Comet, brings up another point.
It was a modern aeroplane, designed properly, and being tested
at R.A.E., but I think that, with any major advancement in design,
there is bound to be some risk involved because you are stressing
materials to a higher level than they have been stressed before.
If you'd waited to do all your exhaustive testing before you
brought it on to the market, you would have lost the market.

191

Sir Alan: Yes, you're stepping into an unknown situation and this, in the Comet case, was the fact that each time the plane went up high it did another cycle of a fatigue test on the body. You can say, with the hindsight, that people should have thought of this and it's absolutely true, they should have done, but it's always easy to have hindsight. I don't see that politicians can do an engineer's job for him, they can only monitor a thing and make sure they've got an adequate engineering organisation and if it is inadequate this will show itself up in a run of failures and then they have to beat the big stick and improve the engineering profession, but they can only work in that sort of way.

ICF4: *We have two questions here that bear directly on this. I'll go through them both if you like - the first one is "What improvements could be made with respect to the extent to which society uses engineers and scientists to define and solve matters of great ecological, economic or sociological impact?" and then, more specifically, "What is seen as the future role of Standards Organisations and Professional Institutions in the rationalisation of the specification of materials and fracture behaviour from a design viewpoint?"*

Sir Alan: On the first question the initial thing you can say is the very general one that the engineering profession in Britain is at a rather low ebb. One can think of all sorts of reasons for it; to my mind the most important reason for it is that industry is at a low ebb, it's had such a battering from the government and unions and all this sort of thing, that it has little self-confidence now. I think this has affected the engineering profession as well and, until one has had several years of a governmental, political and social climate which is much more favourabl to manufacturing industry in the country, I think you won't see the situation improve.

ICF4: *But is it just that? There is also the prestige point of view. We were talking earlier about the proportion of people doing fundamental research.*

Sir Alan: Well I think all that comes in as a consequence of it. I think that if we were going ahead with lots of great engineering ventures, as we started to do in the 1950's the situation would be different. Unfortunately the choices then were rather poor ones but certainly in the early days of atomic energy, the early days of Concorde and some of the supersonic aircraft, they brightened up their branches of the engineering profession enormously. Wit more sensible choices then, we would have now a much stronger engineering profession. I think a lot of the problems raised there would have been solved in that way. So I would put that down to the general low state of morale in the engineering profession which goes back to the low state of industry.

ICF4: *Is that simply a British phenomenon? Is it different on the continent or in the States?*

Sir Alan: It is particularly pronounced in this country.

ICF4: *Have people lost pride in calling themselves engineers in this country more than they have in the States? On the continent the distinction is quite often lost and the engineer is also a*

Professor in an Institute and is highly regarded, and Engineer becomes the title you put in front of your name. Whereas Britain has the respect for science; the excitement about the 50's development in atomic energy was attributed to scientists as much as to anybody else.

Sir Alan: I think this country has partly lost the understanding of a need to work. The whole business of earning a living seems to be no longer a natural assumption in the country and I think many of the things we were just talking about really flow out of that sort of change of outlook. You had another question?

ICF4:   *The second point, yes, was on the role of Standards Organisations and Institutions on the rationalisation of the specifications of materials and fracture behaviour from a design viewpoint.*

Sir Alan: I think this is important, because we're moving in a world where resources will be much tighter than they have been in the past. Populations are going up and we're beginning to exhaust some of the easier ores and energy sources. The result is that if we're going to get by we've got to skate on thinner ice, in our use of materials and facilities. That means that all these things are going to be so much more vulnerable to breakdown, therefore they've got to be quality-controlled and the monitoring and standardisation have got to be so much better in order that one can skate on thinner ice without falling through. The demands will be much higher.

ICF4:   *But I think in rationalisation of the specification of materials as well, it is an odd feature perhaps in the way in which, shall we say, steels, have been developed in this country. There is a whole range of materials with very similar compositions and properties which are covered under a large number of different codes and where people may be working on almost identical materials because they're in different industries with very little contact between each other. It would seem that, if one could rationalise the structural materials that are being used, one could concentrate the research and design effort on to a much narrower range of materials.*

Sir Alan: There's certainly scope for doing this, yes. I think it's a thing we shall see before long: computer data banks for all these things.

ICF4:   *Now we come back to the field of questions concerning brittle fracture and risk analysis and safety and things of that nature, related to the choice of nuclear reactors in Britain. I think that is seized upon as a single example of political thinking in this country, presumably this would be the case in Sweden as well, where a government was voted out on the issue of particular types of nuclear reactors. We would certainly welcome seeing you broaden your answer to that in the sense of relating to brittle fracture in the context of your other more recent expertise in the energy field: in terms of oil pipelines, gas pipelines and so on. You are one of the people who has preached a policy on energy, which I fully endorse, but one of the problems in many of the ways of acquiring and transmitting energy, is that one has to face brittle fracture questions. If you could perhaps broaden your answer it would be interesting to us.*

Sir Alan: I can't talk about the backgrounds of Government decisions of
            course, because of the usual rules about this. The Government
            decision itself, and I mean the most significant decision, was
            the one in July 1974 when the Secretary of State for Energy at
            that time announced the choice of the Steam Generating Heavy
            Water reactor. What he said, when he made that announcement,
            was that the Chief Inspector of Nuclear Installations advised
            that there should be no fundamental difficulties in giving SGHWR
            safety clearance. The reliability and the confidence that we can
            have in a system are of particular importance. So, in the way
            that governments move, it had considered safety very important
            when it made its decision. We do know that SGHWR is a pressure
            tube reactor and that pressure tubes have the fail safe principle
            in that they possess the "leak-before-break" feature in them.

ICF4:   *I think this had followed on, if I remember, from some corres-
            pondence in The Times and elsewhere?*

Sir Alan: Well, I had expressed my views, that's right! That is different
            from a Government decision, of course. I had put the view that
            the thing I did not like about the pressure vessel light water
            reactors was that they did not have "leak-before-break". It
            seemed to me that, where the consequences of a major failure are
            as serious as they could be in a nuclear reactor, then one does
            need a natural safety feature of that kind; and that's why I
            argued against that system.

ICF4:   *Yes, I remember the articles, emphasising, if I recall, the
            need for the most stringent care being paid to fabrication
            techniques and to non-destructive testing.*

Sir Alan: If you haven't got "leak-before-break" then the first thing you
            will know about such a failure is that the whole structure is
            coming apart. Nuclear technology is what someone once described
            as an 'unforgiving' technology: if you make a mistake, then
            it's a bad one! That means that, if you're going to dispense
            with a natural feature such as leak-before-break, then you're
            forced to the utmost precautions in your general standard of
            work. I think the specifications that the Americans have set
            for their water reactor pressure vessels are extremely rigorous,
            there's no doubt about that. If human frailty is able to achieve
            that degree of rigour in practice then they will be alright, but
            you must always have a question mark against human frailty and
            this is the thing that worries me. Whereas, with a pressure
            tube kind of reactor, again you have to be just as good as you
            can be against human frailty: nevertheless if you are let down
            by human frailty you've got a natural back-up - that's where
            the difference is and I still feel strongly about that point.

ICF4:   *Do you think, because of the emotive word 'nuclear', that most
            attention is given to the commentary you've just been going
            through on the nuclear reactor case, than is given in the
            equally worrying ecological case of having large pipelines run-
            ning hundreds of miles across the bottom of the North Sea with
            large amounts of oil running through them where a split again
            could be equally disastrous.*

Sir Alan: Well I think so, yes. My own position on that specific reactor problem doesn't reflect any sort of general position I have about nuclear power. In general, I feel that politicians and the general public are being taken for something of a ride by the environmentalist lobby which has being going very hard against nuclear energy. I feel that this is an extremely unfortunate development because the only assured new major source of energy for the world in the thirty years time or so is nuclear energy, and to turn one's back on that without very, very, good reasons could, I think, be a disastrous step for mankind. I think that the fossil fuel position, certainly in the western world, is really alarming. It's much worse than is said in the newspapers. We, in Britain, are locally in a good position for oil, since the North Sea will give us all we need, certainly for twenty years, possibly forty years; but if you go outside Britain into the rest of the Western world then the position is really alarming and we may already have left it too late. The only way out of the situation is the nuclear one. I think the environmentalists have served the Western world badly with their over-done campaign against nuclear energy.

ICF4: *What I was really trying to get to, from outside, is that a double standard is applied because we're asking the nuclear world to fulfil these criteria of safety and I would thoroughly endorse those. However, it does seem to me that when we talk about laying pipelines along the bottom of the North Sea, where a major split; in terms of our fish, docks and so on; would be almost as disastrous as a nuclear reactor core going up, we're not asking for the same sort of stringent safety measures.*

Sir Alan: Well this is true, and it's true of other things. A highly dangerous source of energy is hydro-electricity. You have the big dams and if a big dam bursts it could not only take out enormous acreages of ground but could drown large numbers of people. On the whole, a big dam bursts about once a year and these, as incidents, are large scale even by the standards of the worst imagined nuclear reactor incident.

ICF4: *The next question is again a very general one but it brings us forward from the last point. What guide lines do you think should be established in public inquiries of major engineering failure?*

Sir Alan: I don't know all the answers to that. On the whole I
            think that, when there is a major failure which gets into the
            public eye and produces a public inquiry, the inquiry is done
            fairly thoroughly, and objectively. The only point I would add
            to that is that it is important to have some experienced, techni-
            cal people on the board of inquiry; not to let it get entirely
            into the hands of the lawyers. I think that these inquiries
            are conducted to a very high standard.

ICF4:   *Do you think the Boards, which are set up essentially, as I
            understand it, piecemeal, in this country are able to react
            quickly enough? In other words do you think there ought to
            be some semi-permanent national, European, or international
            organisation which can be called upon? If an aeroplane crashes,
            we have a standard routine that's gone into, with a group of
            people always waiting to do the job. Do you think now with
            large engineering structures in general, there's any need for
            some sort of world-wide organisation which can leap in and do
            the same job?*

Sir Alan: I would prefer not. I think that for a particular incident,
            you're going to need particular men with particular expertise
            and they may not necessarily be in this group. It seems to
            me that the right way is if one of these big failures occurs
            and you've identified the very broad technical nature of the
            thing, to go to the most distinguished and independent people
            in the field. I think that's the best method of getting
            objectivity and authority into it.

ICF4:   *But, in aviation, there's a nucleus which exists continually
            and which can always have men on the ground who know how to
            take the right sort of pictures and record the data even if
            they can't interpret it.*

Sir Alan: Yes, that kind of thing could be useful.

ICF4:   *Courses which pay attention to failures are very instructive,
            and it does seem that more use ought to be made of these failure
            reports as parts of the engineers education.*

Sir Alan: Yes, they're very good indeed. They really challenge your
            basic knowledge and you realise what enormous gaps there are
            in it, and for practical teaching of the subject, examination
            of failures and diagnosing them is a very good exercise indeed.
            Perhaps it's a thing that is not practised enough in teaching
            departments but it is a very clear intellectual exercise: taking
            what clues there are in the form of a fracture: whether its one
            side of a pipe or another; discontinuous or marked in various
            ways; and deducing from that what's been going on in the failure
            It's amazing how far you can go with a bit of practice in
            diagnostics.

ICF4: *Do you think that that is the sort of intellectual challenge that we are looking for to encourage more people to come into Applied Sciences and particularly into Materials Science and Technology. In other words, is this the sort of case we ought to be showing as an example of our profession when we go and speak to schoolchildren?*

Sir Alan: Well I think it would certainly fire their imagination, to show how one can analyse these things, because often the principles by which you make these deductions are pretty simple.

ICF4: *And you're using sophisticated techniques, like scanning electron microscopy in order to record this sort of data. The thought processes are very similar to those in solving crossword puzzles or crime novels.*

Sir Alan: And medical diagnosis. You have the symptoms and you've got to build deductions on them.

ICF4: *Now let's have a look at the last questions. I think that we're confining it to the fracture field, and it asks what are the main areas that are in urgent need of development from the research point of view and from the point of view of educating senior political advisers? I think that we've touched on some of the research points.*

Sir Alan: I'm not quite sure what is meant by 'educating political advisers'. I think you must mean permanent civil servants, Heads of the Civil Service Departments.

ICF4: *I think probably it's assuming that these people will themselves be doing what in fact the inspectorates are doing.*

Sir Alan: I would say that chief scientific and technical advisers must go straight to Ministers, concerning advice on decisions such as whether they should build an advanced passenger train, or get into the space programme, or whatever it happens to be. That advice should go straight to the decision makers and not be tampered with on the way, because if it goes to non-technical people on the way and is adjusted to their other considerations, then, because they're non-technical people, they will downgrade the technical aspects relative to the political, social and the economic aspects. They are bound to, because that's human nature, but when the advice gets through to the decision makers the technical advice will have been diminished in its importance. The only people who should rightly weigh the technical factors against all the other ones are the decision makers themselves: the Ministers.

ICF4:   *How does the Minister, or the decision maker become sufficiently familiar with the technical arguments, so that he can weight them in a proper manner?*

Sir Alan: He doesn't, he has to trust his advisers; the advisers have to put it into language that he can understand. An adviser doesn't explain everything about brittle fractures etc.; he just says 'well if you build it in this way, there is a real chance of the thing breaking in the first ten years: in that way, the chance no longer exists", or something of that kind.

ICF4:   *Is it possible to get this system working, or does it always get watered down by political and social considerations?*

Sir Alan: Well, I can only say in my own case that I delivered my advice straight to the Ministers. I think I would say one other thing here. As an adviser, you have to envisage all the main decisions or indeed all the main classes of decisions and you have to say particularly what the consequences of each of those will be, because the Minister will make a choice: he may ignore your advice in the end and he may not choose the technically best answer because the non-technical factors are overwhelming. At least, he must know the consequences of each kind of decision that he will make.

ICF4:   *If I may just recapitulate a little on the first part of the question; the main areas that you believe are in urgent need of development? The field that you quoted before was application to the science of materials in service, that is the number one point, and that would then include the whole gamut of not only fracture and fatigue but presumably factors such as corrosion effects and fretting and all of those things as well. As an aid to assessing the relative merits, would you think that the probability type of analysis is of value?*

Sir Alan: I think so, yes, things do fail and you can't design a perfect structure. You've got to accept some possibility of failure.

ICF4:   *If I may just look at one or two "supplementaries" - here is a question on "major new areas with regard to the development of high strength and toughness in materials over the next fifteen years". Do you think from the materials point of view that there is much scope for new materials.*

Sir Alan: I think the answer is more work on composites.

ICF4:   *Then there are one or two questions concerning the state of fracture at the moment as you see it, questions directly on your examination of the contents page and papers of ICF4. Is the subject where you would have expected it to be tens years after leaving active research, in balance as well as content?*

Sir Alan: I think by and large it's at the stage I would expect, I think its gone forward a lot and its becoming much more of an applied science if you like, because the basic principles are increasingly understood and so one is getting down to the details of actual systems and how they work, I think there's one thing I am surprised hasn't been answered yet is why in so many cases of very brittle fractures the measured work of fracture comes out at a few times greater than the ideal value for the surface energy of the material. I would have hoped that with people working away at that for many years we would have had some enlightenment on that question. I think that is one of the few fundamental sore points remaining in the basic part of the subject.

ICF4: *Yes I think the situation has been looked at, in glass as I recall, by Marsh who tried to put up a physical model with movement of little ordered regions around the crack tip, the problem has certainly intrigued me in iron. Until one has really detailed ideas of the potentials and forces around in terms of atomic theory I think it's difficult to give a quantitative answer.*

Sir Alan: Yes, I think the obvious way you approach this question is to think that there is some sort of localised plasticity going on at the tip of the crack, even in the "completely brittle" case, and that somehow this plasticity reverses itself when the crack was gone past so that you see no traces of it at the end of the day. Certainly I struggled with that idea, some fifteen years ago, and it always seemed to me that if you confine it to that geometrically reversible stage then the amount of work you could get out of that effect was really rather small and you couldn't add more than perhaps 50% or so to the apparent surface energy of the material. If you tried to get a big plastic work out of that, say three or ten times the surface energy of the material, then this plasticity has to become far-ranging and would leave some permanent traces in the material.

ICF4: *We still don't have very accurate measurements of surface energy, do we?*

Sir Alan: That's true. If you put in the cleavage values, in general you would be putting in a value three or ten times the ideal value. But we know quite a lot about the ideal value. I think we know it to within a few percent, from pure theory, also from the high temperature experiments where you measure the stretching or contracting of wires with various weights on. These more or less agree with the theoretical value. Also you know the values of liquid metals and you know from general theory that the liquid value can be only say ten percent smaller than the solid value. So you can fix the ideal surface energy within ten percent I think. Those are the sorts of values I've always had in mind when talking about the ideal value.

ICF4:   *If one could move a dislocation a couple of Burgers vectors on either side of the crack tip, this would produce an amount of work which if you took as being the incremental work for the fracture of each bond as the crack advanced, would give you a few times the surface energy. There is an energy hump which is reached after a couple of Burger vectors and what it seems you're doing work against is an image force. Once the crack tip has gone past and the surface unloads, the image force can pull them back out again.*

Sir Alan: The dislocation, in getting back, has somehow got to slip past the next dislocation which is coming forward out of the crack; and this gets into a very messy problem. I think the natural movement of the dislocations is to be repelled further away from the crack, by the other ones coming out. You're then onto full plasticity before you know where you are.

ICF4:   *Yes, once you've got the second dislocation pushing the first one you reach yielding. The only virtue about this is that it can still be confined to within the core radius, as it were, at the crack tip.*

Sir Alan: It may be that you can get away with it on that basis; perhaps if one did some computer modelling of fracture to see what would happen.

ICF4:   *Sir Alan, thank you very much indeed.*

POLITICAL AND SOCIAL DECISION MAKING IN RELATION TO
FRACTURE, FAILURE, RISK ANALYSIS AND SAFE DESIGN

Max S. M. Saltsman*

The main theme of this meeting is purely technological but it is a sign of
our times that the proceedings should include consideration of the theme in
political, social and educational terms. Politics and public opinion now
play a major role in deciding the level of support that can be given to
technology, and while this may be somewhat unpalatable to research engineers
and scientists, it is not wholly unreasonable that he or she who pays the
piper should call the tune. At any rate, for better or worse that is the
reality of politics.

It is proper therefore, that I should address the question in my capacity
as a politician. It is not my purpose to be political but rather to try to
give you a politician's view on technology in general, and fracture in
particular.

First a disclaimer, like the disclaimers put forward by writers in their
prefaces of books. They thank those whom they have consulted, but at
the same time, assure the reader that they and they alone bear the responsibility for error. My preface is somewhat similar, but in reverse as regards
this paper.

Lest you come to the conclusion that politicians have learned too much about
science, I must say that most of the factual information was put together
by Dr. F. R. Thurston, Director of the National Aeronautical Establishment
of the National Research Council, to whom I ran in humble supplication
after committing myself to speak and only then realizing what I had done.
A kind of fracture strikes the politician confronted by facts. Dr. Thurston
helped me out.

And to show my gratitude to Dr. Thurston for his generous response, I hold
him totally responsible for all errors of fact. The political parts are
my own and I accept their paternity, particularly since they can't be verified or checked out on a slide rule.

In Canada, fracture is always on our minds. Even the Prime Minister when
addressing a joint session of the U. S. Congress and Senate, used the word
to describe the Canadian dilemma. Canada has lived with the problems of
fracture from its very beginnings.

Not just the fracture of politics of English-French, Catholic-Protestant,
farmer-manufacturer, but the fractures that we have had to overcome from
the nature of the Canadian geography.

We have been essentially a resource based economy, living off the bounty
and hazards of nature, and to conquer nature, we have fractured nature

---

* Member of Parliament for Waterloo-Cambridge and Adjunct Professor,
  Department of Management Sciences, University of Waterloo

herself and learned to overcome nature's angry responses. The mountains
were fractured to build the railways - and mountains wait to break the
tracks and the vehicles that race the wounded passes. The earth was fractured by drills and shovels to move the Antediluvian sleeping hydrocarbons
to machines that are then operated by the oil, gas and coal that we had
extracted.

In return, the heat, pressure and conveyances to accomplish these purposes
are a Damoclean sword over our impertinent scientific heads. Some of these
energies are used to further fracture the earth for its riches of copper,
potash, nickel, iron, etc. The very chemicals in the air are separated or
transformed in the course of turning the rich rocks into tools, shelter and
machines for man.

We live surrounded by an ever ready retaliation for our acts and just as
disease moves man to value his physician, Canada needs and cannot live
without its fracture specialists.

As an example of current concern, Mr. Justice Thomas Berger, conducting
an inquiry for the Government of Canada, had this to say: "One of the
important technical problems that needs to be resolved, is designing pipelines that are capable of withstanding frost heave. The Russians have been
able to avoid permafrost areas in building their pipelines, while the
Americans elevate their northern pipelines. The gas pipeline that Arctic
Gas and Foothills Pipeline propose to build, on the other hand, because
of costs, will be buried. Chilling the gas, because permafrost areas are
discontinuous, creates the problem of frost heave. Arctic Gas says that the
heave can be negated by covering the pipe with mounds of earth. But dis-
senters assert that the amount of earth needed for this is too great to
be feasible. The concern here is that the largest civil engineering project
ever to be considered for Canada has left a fundamental scientific/engineering
question unexamined by any department or agency of government. It is not
appropriate that such questions are left entirely to industry."

In elaborating on this problem, Mr. Justice Berger emphasized that Canada
is a northern country, that the North is part of the Canadian psyche, that
the North is a kind of window "opening on to the infinite". Despite this
and despite the obvious need for the sort of scientific work which would
have resolved the frost heave question, no agency of government has under-
taken or supported the necessary research.

As industrialization moves farther and farther into the North, we will
encounter other unique scientific and technical questions that Canada
should be looking into now.

Another example was given by Dr. J. J. Shepherd of the Science Council of
Canada. He said, "an emerging major programme is the transportation by marine
mode of gas and oil from the Arctic to our eastern seaboard. Northern
based control systems, ship and containment design and construction, *en
route* and terminal navigation systems - all pose special problems." He
might have added special problems in *fracture*.

The worry about resource energy shortages and escalating costs calls for
the preaching of a doctrine of greater durability, safety and conservation.
History can teach us many lessons and perhaps one of the best examples of
a long life structure is to be found in Egypt. While we can all marvel at
the long life of the Pyramids and still admire their beauty and proportions
after so many generations, we must question their cost and purpose. I doubt
that they did much to improve the quality of life of the slaves who built

them although undoubtedly they did a greal deal for the quality of death
of the Pharaohs. Equally certain, they must have had a big impact on the
unemployment rate.

The great difference between the Pyramids and modern construction is that
the Pharaohs' builders didn't have to worry about zoning by-laws, spills,
radioactivity, explosions or what the effect of these disasters would have
on the environment or the people about them.

Undoubtedly in very many cases, durability and safety can be bought; in
other words we can have long lives and high reliability if we are prepared
to pay the added costs. The questions now are what should our proper goals
be in both product lives and costs. These are very important and difficult
questions to resolve and, in fact, they can make or break an industry. If
either the life is too short or the price too high there will be no public
acceptance and no sales. We may also enter the realm of planned obsolescence
in which a manufacturer deliberately designs in a limited life to his product
in order to obtain repeat orders. In such a case, the benefits of increased
employment must be carefully balanced against the costs to the consumer.

The price that we are prepared to pay for durability and safety will depend
on both our ability to pay and on the consequences of failure. The risks
that we can accept in consumer products are many orders of magnitude greater
than those that we can even contemplate in an atomic power plant. In fact,
to a politician, to the general public, or to the scientist, there is no
"acceptable" risk of failure of an atomic power plant if it would result
in the release of radioactive material. This is a major dilemma in our
growing needs for energy.

It is, as I am sure you will all agree, impossible to design anything or
develop any material that will not crack, degrade or otherwise fail under
some set of circumstances. Unbreakable houseware is a salesman's illusion
and as fictitious as the irresistable force meeting the immovable body.
This is why it is, in my view, essential that materials scientists and
engineers work closely together at meetings such as this. The materials
man must be able to say clearly when, how and why his material will fail
and the engineer and designer must be able to produce designs that will
minimize or eliminate the consequences of the failure.

The spread of information about failures within the technical community
is essential and it is a sad reflection on society that legal and political
constraints are doing so much to prevent this spread.

History contains many excellent examples of the benefits of freedom of
information about failures. In Canada, we can have cause to be proud of
the Report of the Royal Commission on the failure of the Quebec Bridge
that was published in 1908. This report, which I might add was produced
by a Commission of three Engineers and no lawyers, contains a frank and
exhaustive analysis of the failure, which was attributed largely to the
then poorly understood buckling failure of a compression member. The
work of the Commission led to a large number of experiments and studies
that, to quote Steinman and Watson in their book "Bridges and their
Builders" - "-more than any other occurrence in the evolution of bridge-
building, revolutionized the art by bringing it to a new high level of
scientific analysis and design".

Similar benefits have accrued in more recent times from the information that was released in the United States following the Polaris and the F-111 difficulties.

The subject of freedom of information on failures inevitably leads to a discussion of legal issues. While there can be little question of the technical benefits of making information on failures freely available within the technical community, this is only possible if the legal profession and the laws of the land take an enlightened attitude. No manufacturer likes to take the risk of exposing himself to lengthy court cases and possibly punitive damages because of a genuine error on his part. By the same token, no regulatory agency would willingly expose itself or a manufacturer to this risk. There are, of course, many legal defences available; notably that the product was designed and manufactured according to the state of the art - the legal doctrine of the "reasonable man". Unfortunately however, the mere act of defending oneself against court charges is expensive and will tie up technical experts over a lengthy period in which they might well be better occupied with more profitable exercises. The principal difficulty is that of interfacing between the legal and technical communities to ensure that technical matters are fully understood by lawyers before lawsuits are commenced. Indeed, with all due respect to the legal profession there is in my view, a need for more lawyers with a technical background. These problems can only become more acute as technology advances and perhaps we should be taking a fresh look at the whole question of the interactions of the law and technology - if we have accepted the idea of a Family Court to deal with one class of legal problem, why not a Technical Court for another?

What do we mean by durability? This is an all-encompassing term that simply implies that a product must be capable of performing its designed function for an appropriately long time. While we recognize that nothing will last forever, a product must serve a function that justifies its initial cost and, perhaps equally important, justifies the costs of maintaining and repairing it. We should then examine the factors that limit the lives of products.

First and foremost, a product must be capable of sustaining all of the loads that it sees in service without breaking. This implies that the designer, and frequently the customer, must have a complete knowledge of both the strength and of the loads and environment that the product will see in service. The strength will be determined by technologies such as those discussed at this meeting, by considerations of fatigue, fracture toughness, creep and other forms of what are perhaps improperly referred to as material failures. The loads are usually estimated from a knowledge of the task that the product must perform and on a judgement based on past experience, of future use. The environment is all too often ignored but a wise designer will always attempt to take into account the effects of extremes of temperature and of corrosive surroundings.

From the titles of the papers to be presented at this meeting, I would have judged that you have made a great deal of progress in understanding the processes of material failures. I am sure, however, that much remains to be done and it is regrettable that the level of support for research in this important area, in Canada, is so low. In spite of your progress I am concerned that the fruits of your efforts do not always reach the audience that they should. The literature is full of examples of costly design errors that could have been avoided had the designer been properly

aware of the current knowledge. A good Canadian example was the failure
of the Duplessis Bridge at Three Rivers in Quebec in 1949. This failure
occurred when the temperature was 230 K and was largely attributed to the
use of a steel which became brittle at such temperatures, a fact that should
have been known to the designers. It is almost astonishing to learn that
in July 1962, the Kings Bridge at Melbourne, Australia, also suffered a
brittle failure, albeit at a much higher temperature. In this case it was
reported that no adequate impact tests were made on the material. The
lesson had apparently still not been learned.

Impact tests are easily understood and there can be little excuse for them
not being required by designers in this day and age. There are more sophisticated tools, such as fracture mechanics, that pose an entirely new set
of problems. These tools all too often demand of the designer a high level
of knowledge of mathematics and I suggest that a major problem confronting
research workers is that of simplifying their concepts so that they are less
demanding in expertise and can become everyday tools.

I have referred to the needs for knowledge on loads to enable designers to
produce goods with the appropriate durability. It is surely not good enough
for a manufacturer to defend the failure of his product by the claim that
it was abused. It is almost an aphorism that if a product can be abused it
will be. One might cite as an example the recent introduction of high
performance glass or carbon fibre reinforced composites into the manufacture
of fishing rods. The stiffness and strength of these are ideal for the
sporting fisherman but the strength when loaded across the diameter of the
tubular rod is lamentable and what fisherman has not accidentally sat on
his rod?

The hazards of an incomplete understanding of loads and environment were
demonstrated only too clearly by the unfortunate experiences of Rolls Royce
in the U.K. You may remember that this company was responsible for the
development of turbine engine carbon fibre composite compressor blades that
had remarkable strength and fatigue properties. Unfortunately, their resistance to impact and erosion damage was inadequate and these facts were not
appreciated until far too late in an engine development programme. The
economic consequences played a major role in the subsequent bankruptcy of
the company, and had profound political repercussions.

So far in this discussion, I have referred to fracture as though it implied
failure. I am sure that many of my audience would wish to correct this
impression and I hasten to do so myself. Fracture is of course, an essential
part of any manufacturing process as many forms of cutting are nothing more
than controlled fracture. Thus, there are times when the need is to produce
fractures rather than prevent them. We might even refer to the production
of jewellery in which the fracture surfaces of precious stones provide one
of the oldest art forms.

In terms of our current needs, I would draw your attention to the enormous
expenses that are incurred worldwide in tunnelling and in excavating. The
search for oil and gas is paced by man's ability to bore holes in the ground,
a technology that has changed little in past decades. The congestion of
our cities can only be relieved by underground tunnelling to provide route
for rapid transit and in many cases this is prohibitively expensive. I
suspect, too, that as the energy crisis grows, we shall find that we would
be better to revert to being troglodytes by burying our housing underground
where we can find more stable and comfortable temperatures than we have on
the surface.

While research workers have protested at the lack of adequate funding for their efforts, they have done little to provide quantitative support for their case. Most of the protests that have been heard by the politician are based on arguments to the effect that support of research activities should be based on some fixed relationship with the gross national product or that there should be some other international comparison. On any of these grounds, support of research in Canada is clearly inadequate but it is my impression that the significance of this inadequacy has not been properly brought to light. In preparing this paper, I have tried to obtain statistics on the cost of lack of knowledge on durability that might be used to provide arguments for or against greater support for research on fracture or on durability. Appropriate statistics seem to be lacking. To quote from H. E. Morgan from National Bureau of Standards Publication 423 "Comprehensive statistical data on the economic cost of mechanical failures do not exist. Sellers do not wish to publicize their mistakes; this may account for the fact that we have only anecdotes rather than aggregate data on failures". Morgan then goes on to discuss personal injury and fatality rates in the U.S. due to accidents and suggests that the total cost of personal injuries in 1963 was of the order of $9.9 billion. Fishlock in his book "The New Materials" quotes a British chemist as having estimated the cost of corrosion to the British economy as being 600 million pounds a year. A related statistic given by Jost in the August 1975 issue of "Mechanical Engineering" is that increased attention to tribology could save the U.S. economy as much as $16 billion per annum.

These are all very large sums of money and, as a politician I am particularly anxious to know more about them. I think that you will agree that as they have been presented, they are not particularly useful. We need to know not merely the cost of corrosion, but also the costs and effects of preventing it. In this context I must remind you that costs cannot be measured simply in dollars, we also need to know the effects on things like employment and on energy conservation. As far as accidents and personal injuries are concerned, we must accept that all accidents are preventable, human beings are fallible. What we need are reliable figures for the costs of preventable accidents and the costs of minimizing these.

The politicians and society at large, are very well aware that the modern world is critically dependent on technology, but they are not prepared to buy technological solutions to comtemporary problems at any price. There is a prevailing tendency to accept solutions that may be socially retrograde, unless proposed technological solutions are proved to be economically and socially superior beyond all reasonable doubt.

It is in this context that conferences such as this can serve a major purpose. Although the primary benefit is to the scientific community dedicated to the study of Fracture, the politician hopes that he will obtain some authoritative guidance in terms of the durability, cost and safety of consumer products and of certain engineering works costing millions of dollars.

THE TEACHING OF FRACTURE IN UNIVERSITIES

C.K. Knapper[*] and D.M.R. Taplin[**]

ABSTRACT

It is argued that the teaching of fracture as an undergraduate engineering subject is especially important because of the potentially disastrous consequences of the failure of materials in engineering structures. Some contemporary approaches to teaching fracture are discussed on the basis of results from a questionnaire survey of delegates to ICF4, and the pedagogical implications of different teaching methods are examined. It is concluded that fracture is now, in 1977, a "discipline" and that a reappraisal of its place in engineering education is required.

1. INTRODUCTION

Fracture is a phenomenon so ubiquitous that it is a daily concern of such diverse specialists as engineers, physicians and kinesiologists as well as the public at large. Although the topic of fracture is relevant to most branches of engineering, it sometimes occupies a less important place in the engineering curriculum than seems desirable - especially when it is considered that a major concern of engineers is building and designing useful products and that such products often fail by one or more fracture processes. One reason to argue that fracture should be a central component in engineering education derives from the fact that the technical and social consequences of some types of fracture are immense. This presents considerable challenges to those experts in fracture who are involved in education.

A good example, among the many that might be cited, is the explosion at Flixborough described by Kennett [1]. Twenty-nine people died and more than 100 were injured when a pipe in a petrochemical plant leaked and caused an explosion that devastated the plant and was heard more than 30 miles away. The basic cause of the failure, according to Cottrell [2], was the rapid development of intergranular cavities in the steel pipe that was operating in the creep range of temperature. Failure was thus due to creep cavitation fracture. The explosion brought into question the wisdom of building large-scale factories or other massive engineering structures, and the disaster eliminated the only source in Britain of an essential raw material for the production of nylon. In the Open University course on Materials Failure a radio interview was held with Sir Alan Cottrell, a leader of the official Flixborough enquiry, and the tape recording of the interview was used as a teaching aid. This tape, as well as other real life examples of failures and their political and social consequences, is also used at the University of Waterloo to provide an added realism to the teaching of fracture analysis.

Society faces the possibility - however remote - of even more catastrophic environmental disasters than Flixborough, for example in nuclear reactors

---

* Teaching Resource Person, University of Waterloo, Waterloo, Ontario, Canada.
** Professor of Mechanical Engineering, University of Waterloo, and Vice-President of the International Congress of Fracture.

and in operations of the petrochemical industry such as offshore oil-rigs. These examples are especially politically sensitive. Month by month significant material failures cause major problems, and headlines such as "Jet Engine Fails due to Metal Fatigue", "Cracks in Prestressed Concrete Skyscraper cause Evacuation" are not uncommon around the world. It thus seems imperative that engineers be thoroughly educated to deal with such problems.

But not only is it necessary that scientists and engineers be educated to recognize, prevent and rectify fracture problems, they must also be able to do this within the wider political and social context of contemporary society. For it is to these experts that the public and the political decision-makers must turn to advise upon and to resolve, such crucial issues [3]. Thus it is important that engineering students have a practical understanding of the political and decision-making processes in society. Just as, for example, students of sociology, humanities and law need some understanding of basic technology, the converse is equally important. Furthermore, society as a whole must also be educated to understand some of the important technical consequences of large-scale fracture, so that laymen may respond intelligently when called upon to make political decisions that have far-reaching consequences for the use of technology in society. The aim of this paper is to provide a preliminary examination of how the subject of fracture is presently handled in universities, focussing particularly on innovative approaches, in order to determine how fracture may be taught in a way that is most relevant to the changing needs of society at large.

2. THE PLACE OF FRACTURE IN THE UNIVERSITY CURRICULUM

In an attempt to discover something of the present status of fracture in the university curriculum, in April 1977 a short questionnaire was sent to each of the registered delegates to ICF4. Of over 400 questionnaires distributed, 80 were returned, from 60 different institutions in 17 different countries. The complete returns by country were: United States, 25; United Kingdom, 14; Japan, 9; Canada, 7; Finland, 3; France, 5; Germany, 3: India, 2; Netherlands, 2; South Africa, 2; Sweden 2, Argentina, 1; Denmark, 1; Italy, 1; Israel, 1; Switzerland, 1; Yugoslavia, 1.

Only 10 respondents did not teach at a post secondary institution, and the great majority were affiliated with a university or its equivalent. The highest degree offered by such institutions was almost invariably the Doctorate (6), usually the Ph.D., but occasionally a D.Eng., D.Sc., or D.Tech. In 3 further cases the institution concerned offered only a Master's degree. The departments represented were Materials Science/ Metallurgy - 28; Mechanical Engineering - 23; Engineering (in general) - 9; Aerospace Science/Aeronautical/Naval Engineering - 4; Solid Mechanics - 3; Mining - 2; Physics - 2; Chemical Engineering - 1. Most of these departments themselves gave the Doctorate (55), with 4 departments giving only a Master's degree in their subject.

Graduate Courses on Fracture

Respondents were asked whether graduate courses specifically on fracture or failure analysis were offered in their department or another department they knew of. Responses indicated considerable differences between European institutions and those in North America. In the case of the United States and Canada, no fewer than 30 separate graduate level courses were mentioned by respondents, offered primarily in departments of metallurgy, mechanical engineering, or materials science (the courses concerned invariably included the word "fracture" in the title); 1 institution reported that fracture was covered as part of another graduate level class; and 2 respondents reported that a course on fracture was offered on an irregular basis or at the inclination of a particular lecturer. For the European universities represented, only 7 graduate courses in fracture or failure analysis were reported; though in the case of 2 institutions such courses were sometimes given; in 3 other cases fracture was covered as part of another course (for example, the mechanical behaviour of materials).

Undergraduate Courses on Fracture

Respondents were asked whether the subject of fracture was an officially designated part of the undergraduate curriculum - for example was there a course entitled Fracture or was fracture a specified part of another course? Analysis of responses here did not reveal the differences between countries apparent in the case of graduate level courses. In fact only 12 fully fledged courses on fracture or failure analysis were reported in total - 4 in Europe, 3 in the U.S.A., 2 in Britain, 2 in Canada, and 1 in Japan. In the case of 33 other institutions, fracture was covered as part of a course or courses, generally in materials science or mechanical metallurgy.

Fracture as a Compulsory Part of the Curriculum

Respondents were asked if the subject of fracture was part of the core (mandatory) curriculum of the undergraduate engineering program. Only 1 department replied affirmatively, with another 1 department reporting that the subject of fracture was compulsory for some students, depending on the option they took. A further 17 reported that the topic of fracture was mandatory in the sense that it was part of a wider course or area that itself was part of the core curriculum (typically a course on materials): 4 institutions in England, 3 in Canada, 2 in Finland, 2 in South Africa, 2 in the United States, 1 in France, 1 in Japan, 1 in Sweden, and 1 in Switzerland. This contrasts with a total of 36 institutions which indicated that fracture was definitely not a mandatory part of their degree requirements at the undergraduate level - except as a rather small part of a general materials course.

Novel Approaches to Teaching Fracture

Respondents were asked if the topic of fracture was taught in their institution in any special or unusual way - for example by means of a "case study" approach, through the use of extensive project work and so on. In all, 20 affirmative responses were received, including 5 from universities in the U.S.A. and 4 from British institutions. The majority of respondents mentioned project work or case studies, occasionally done in a real life industrial setting. The following comments were made by individual respondents,

- "The topic is the subject of a full year's research for the final year student in some laboratories in our department." (Yokohama National University, Department of Mechanical Engineering)

- Fracture is taught by the case study approach. (University of Southampton, Department of Mechanical Engineering)

"Apart from standard lecture units on various aspects of Fracture in courses such as 'Mechanical Properties of Solids', we use one less usual type of 'case study' approach. I introduced this method when I joined the staff ... and it is still proving useful. Students are each given a different article (i.e. a metallic component) which has failed in service. These components can vary from bearings for large rolling mills to rocker-arms from motor engines. The student is given a period of laboratory time to determine the cause of failure, prepare a detailed report (in a form comparable to a report from a consultant) and then face an oral examination for about an hour with a member of staff. This approach not only gives them practice at report preparation and orally supporting their work, but provides an excellent check on their understanding of metallography, fracture modes, etc., and forces them to become familiar with 'material specifications' and 'codes of practice', etc. It certainly brings home to them the gulf between academic lectures and the realities of service failure examination." (University College of Swansea, Department of Metallurgy and Materials Technology)

- "A small service failure analysis project is attempted." (University College, London, Department of Mechanical Engineering)

- "One of the undergraduate laboratory courses has an experiment on fracture using compliance methods." (University of Rochester, Department of Mechanical and Aerospace Science)

- Fracture is integrated with the research work of some staff, and an attempt is made to relate fracture to fundamental deformation mechanisms. (University of Cincinnati, Department of Materials Science and Metallurgical Engineering)

- "Yes, in the failure analysis course fracture is taught as both a 'case study' subject and as a real case experience, with students working with four industrial concerns on problems selected by them." (University of Michigan, Department of Mechanical Engineering)

- "We frequently use case studies dealing with fatigue and/or fracture from the Stanford/ASEE case library." (Wichita State University, Department of Mechanical Engineering)

- "We often use outside lecturers for specific expertise." (Carnegie-Mellon University, Department of Mechanical Engineering)

- The case study approach is used. (McMaster University, Department of Mechanical Engineering)

- Fracture is taught by a combination of lecture, laboratory, and research project. (Carleton University, Department of Engineering)

- Fracture is taught as a research project as part of the study of reactor materials. (Technical Research Centre of Finland)

- "Sometimes we are able to co-operate with industry when they ask us to investigate damage or similar problems." (Technical University of Aachen, Faculty of Mining and Metallurgy)

- "Not yet, however the Department of Mechanical Engineering is going to start a 'case study' approach by way of an experiment." (University of Witwatersrand)

- "We are developing a method of analysis for nuclear structures taking into account fatigue, failure and plasticity for a tri-dimensional case." (Ecole Polytechnique of Paris, Department of Solid Mechanics)

- Fracture occasionally appears in final year project work, often done on an industrial problem. (Chalmers Technical University of Sweden, Department of Physics)

The Place of Fracture in the Undergraduate Curriculum

No fewer than 46 respondents replied to a question which asked them what place courses on fracture should have in the undergraduate curricula for engineers. A majority felt that it indeed should occupy a fairly prominent place, though most of these stopped short of saying that fracture merited treatment in a special course at the undergraduate level. A further 5 respondents had considerable reservations about (or were completely against) the teaching of fracture to undergraduate engineers.

The following comments are selected from those made by individual respondents.

- "Fracture should be taught as a topic (general approach) in the course on Materials Science."

- "There should be a mandatory one-semester course for all engineers; it is advisable to have a part of this course based on 'case histories'."

- "Contributions to fracture mechanics by metallurgists are very much needed. Unfortunately, however, the status of the subject 'Fracture' in metallurgical engineering is of secondary importance, at least in this institution. It appears that fracture is more important to users than to producers. I think that the theory of fracture should be developed more systematically from the point of view of physical metallurgy: then fracture will attain the status of an independent subject within the field of metallurgical engineering."

- "Fracture is too sophisticated for undergraduate students to understand completely; so it may be enough at the moment to give some basic concepts."

- "In a general engineering course fracture could occupy 5% of the curriculum, or about 27 lectures (say 1 lecture a week for a year; in mechanical and electrical engineering up to 10% of the curriculum could be devoted to fracture (in the case of electrical engineering that concentrates on nuclear applications, turbines, etc.); in civil engineering internal fracture and structures should occupy about 10%; materials engineering also needs about 10%; aeronautical engineering needs at least 10% on fracture in relationship to structure, and somewhat less than 5% on fracture in relation to engines; in the case of electronic and management engineering fracture deserves less than 5% of the curriculum."

- "Fracture is an essential part of all undergraduate curricula in engineering, at least as part of a course on materials."

- "A permanent place, but at present fracture should be dealt with within materials and mechanics courses rather than separately; fracture should be part of courses on mechanical behaviour of materials and design."

- "It is essential that design for structures containing defects be included in all courses in mechanical engineering to an extent which recognizes that most engineering failures are of this type and not failures of the traditional 'stress analysis' method which occupy so much of present courses."

- "Fracture should be taught in the core courses of materials science as well as in the sequence of machine design courses; in particular, the importance of fracture control in design should be emphasized."

- Fracture should be taught as part of a mechanical metallurgy course - more time should be spent on fracture ("a problem of general interest") and less on rolling etc.

- Fracture should be an elective course, but encouraged for metallurgy, mechanical engineering, chemical engineering and civil engineering, and should be taught at an advanced undergraduate or graduate level.

- "Fracture should be introduced early in the curriculum for most engineers and should have at least as much time devoted to it as the traditional mechanics type courses such as strength of materials, in which fracture is barely touched upon."

- "Fracture should not be taught as an exercise in mechanics - this needs a proper base in materials and plasticity. Rather, fracture should be presented as a culmination of response to loads, and related to service environments and structures."

- "Fracture is one mode of failure, which should be covered in some undergraduate curricula - I do not believe undergraduates really comprehend fracture or can learn about it in the period of time I can devote to it (about two weeks). Fracture should be introduced and some idea given of the importance of the topic; however there is no time to treat the topic in depth in the undergraduate curriculum."

- "I believe fracture topics should be integrated into courses in mechanics and in materials."

- "There should not be a full course (i.e. a semester-long course) on fracture, but it should be covered as a part of a course sequence in materials and mechanics."

- "Fracture should be a specifically designated part of the core curriculum for engineers."

- "Fracture as a mechanism should be taught both under 'strength calculus' and 'materials' courses."

- "In Finland we should have more fundamental and mandatory courses in fracture."

- "In our system these courses should be given at the graduate level."

- "Fracture should be part of a course on 'mechanical properties of materials'; I do not consider that fracture can be separated from other mechanical properties."

- "Fracture should be taught at a later stage, following courses in strength of materials, elasticity and basic materials science."

- "The topic of fracture should be briefly outlined as part of the mandatory curriculum; it should be dealt with in more detail as an elective and/or graduate course."

- "There should be more interdisciplinary co-operation in the teaching of fracture. Very often the development of fracture is prohibited by the sharp limit between mathematicians and materials scientists, i.e. the two groups have large difficulties learning from each other and work in isolation. Furthermore, I think there has been a stagnation in the field since the major developments between 1957 and 1970."

- "In my opinion fracture should not be treated separately from other mechanical properties. In the case of steel we have always to look for the strength *and* fracture properties. Therefore we always try to find correlations between the structure and fracture appearance. That is the reason for treating fracture as a part of physical metallurgy of steel in our courses."

- "In our teaching course *'Fractology'*, the first special way is to understand the fundamentals and the methodologies of fracture by an interdisciplinary approach, the second by the development of the comparative science of fracture."

3. TEACHING FRACTURE AS A FORENSIC SKILL

Although the evidence for successful innovation, based upon the results of this survey, is somewhat sparse, distinguished experts in the field of fracture have not fought shy of calling for radical changes in the way the topic is treated in university education. Cottrell [3] calls for a "science of materials in service", with a stress on applied aspects of materials and materials failure. He points out that universities presently tend to emphasize unduly the training of research workers, and argues that a more applied approach is needed with more integration of university and industrial research. This indeed is a basic thrust of the University of Waterloo in its co-operative approach to education, and it has been actively pursued, (if, perhaps, less successfully) in Britain.

For Cottrell, this hiatus is indicative of what is wrong with the way fracture is often taught in universities, and he argues that the subject should not be taught as pure science, with the aim of producing a "good research man", but that what is needed is a qualitative, illustrative and applied approach. A similar point has been made with regard to the teaching of science and technology to non-science majors. It is felt by some critics that stressing theoretical and technical detail (analysis) only serves to confuse the student with unnecessary, complicated information; what is needed is an emphasis on the more qualitative general issues, with provision of concrete examples and exploration of the various implications of modern science and technology.

Reid *et al.* [4] argue in a similar vein when they comment on the tendency of the traditional university teacher to tell his students what he knows, regardless of its relevance to their (and society's) needs. Cottrell regards engineering as essentially a profession for the "generalist". One implication of this is that engineering education must train its students to know when they are out of their depth with a particular problem and when to call in a specialist. This implies familiarity with learning in very practical situations, and in the case of fracture Cottrell recommends the examination of real or simulated failures, and practice at diagnosing them. This is very similar to the "forensic" or trouble-shooting approach that Reid *et al.* report as being used at the Open University. The technique typically involves what is generally called a "case study approach", in which students are set an empirical task for which there is not necessarily any single "right answer". (This is in contrast to the situation to be found in more traditional laboratory work, where there is often an undue emphasis on producing "the correct results" rather than on the process of learning and discovery leading to the conclusions.) To help them with their task or problem students are typically provided with certain factual data which may be more or less relevant to the problem's solution. This material may come in a number of forms, ranging from written documents to physical specimens, and the possibilities for using audio-visual aids are considerable, as is exemplified in the approach adopted by the Open University, which has made use of written reports, photographs, recorded interviews with experts, and even a home experiment kit.

The use of case studies in this way has the advantage of presenting the student with a problem that is very close to real life - indeed, genuine examples taken from industrial settings may be selected, and a student's manner of arriving at a solution may be compared to those of established experts in the field. Where co-operative programs exist there is the possibility of carrying out such projects in an authentic setting, perhaps with the student acting as a member of a team rather than an individual.

The case study approach has some severe limitations - in providing a collage of cameos, for example - but it does appear to be extremely valuable in exposing students to reality-based contingencies where practical remedies are called for. The pedagogical advantages of this type of learning are that the student acquires knowledge and skills by direct (as opposed to second-hand) experience, and hence, it is believed, is better able to transfer the skills he has learnt to the work situation.

The case study approach has a close relationship with the "heuristic" technique pioneered by Armstrong in the late nineteenth century [5]. This approach eventually developed into the Nuffield Science Teaching Project for schools [6],[7]. Indeed university science and engineering professors have much to learn from the science educationalists who focus on teaching in high schools (14-18 yr olds): an example would be the work of Jenkins and Whitfield [8].

Heuristics derives from the Greek *heurisko* meaning "I discover". Van Praagh [5] first developed the technique in full for the teaching of "Chemistry by Discovery" in England and later went to lead the Nuffield Chemistry Team developing and spreading the basic method. Reynolds [4] was an exponent of this technique for studies of materials failures and provided much of the original "heuristic" thrust for the Open

University course. Martin [9] has also done a great deal to bring the study of materials, strong solids and fracture into the secondary school curriculum.

4. ENGINEERING EDUCATION AT THE UNIVERSITY OF WATERLOO

The University of Waterloo is *de facto* and by deliberate design a technological university with considerable emphasis on professional programs in Engineering, Mathematics and Science. It grew quickly to a place of international prominence in these areas and it has a particularly high reputation within the engineering community in Solid Mechanics and Fracture. Indeed there are some 40 researchers, including over 20 faculty members, working in the field of fracture. Much of this research is of a directly applicable "mission-oriented" nature and it is reflected in the teaching on the Faculty of Engineering. An important tenet of Engineering education at Waterloo is that Engineering education has an important responsibility to train students who will be able to use their knowledge in a practical way to solve very practical problems. One way to expose students to the industrial environment they are likely to encounter in their later careers involves the notion of co-operative education, and this co-operative approach was a major element in the founding of the University of Waterloo 20 years ago. It had the aim of providing an educational system that would formally integrate a student's academic study with suitable work experience, or as Wright [10] noted, to produce an engineer who could function in the modern role of "manager technologist". The advantages for students were thought to be considerable: not only would they achieve some level of financial solvency, but they would benefit academically, experimentally, and in terms of eventual career opportunities. Presently at the University of Waterloo there are over 5,800 co-op students (and 900 employers) throughout all six faculties of the institution - with some 2,700 students in Engineering and over 1,700 in Mathematics. This makes Waterloo's co-operative program the largest in Canada and the second largest in North America (the largest being at North Eastern University, Boston); in fact *just under half* the total undergraduate enrolment is engaged in co-operative programs.

The mechanics of the scheme now run exceedingly smoothly. The Department of Co-ordination and Placement, staffed by over 30 co-ordinators (all with academic *and* practical backgrounds in their particular programs) arranges student placements in carefully selected positions after interviews with employers and with each individual student. The co-ordinators also visit students on the job and help with the general assessment on an ongoing basis. A co-operative student typically requires 6 work terms and 8 academic terms to graduate with an honours degree. Work reports (a professional level paper or research report on a topic of interest to the student and his employer) are required, and these are graded by the employer or by a faculty member; in addition, the employer must complete an evaluation of the student each work term. Cohesion among students is partly achieved by a system of streaming, whereby the same group of students proceeds through its academic career together, being enrolled in the same classes for the same academic terms. This is a fairly uncommon procedure nowadays in North American universities, and it would be interesting to determine empirically whether or not the plausible benefits in terms of morale do indeed exist in comparison with students who have no sense of class identity. There is a feeling that cooperative structures lead to rather less friendly and humane places than traditionally structured courses, perhaps due to the complex and changing work schedules involved.

Additional features of the co-operative scheme as it operates at Waterloo include adjunct appointments from industry and government, a student advisory council, and feedback sessions with students at the start of each academic term. There also exists an Industrial Advisory Council of 24 engineers and scientists from Canadian industry and government agencies that meets twice a year on campus with the Department of Co-ordination and Placement and faculties of Engineering and Science to provide input and advice on directions that scientific and technical education at the University of Waterloo should be taking.

As can be inferred from the above information, the University of Waterloo is extremely anxious to continually monitor the success of its co-operative program. To achieve this, an annual questionnaire survey of all co-operative students is carried out, supplemented from time to time by surveys of former graduates of the program. Results show that students overwhelmingly (over 94%) feel that co-operative education is a valuable experience, although surveys of past graduates reveal considerable criticisms of the relevance of the undergraduate curriculum to the jobs they eventually occupy. In terms of some other criteria co-operative education appears to have been markedly successful for Waterloo: those departments that offer co-operative programs have seen a considerable rise in enrolment, and an increase in the calibre of students applying. There is also some evidence that such students are preferred by employers compared with graduates of other equally prestigious, but non-co-operative schools. Certainly if the remarks made above about the pedagogical value of practical experience for the engineer have any validity, then it might be expected that students educated in co-operative programs would be in a better position to transfer what they learn in the classroom to the tasks they eventually face as professional engineers or scientists.

5. PEDAGOGICAL IMPLICATIONS

One theme that can be seen to run throughout the above discussion - but which has not been identified specifically - is the psychological concept of "transfer of training" [11]. This refers to the idea that what is learnt in one situation will be in some sense transferable and applicable to other situations. Transfer is an implicit assumption underlying university education - and indeed, all education. Assuming general agreement that it is desirable to teach for such transfer, the question remains as to how best to do this. For example, is it preferable to go from the general and the theoretical to the particular and applied, as in the traditional notion of discipline-oriented university teaching, or is the reverse precept a better strategy, as is claimed by those who favour a case study or problem oriented approach? Most people would agree that the closer the learning experience to the practical ("real life") situation, the easier it is for transfer to take place, and on this basis it might be expected that co-operative education would be particularly effective. However, this would clearly only be the case if students are given work experiences relevant to experiences they will encounter in their later careers, and if the program of academic training is closely geared to the practical situations they encounter in their work terms.

This points up the necessity - in co-operative programs as elsewhere in education - to specify precisely the objectives for learning [12], at the level of the individual course as well as the entire program. (In passing, it is interesting to note that the stress upon writing objectives and then testing to see whether they have been attained, which is now ubiquitous among educationalists, originally derives from systems engineering.) Unfortunately it is often more difficult to specify precise learning objectives than it sounds in principle, especially where less tangible skills are required. However, as work at the Open University has demonstrated so effectively, learning that is based on solutions to specific practical problems lends itself particularly well to the "objectives approach", and it is often possible to see quite clearly the degree to which students are able to achieve the desired behaviour and in what ways they failed to do so [13].

One other pedagogical advantage of the integrated case study approach that is claimed by its advocates is its stress on the process of synthesis as opposed to the analysis that is often the major concern in more traditional, discipline-oriented approaches to learning [14]. Common sense would suggest that this is the case, but to guide students through the sophisticated process of synthesizing solutions to a problem is extremely demanding on faculty time and departmental resources. Not only do suitable problems have to be delineated (and the same problems cannot necessarily be used from one year to the next), but experience at the University of Waterloo has shown that the teacher often needs to be available to small groups of students seeking advice on a daily basis: demand on faculty time is not only great but unpredictable. Furthermore, not all university teachers are able to respond very well in this type of learning situation, which demands very different skills from the didactic approach normally taken in a lecture. One solution to this problem that has been used at the Open University is team teaching (both to prepare the course materials as well as to teach them) and the division of responsibilities amongst teachers according to particular skills in style of *teaching* as well as *topic*.

The reference above to the need for careful selection of problems is oversimplified. It is now well recognized that to be a successful engineer involves much more than solving problems within technological constraints. There may be financial, moral, social and political influences and restrictions on decision-making. Wright [10] noted that the rapid advance of technology means that the old "empirical" solutions to problems are no longer adequate. He felt, writing over a decade ago, that rigorous analytical procedures had perhaps been overdeveloped at the expense of skills involving synthesis as well as analysis. One consequence of this was the development of a separate and very successful Department of Systems Engineering at the University of Waterloo. Wright - the first Dean of Engineering at Waterloo - saw the task of the future engineer to be one of "managing and planning an industrialized society in which economic and social factors are no less significant than technical factors". Ironically, co-operative education at the University of Waterloo may make it somewhat harder for the motivated student to obtain exposure to classes outside the faculties of science and engineering because of the limited range of electives offered during the summer periods when he is frequently on campus for his academic work. It is encouraging, however, that the surveys of co-operative students show many are aware of the need for knowledge about the social, psychological and political aspects of technological decision-making. In fact what is probably needed is integrated courses for engineers on "Fracture and Society", for example, rather than an odd course in politics or sociology.

Just as the idea of university education has changed considerably in the last 30 years - due to changes in the subject matter, new challenges from the outside world, and a different range of student entrants - so the concept of the engineer/technologist is also changing [16]. For example, one compelling need in contemporary society is for an engineer who can communicate what he knows to those without technical training. Another is for the engineer/scientist of the future to be able to straddle disciplines - and nowhere is this brought home more clearly than in the case of fracture, which is not only a discipline but also a phenomenon that, to understand it, requires skills from a variety of traditional disciplines. It is already being recognized that the range of skills in such problem-solving has to go beyond traditional scientific fields and encompass information and insights derived from social science. Soon it will be necessary to go even further afield into questions of philosophy and aesthetics. Inclusion of such topics may cause the academic engineer to balk at curriculum planning but guidance exists in the literature e.g. [16,17].

But if pedagogical guidelines in this area have yet to be worked out, there is one educational axiom that remains true, and it is one that takes us back to the concept of transfer of training, mentioned repeatedly in the paragraphs above. This is that the knowledge and skills taught to university students will quite certainly change even more rapidly over the careers of the present generation of students than they did during the past three decades. This means that it will no longer be adequate for university educators to rely on subject matter expertise and traditional teaching methods that view students as passive receivers of transmitted information. Rather, students will need to be taught that most fundamental skill - the skill of learning how to learn [15].

The engineering educationalists who debated the issue at ICF4 have provided considerable data that may now be explored further to produce some integration which may lead to useful curricular changes. Some may see Professor Rice as a purist and traditionalist and Reid and his colleagues [4] as practical and radical. Yet these two approaches are certainly not as diametrically opposed as might appear at first sight.

The purpose of the present discussion has been simply to raise some questions, with the object of enhancing the quality of education in the area of fracture. However, in 1977 it can be fairly claimed that Fracture is a *discipline* in its own right in the sense, for example, that the relevant subject matter is "available knowledge organized in such a way that it is suitable for learning." As Whitfield [8] put it:

> "A discipline is thus the flexible conceptual structure,
> a community of concepts, which contains the raw knowledge
> and experience of particular fields of enquiry. Dis-
> ciplines are concerned with particular domains of
> experience; they have a history and a heritage of litera-
> ture; they have developed their own distinctive public
> criteria, conceptual frameworks and modes of investiga-
> tion... They generate a communicating community of men
> who have been initiated into the domain of experience and
> they all embody some expression of the human imagination."

In 1977 *Fracture* surely fulfills this definition and requires a full reappraisal of its place in the core and elective engineering curricula of educational institutions, and in particular with regard to new integrated courses under the title *Fracture and Society*.

REFERENCES

1. KENNETT, F., The Great Disasters of the Twentieth Century, Marshall, Cavendish, London and New York, 1975.
2. COTTRELL, A.H., A BBC-Open University Interview, Open University Press, Milton Keynes, 1975.
3. Contributions to *Fracture and Society* Panel Discussion by A.H. Cottrell and M. Saltsman, Fracture 1977 - Advances in Research on the Strength and Fracture of Materials, ed. D.M.R. Taplin, Vol. 4, Pergamon Press, New York, 1977.
4. REID, C.N.,NEWEY, C.W.A., REYNOLDS, K.A., WEAVER, G., and WILLIAMS, K., A Case Study Approach to the Teaching of Fracture, Fracture 1977 - Advances in Research on the Strength and Fracture of Materials, ed. D.M.R. Taplin, Vol. 1, Pergamon Press, New York, 1977, page 429.
5. PRAAGH, G. van, Chemistry by Discovery, Science Press, London, 1948.
6. HAYSOM, J.T., and USTTON, C.R., Innovation in Teacher Education (Science Teacher Education Project), McGraw-Hill, Maidenhead, 1974.
7. TAYLOR, L.C., Resources for Learning, Second Edition, Penguin, Harmondsworth, 1972.
8. JENKINS, E., and WHITFIELD, R., (eds.) Readings in Science Education: A Source Book (Science Teacher Education Project), McGraw-Hill, Maidenhead, 1974.
9. MARTIN, J.W., Elementary Science of Materials and Strong Materials, Wykeham, London, 1973.
10. WRIGHT, D.T., Co-operative Engineering at Waterloo: The First Seven Years and Prospects for the Future, Address to the Industry-University Conference, University of Waterloo, October 1, 1964.
11. ELLIS, H., The Transfer of Learning, Macmillan, New York, 1965.
12. BLOOM, B.S., Taxonomy of Educational Objectives, Handbook I: Cognitive Domain, McKay, New York, 1956.
13. KNAPPER, C.K., GEIS, G.L., PASCAL, C.E., and SHORE, B.M., (eds.) If Teaching is Important: The Evaluation of Instruction in Higher Education, Clarke Irwin, Toronto, 1977.
14. BLIGH, D.A., What's the Use of Lectures? Penguin, Harmondsworth, 1972.
15. CROPLEY, A.J., Lifelong Education: A Psychological Analysis, Pergamon Press, London, 1977.
16. FLORMAN, S.C., The Existential Pleasures of Engineering, St. Martin's Press, New York, 1976.
17. PIRSIG, R.M., Zen and the Art of Motorcycle Maintenance, Corgi, London, 1974

APPENDIX - Questionnaire Sent to all ICF4 Delegates

Two panel discussions have been organized at ICF4 under the general heading *Fracture and Society*. The first of these is scheduled for Wednesday 22 June, 1977 and is designed to provide a basis for the development of clear educational objectives with respect to *Fracture* as an Engineering subject. It is expected that the discussion will be very lively and several speakers have been lined up. If you wish to contribute to discussion on this topic, please so indicate below. It would also be valuable if all participants would complete and return the remainder of this short questionnaire.

1. Do you wish to contribute to or participate in the discussion on
   Fracture Education and Society? Contribute Yes____ No____
   Participate Yes____ No____
2. Do you teach at a post-secondary institution (university, technical
   institute, etc.? If so please give the name of the institutions,
   and the highest level of qualifications offered (e.g. Ph.D.)._____

3. Please give the name of the department or academic unit in which you
   work, and the highest qualification it is possible to obtain in your
   own department._____
4. Are graduate courses on Fracture or Failure Analysis specifically
   offered in your department or another department? Please give
   details._____

5. Is the subject of fracture an officially designated part of your
   undergraduate curriculum (e.g., do you have a course entitled fracture,
   or is fracture a specified part of another course)? If so, please
   give details_____

6. Is the subject of fracture part of the core (i.e. mandatory) curri-
   culum of the undergraduate engineering programme (if you have one)?
   Yes ____ No ____
   Please give details._____

7. Is the topic of fracture taught in your institution in any special
   or unusual way - e.g. by means of a "case study" approach, through
   the use of project work, etc.? If so, please give details._____

8. What place should courses on fracture have in undergraduate curricula
   for Engineers?_____
9. Please add any other comments on this topic - overleaf.

Return to Professor D.M.R. Taplin, Department of Mechanical Engineering,
University of Waterloo, Waterloo, Ontario, Canada, N2L 3G1.

FRACTURE AND SOCIETY

D.M.R. Taplin

Two Plenary Panel Discussions were organized at ICF4 under the general heading *Fracture and Society*. These were conducted from approximately 14.30 - 17.00 hours in the Humanities Theatre of the University of Waterloo on the Wednesday and Friday afternoons of the Conference (June 22nd and 24th, 1977). The Discussions were open to the general public and were widely reported upon in the Canadian Press. There was also a good deal of effective television coverage of the Conference, including these Panels.

The object of these panels was to explore and integrate our thinking as experts in fracture and delegates to ICF4 with the wider purpose of the society in which we all live on this planet, Earth. Specifically, the overall aim was to provide a formal framework within which we could examine the broader implications of our technologically based fracture research - be they sociological, philosophical (ethical, moral, aesthetic), education- al, political, economic - to perhaps bring us down to Earth from our "lofty" mathematical and "hidden" microscopic studies. For, surely, it is those of us who are educated in the problems of fracture who must ensure that intel- ligent social and political decisions are arrived at in our areas of competence. Much is spoken of the need to educate the general public in science and technology but a more crucial issue in education is the educa- tion of we engineers and scientists and our students in the practical and philosophical facts of social and political life and decision-making.

The two other major innovations in the topics addressed at ICF4 relate to this innovation. They are firstly, an emphasis on the *control of fracture in large-scale engineering structures* - Risk Analysis (Tetelman), Ships and Tankers (Burns), Nuclear Reactors (Nichols), Railways (McClintock), Pipelines (Hahn), Aircraft (Nemec) - clearly such technological structures are, along with the fracture of the Earth itself, the most pressing concerns of society today in considering fracture problems. But the overriding innovation at ICF4, perhaps, was to focus upon an *integrated* approach to fracture. This quite clearly included these panels here being introduced but it also included the aim of integrating the micromechanistic and mechanics approaches both through commissioning broad papers, such as those by Ashby (Maps) and Knott (Alloys) and the judicious juxtaposition of papers from the different disciplines. Furthermore, as Max Williams is often quick to point out, Fracture is an interdisciplinary topic. Perhaps it can best be approached as a *phenomenon* which has many consequences and implications and which needs the application of many disciplines for its understanding and control - including those outside science and technology.

The Conferences in the present series are, indeed, the *Olympics of Fracture*. We must therefore surely endeavour to encompass the study of it in all its aspects - and yet produce a vigorous, integrated theme, philosophy and science. Our purpose at ICF4 was to encompass all nations and all aspects - but, I hope, forward-looking, *critically* and with the application of high

standards. Not all things to all people but, at least, a significant land-
mark and some new departures.

Accordingly, these Panels on *Fracture and Society* represent just a beginning.
What follows in this Volume is an edited transcript of the taped record of
the two Panels. This written record goes somewhat further than the actual
discussions themselves - it is essentially the "book" of the "play" - and,
I believe, several aspects of the topic addressed are revealed more clearly
in this written transcript. It is surely revealed as a topic worthy of
considerable attention - for the collaboration of researchers from several
disciplines and a responsibility for ICF and ICF5 to continue.

The first Panel, on June 22nd, 1977, focussed upon *Fracture, Education and
Society* and directly followed in the Programme, Dr. Reid's presentation on
the Teaching of Fracture. The meeting was turned over to the co-chairman
for this session, Professor Ronald W. Armstrong. Dr. Armstrong was at that
time with the United States National Science Foundation as a Programme
Director in the Science Education Branch. His permanent post is in the
College of Engineering at the University of Maryland. The panellists were
drawn in the main from prominent engineering educationalists both in Canada
and elsewhere, with some wider representation to add a little spice. The
full list of panellists is recorded below with their affiliations. Two
particular papers were examined as background information - the paper by
Reid et al and the Cottrell interview - along with the results of a survey
on education. The idea for this survey derived from discussions with
Professor Alan Tetelman and it was conducted by Dr. Knapper and myself.

The second Panel on June 24th, 1977, directly followed Professor Bruce
Bilby's broad survey of Fracture. Bilby is such an eloquent speaker and
penetrating mind that this proved to be, as many expected, perhaps the
major single technical highlight of the Conference. Even though, by this
time, delegates were themselves somewhat fatigued for, apart from the long
and full technical sessions, the evening before this final day was the
night of the Conference Banquet and Cabaret. The team of Bruce Bilby,
Roy Nichols and John Knott proved themselves to be great cabaret performers,
indeed rivalling our main Cabaret star, Dinah Christie. Roy Nichols'
poetry, Bruce Bilby's apologia to the J-integral and the Duke of Wellington
and John Knott's purple bailiwicks will be long remembered - long after we
forget science, perhaps.

This final session of the Conference was conducted by Dr. R. W. Nichols.
Apart from being President of ICF and Chairman of the International Council
on Pressure Vessel Technology, Dr. Nichols is an Engineer, Manager and
Designer concerned directly with perhaps the most crucial and certainly
most politically sensitive industry   -   Nuclear Power Technology.
Once again, two pre-published papers were available as a starting point
for discussion - Sir Alan's interview and the paper by Max Saltsman, the
Member of Parliament for Waterloo-Cambridge. Max Saltsman M.P. is well
known in Canada for penetrating political analysis and debate and he has
been the parliamentary representative for this constituency for many years -
as a member of the New Democratic Party. He is also a Professor in the
Faculty of Engineering at Waterloo, teaching courses in Management Science.
The panellists were selected to cover a broad front of the interface with
the public at large and their names are recorded below with their affiliations.

Naturally, the names of many eminent and well-qualified people are not
listed on these panels. For example, Dr. Alan Tetelman contributed to the
thinking behind both panels and Professor Mike Ashby unavoidably had to

leave early - many others contributed from the floor and via informal discussion on questions raised. One scientist and engineer whom we should here record as a crucial contributor to our discipline is the late Dr. A. A. Griffith FRS. Griffith is surely the father of the science of fracture mechanics and in his honour we have named the unit of fracture toughness the *Griffith*, where $1 \text{ Gr} = 1\text{MPam}^{\frac{1}{2}}$. This should prove to be a useful standard designation amongst the community and it may well achieve formal recognition as an SI supplementary unit. The major part of Griffith's working life was not spent on fracture mechanics. This was a consequence of an unfortunate accident during his experiments on glass. Armstrong (1) records the fact that Griffith's assistant caused a fire by leaving the glass-melting torch on overnight. The work was subsequently scrutinized and the Committee of Scrutiny decided the research was not worthwhile. Thus Griffith turned to other matters, and developments in the science of strength and fracture were halted for a period. As Mrs June George (2) pointed out in an interesting letter about her father, he worked mainly on the Jet Engine. Mrs George's letter is worth quoting as it provides a useful basis for discussion on the topics we now address:

"Sir, Your readers, when trying to decide about the arguments for and against Concorde, might be interested in the story of my father, Dr A A Griffith. He was an aeronautical engineer concerned in most of his working life with the development of the jet engine.

In the mid 1930s, convinced that there would be another war, my father, then working at the Royal Aircraft Establishment in Farnborough, tried to get money from the government to develop the jet engine. I remember, as a child of five or six years, the excitement in my family, and his anger and disappointment when he returned from London having had his request refused. According to the government there would not be another war.

In the early 1950s my father had another idea. By this time he was working for Rolls-Royce, where he could get money to develop his projects. The new idea was for a supersonic airliner which would dramatically reduce flying times between cities. Again the family shone in reflected glory and hoped that this time he would make his fortune.

This time he gave up the project himself, having decided that the disadvantages, particularly those of noise, far outweighed the advantage of increased speed. He went on to develop the "Flying Bedstead", the first vertical take-off machine with a jet engine. He was always interested in the idea of developing this commercially, but again was very concerned about the problem of noise.

My father predicted that the supersonic airliner <u>would</u> be developed at some time in the future at enormous expense and mainly for its prestige value.

Towards the end of his life he became more and more concerned with environmental problems. He believed that if noise reduction had been also of commercial value then it would have happened much sooner and that it was technically very possible.

Perhaps the lesson to be learnt from his advanced thinking is that we should listen carefully to such men, for they do not always shout the loudest. There are also implications in this story for the education of future scientists; that they should be concerned with the human problems associated with their inventions. My father had an arts education before he became a scientist.

Yours sincerely,
June George

Clearly much debate on this overall topic will continue under the auspices of ICF and elsewhere. Hopefully the start made here will be developed further at ICF5.

REFERENCES

(1) ARMSTRONG, F.W., The Aero Engine and its Progress - Fifty Years after Griffith, J. Roy. Aero Soc. December 1976, 499-520.

(2) GEORGE, J., Letter to the Editor, The Times, January 24, 1976.

PANELS - FRACTURE AND SOCIETY

Chairman : D.M.R. Taplin

1. FRACTURE, EDUCATION AND SOCIETY

Chairman : R. W. Armstrong   National Science Foundation, U.S.A.
             J. D. Embury   McMaster University, Canada
             D. Francois   Université de technologie de Compiègne, France
             R. W. Hertzberg   Lehigh University, U.S.A.
             E. Hornbogen   Ruhr-Universität, Bochum, West Germany
             C. K. Knapper   University of Waterloo, Canada
             J. F. Knott   University of Cambridge, England
             H. Liebowitz   George Washington University, U.S.A.
             A. J. McEvily   University of Connecticut, U.S.A.
             C. N. Reid   The Open University, England
             A.R.C. Westwood   Martin-Marietta Laboratories, U.S.A.
             T. Yokobori   Tohoku University, Japan

2. FRACTURE, POLITICS AND SOCIETY

Chairman : R. W. Nichols   U.K.A.E.A., England
             B. L. Averbach   M.I.T., U.S.A.
             B. A. Bilby   University of Sheffield, England
             G. T. Hahn   Battelle Laboratories, U.S.A.
             J. F. Knott*   University of Cambridge, England
             D. Mills   Ontario Hydro, Canada
             M. Saltsman   House of Commons, Canada
             T. Yokobori   Tohoku University, Japan

Chairman from 16.00 hours

FRACTURE AND SOCIETY - PART 1

Edited Transcript of the First Panel Discussion on

FRACTURE, EDUCATION AND SOCIETY

Armstrong : The schedule for the conduct of the meeting is as follows. First, Dr. Knapper will tell us about the questionnaires on the teaching of fracture that the participants at ICF4 have returned. Second, Dr. Weaver will show a film relating to Dr. Reid's plenary lecture showing us more about the Open University Course *Materials under Stress*. Then, we will turn to the panellists and, beginning with Professor Embury, each panellist will give a short presentation on the topic we address. Following this, we will open the meeting to further discussion, possibly from the panellists themselves, but preferably from members of the audience.

Knapper : I will talk only briefly about the results of the survey since this is recorded separately. I am a psychologist and my job title here at Waterloo is "Teaching Resource Person". Dr. Taplin invited me to collaborate with him in this enquiry into the place of Fracture in Education, as an educationalist as such and I make no claim whatsoever to any expertise in either engineering or fracture. Fracture is quite clearly something that people should be educated about. This has educational implications not only for those involved directly in fracture research and the teaching of fracture, but also for society as a whole, which should know rather more about potential problems that can arise. This is particularly true when these problems require decisions of a political nature. Experts in this field have a very special responsibility, not only in educating students, but in educating students to be able to communicate with the public at large.

About 80 responses were received to the 400 questionnaires sent out by Professor Taplin. It may be assumed that these are not a wholly reliable indication of the teaching processes going on around the World in the 60 institutions represented, but presumably they do represent the responses of those who are most interested, and it is a good psychological precept that people who respond to questionnaires are the people who have a definite point of view.

Let me extract just a few of the findings. Everybody of course agreed that fracture was a very important subject for people to know about and presumably most of those who responded were actively involved in teaching this subject. Nonetheless the number of formal courses, particularly at the undergraduate level, for the teaching of fracture, was rather small. Less than a third of those responding mentioned a formal course. We were more interested, however, in reports of innovative teaching. The innovative method usually boiled down to a case study or project approach, perhaps along the lines that Dr. Reid and his colleagues have written about.

It should be noted that the proportion of reported innovations is relatively small in relation to the number of institutions that we surveyed and of course we have no good indication as to how successful they were. But there are examples there and several detailed descriptions : these are summarized in a short report.

I would like now to mention the contributions by Dr. Reid and his colleagues from the Open University and the interview with Sir Alan Cottrell. It does seem to me that there are certain themes which run through both these presentations. One is a dissatisfaction with the teaching of fracture and, in fact, you could take it beyond fracture to the teaching of engineering as a whole, in the standard discipline-oriented research way that is characteristic of traditional universities. Instead these presentations are arguing for a much more practical approach to understand fractures, an approach which is rooted much more firmly in real-life situations. There are a number of reasons why this is the case and is true for the Open University for instance. This stems particularly from the type of clientele that they have - very different from the sort of students that many of us would find at more traditional universities. In the case of Sir Alan Cottrell this feeling seems to emanate from his experience outside of universities dealing with real world problems. But I feel quite strongly that this dissatisfaction is appropriate for a topic like fracture. I say this is a "topic" although Professor Taplin calls it a "phenomenon" rather than a "discipline". Thus, fracture does seem to be a rather central topic in this regard because it does not particularly lend itself to teaching in the normal "discipline-bound" ways. You could read about this discipline-bound approach much more eloquently in Sir Alan's interview, where he says, "look, we do not want to produce any more researchers, we have got enough of them anyway - it is just not good enough to train graduate students who are going to be training other graduate students, and so on."

I give fracture as an example of an instance where it does not emanate directly from a discipline in the same sense as physics and where at the same time, people need to know about fracture in order to solve very immediate and worthwhile crucial problems. So you have those twin concerns. How do you do this? That really is the point. That is what presumably some of our panellists and I hope many more of our audience are going to address themselves to today. One approach is the approach you will see demonstrated and it is a case study or project approach used at the Open University. Another approach which I talked about at length in the written version of our paper is the attempt at universities like Waterloo to involve the students in a large measure of co-operative education. I will not talk about either of these efforts but will just end by raising a number of issues that are psychological more than anything else and which perhaps the other panellists might keep in mind when they are talking about training or educating people in the subject of fracture. If there is one point above any, that I would mention as a psychologist, it is the notion of *transfer of training* or *transfer of learning*. This is the basic underlying assumption of all education and that is that what you learn in one situation is applicable in another situation and this is something which I think as university teachers we tend to forget. There are many points which emanate from this idea of transfer of training. Let me just say the psychologists find that transfer of training is not something you can blindly assume will happen. It only happens in very specific situations. Particularly it requires a fairly close match between the learning situation and the situation in which you are to apply that knowledge. Secondly, related to the notion of *transfer of training* is the whole question of what particular skills you, as engineers, will want to teach under the topic of fracture to the students who pass through your hands. The point here is to bear in mind that the very nature of what they are going to be doing is likely to change even more radically in their career span than it did in yours. Various educators have called for a quite new approach to teaching which, instead of teaching a body of skills and information, teaches one single skill, perhaps, the ultimate one, and that is the ability to learn how to acquire information or to learn how to learn.

Armstrong : As Professor Knapper has mentioned the interview with Sir Alan Cottrell, this may be an appropriate moment to make a few remarks of my own relating to the educational aspects of the Interview.

The first remark is concerned with the point, mentioned again, which was mentioned in the Cottrell interview to the effect that everyone is concerned, even from a very early age, with why things break. In fact, there is an instructive comparison to be made in this regard between fracture technology and space technology. The thesis is that the everyday experiences of a person with common fractures in respect of fracture technology are somewhat analogous with the counterpart everyday experiences of a person, say, with common flying objects relative to space technology. For both subjects, the principles are largely the same in everyday experience and in these significant engineering accomplishments. The interesting consideration is that an understanding of space technology involves observations on the largest scale imaginable whereas understanding fracture involves observations in the opposite direction on the microscale. The problem of extrapolating either way - for most people - makes each subject a bit difficult.

The second remark is a reminder of the pioneering emphasis given by Orowan to the effect of fracture being a model- or mechanism-sensitive phenomenon. This observation is fundamentally important to differentiating between physics and engineering educations in many places in the World even today. In this perspective, there is reasonable educational difficulty involved in connecting the physics of fracture, with its model understanding which is imperative, and engineering considerations of fracture, extending themselves from abstract applied mathematics for purposely poorly defined continua to empirical analyses of service failures.

The third remark is concerned with the fact that fracture is a many-material-parameter subject - and this, too, adds special difficulties for the educational process. Even with the few experimental quantities which are determined, often with difficulty, in fracture mechanics experiments, it is important to understand that these quantities must encompass, at least, all of those parameters which are involved in understanding the strength properties of crack-free materials. This result is a consequence of the theoretical requirement that plastic flow must always be initiated in some limited sense for fracture to occur because of the growth of a single crack or a number of them.

Now I would like to move on to Dr. Graham Weaver of the Open University, who, rather than Dr. Reid, will introduce the OU film.

Weaver : The title for the whole of this discussion, *Fracture, Education and Society,* is clearly much wider than just talking about the sort of teaching one can do with television. It is surely important to produce an understanding among the general public of the issues which can be treated by fracture mechanics and, insofar as Open University television programmes are available to the general public because they are broadcast on the ordinary television channels in Britain, perhaps we are making a start in this regard.

The course on failure is designed to take about 200 hours of student study. Most of this time is spent studying the written work. There are only about 7 hours of television available so it is a scarce resource. We have to make sure that we use it with some care. The extracts we have to show you constitute about a 20-minute sample. The selection is a number of short extracts with which we hope to illustrate how we use television to teach. I should like to say a few words at various places in the middle of the film about what is next to come. You will probably notice some sort of a story line and just in case you do not I will tell you about it.

The first extract deals with designing against failure, without any reference to the presence of cracks or defects at all. It concerns a man-powered aircraft wing. Then we go on to talk about cracks being introduced into a welded structure during its manufacture. Actually this extract is a sequence about welding and then weld testing. The welding operation introduces cracks into the structure but you are not sure where they are. You need to detect them and so there is a sequence about non-destructive testing. Then some of the later sections talk about fracture mechanics and defects actually growing in service. So the sequence is in fairly logical order in terms of the subjects which are presented. I suppose it is no surprise that the presentation roughly corresponds to the order in which the subjects are dealt with in the written part of the course. But that is not the rationale for this selection. As I say, I want to show you how we use television as a medium for teaching. Our written work can give students contact with problems via all sorts of examples which they can try to work out for themselves with direct instruction on theory and its application. There are several things that written work cannot do. It is very difficult through the Open University system to give practical experience. The only practical experience we can give to our students, at least in this course, is with a small home experiment kit which some of you might have seen in the publications room in the other building. So one of the things we would like to do with television is to demonstrate the use of laboratory equipment and the kinds of experiments which it is necessary to do in order to establish evidence for the sorts of theories which are being put forward or in order to establish evidence for taking action in an industrial context. This reinforces learning if you like contact with practical methods. In this limited way, we can put the student in contact with practical applications.

Many of our films serve the purpose of actually taking the students out into industry on location. In the two instances that I have examples of in the film we are talking about fracture mechanics. They are applications of fracture mechanics, in one way or another, or examples of situations where fracture mechanics can be applied. Another thing which we can do is to give students contact with experts in the appropriate fields. I think residential students at universities are often spoiled for this. All the staff are experts in that most of the staff have active research interests, and it is fairly natural that the student receives a good deal of motivation and encouragement from seeing where intelligent study can lead him. But our students, working at home, part-time, trying to hold down a job as well, do not have anything like those opportunities, so television is one way in which we can bring the expert and his opinions and his professional apparatus in front of the student. Well, those are three principle areas that we try to give the student, and they are dealt with in the first three extracts which I shall show you. The remaining parts of the film are just short snatches which actually reinforce these first three items but also show you some of the television tricks of the trade.

The first extract is an example of practical work. It is a demonstration done in a laboratory; actually, the laboratory is the television studio. All the apparatus was taken there and the person who is presenting it is one of our authors from the course team. In one way it is quite a lavish experiment. The students at this time in the course have been studying beam theory and have been following a step by step design of the wing for a man-powered aircraft. This is a situation which was carefully selected because both weight and efficiency of use of materials are important. A man-powered aircraft only has a 300 watt power source which is rather small compared with mechanically-powered aircraft. During this extract you will hear two references which might puzzle you. One is OUMPA (this is merely the acronym for "Open University man-powered aircraft") from which you should not deduce that we have built a man-powered aircraft. We have gone so far as to build a section of the wings of one of these designs that the student is being led through and the programme shows this design being tested. There is also a reference in the extract to "broadcast notes" which I feel I should clarify. Most of our television programmes have associated with them some short written notes which can serve several purposes. Often they are no more than a resumé of the content of the programme which really acts as an aide-memoir to students after they have seen the programme, even a long time after they have seen the programme, because one of the disadvantages of television is that it is a very transient medium. Once you have seen the programme it is all over and done with and there is not much opportunity with the limited Open University schedules of seeing it again. Then, again, the broadcast notes can contain questions for the student to try to answer things based on the lessons of the programme and so he can test the extent to which he has understood it. In this particular programme the "broadcast notes" also contain data which the student needs in front of him actually to use during the time he is watching the programme.

Can I now have the first clip of film, please? It runs about four minutes and there is a blank where I want the projectionist to stop.

FILM   (Sound-track transcribed in italics)

We have seen the tests of individual materials and we must now test a section of OUMPA's spar design, the balsa and spruce main spar. I have set it up here in the studio as a simply supported beam, supported here and at the other end, with a load applied by a hydraulic jack in the centre. I am reading force via a pressure gauge directly in kilonewtons. The displacement in centimetres is registered on this scale here. You can see from your broadcast notes that we have a required stiffness of 0.24 meganewtons per metre for this structure. So if I allow a centimetre deflection of the spar, this should give me 2.4 kilonewtons deflection on the displacement scale. If I achieve that I will have predicted OUMPA's main-spar stiffness adequately. Now we are testing this spar as a simply supported beam and this does not precisely represent the conditions of a spar beam in flight because there it would be cantilevered. We are in fact going to have a higher shear force in the present test than we would normally expect. For the design of OUMPA's main spar, a 5,200 newtonmetres bending moment will occur at the centre of the beam when the force on the scale reads roughly $7^1/_2$ kilonewtons. So, let's begin the test. Ian, can you start pumping? I will call out when we have a centimetre deflection ..... Now! Thus we have just about achieved the required load. This is a pretty successful prediction of bending stiffness and I think we can be pretty proud of that; however the theme of the course is "failure" and we ought to fail this beam now to see whether we can learn anything from *this particular design. So, Ian, can you carry on with the pumping? We are building up the force ..... I am getting some funny noises too. It is starting to crackle but taking more load. There she goes! Now you can see, in fact, if you observe here, that local shear has caused failure of this beam and we pulled out part of the spar from the glued surfaces. So we have not failed at what is something like over 5 kilonewtons. We have not failed the spruce or the balsa but we failed the glued joint. This would lead us to conclude that we ought to reconsider our beam design to have a somewhat larger area between the glued surfaces. However, the shear as I said before in this type of test is quite high and so I think that this is in general a pretty good design for OUMPA's main spar.*

Weaver : Well, that beam cost about $500.00, so we only had one of them. I think it says something for the quality of the presenter that the failure was unrehearsed and he was able to make pertinent comments on it even though it did not fail in quite the way it was expected to. The next excerpt is on location at a pressure vessel works. The main point of this excerpt is the industrial location. You will note that some of the film has the commentary dubbed on to it because it is a noisy factory and it would be rather tiring if all the commentary was backed up with factory noise. However, part of the commentary is deliberately spoken against the factory noise. It is part of our purpose to give an idea of the atmosphere of the place.

FILM

The manual-welding process is used in this company generally for welding nozzles to shells or internal attachments. The circumferential seams and longitudinal seams of the pressure vessels are generally done by automatic processes. In one process the idea is to feed a continuous coil of wire into the electrode over the work piece. Granulated flux is fed round the wire and the arc is struck underneath, melting some of the granules to form a protective blanket. Unmelted granules can be recovered with a kind of vacuum cleaner for further use. The fused slag breaks away easily after the weld pool is solidified and gives a smooth even surface to the metal, reducing the possibility of slag entrapment, which is a big danger with manual welding. With this, the submerged arc process as it is called, very large volumes of welding material can be deposited without stopping, by simply keeping the workpiece rotating. But how can checks be made on weld quality? Sample welds are taken for mechanical testing and sections can be etched in the laboratory so that the grain structure of the welded material can be examined.

Here is a close-up of an etched section of a multi-run submerged arc weld. What does it show? First there is a difference in structure between the weld and the parent metal on either side. The columnar crystals reveal the successive weld pools which are all well-fused in this case with no obvious slag entrapment or porosity. The darkly etched narrow zones immediately surrounding the weld deposit are the heat affected zones of the parent metal. Compare the structure with this next one, which is a section of nozzle, that is the vertical member, manually welded to a thicker horizontal shell. Again you can see the individual weld pools but notice that the root run of the weld on the right-hand side is not as well fused as the later welded zones. This zone could be the starting point for cracks to be initiated. So something has to be done about this. Notice also on the outermost welds that there are depressions between the weld pools and the horizontal surface of the parent metal. These could also act as stress raisers. But neither of these defects is left unattended

*After the nozzle welding is complete, the outsides of the welds are pains-
takingly dressed to produce a smooth contour and thus, leave the surface
free from stress raisers. For the smaller diameter root welds a similar
operation of grinding the irregularities can be carried out. But what
about the problem of the larger nozzle to shell welds? These parts are
only tack-welded in position. They are about to be welded and we know
that the first weld run, the root run, is not likely to give such good
fusion with the parent metals as the later ones, and this is where there
are likely to be faults. The engineers can take care of this provided
that it is realized at the design stage. We can account for it by leaving
surplus metal on the inside of the nozzle. We use this to form the root
run and, later, when the welding is finished, we machine away all of the
metal which is surplus, at the same time taking along with it the faults
involved in the root run.*

Weaver : In the next extract we shall again be out on location, this time
at Hartlepool on the North Sea Coast where an oil production platform is
being built. This sequence starts with an explanation of a practical non-
destructive testing method and, then, we meet an expert who in this case
is the chief engineer on that site for the non-destructive testing sub-
contractors. He explains some of the difficulties of the job through a
recorded interview. "A" in the following is Mr. K. A. Reynolds, a major
author of the course - unfortunately, he could only be present at ICF4 on
film.

FILM

A : *The job is still some months away from completion but already you can
see the size of this enormous structure which is being built. Parts are
being brought in by sea, whereas others are being built on site and
assembled to make the platform. You must remember that the model is only
of the jacket. This which we are looking at is about half complete and
on top of that is going to go the superstructure. You can see that the
large cylinders at floor level are about half their final length and
already the K-frames are being erected. From this you can gain an impression of the size of the thing. That particular fabrication is about half-
way between the top and bottom of the platform. There is going to be
over 35,000 tons of steel used in manufacturing this structure. It is
being joined together by welding with something like 40 miles of welds
being involved. In a few months' time it will be complete and it will be
taken out into the North Sea where it will finally take up its position.
Just imagine the problems of inspecting all of those welds, because when
once it leaves this basin there is no hope of being able to get to it and
remake any faulty work.*

*How does one radiograph welds in a structure like this? Access is so
varied and generally difficult that an X-ray tube with its associated
power source is just not feasible. So a portable radioactive isotope
source is used, emitting gamma rays. This is mounted on a capsule which
includes suitable shielding and is taken to the job in order to make the
exposure. Film is laid over the area of the weld to be examined. As
many of the sections to be welded are tubular, it is convenient to place
the gamma ray source at the centre of the tube which sometimes necessitates
drilling a hole and positioning the isotope at the centre. This allows
the whole seam to be photographed at one exposure. The isotope is fed
into position through this cable from its storage box. The hole in the
tube will have to be welded up again afterwards, of course.*

B : *That is a 36 inch diameter joint, 5/8" wall thickness and it is a panoramic technique. The exposure conditions would be about 80 curie minutes. If we use a 10-curie isotope we would be using an eight-minute exposure.*

A : Panoramic photographs allow us to look at the whole weld at one shot.

What other types of technique could you use?

B : Well, there are single shot techniques where you only want to look at a specific piece of the weld and if you cannot get access to the inside of the joint you can do what is known as a double wall single image technique where you put the isotope on the outside of the joint and the film on the outside of the joint also but at the opposite side. You then penetrate two walls and look at the images of only one weld.

A : Is this a typical situation or is this an easy one?

B : This one is relatively simple for this type of work. It is only about 30 feet up in the air. It is well-scaffolded from ground level. The difficult ones are when you start going up into the 200 feet level. Again scaffolding is there, but access is quite difficult at 200 feet.

A : What are you actually looking for in this examination?

B : We are looking for defects in the weld and, basically by using gamma radiography, we shall be looking for volume type defects, that is slag, porosity and inclusions of any nature, and we are not really in a position to be critically examining for cracking. But the reason that we do this by gamma radiography rather than, say, ultrasonics, which may be more useful for detecting cracks and crack-like defects, is that past experience of structural welding of this nature has led us to believe that cracking is a relatively rare occurrence and volume defects are the major problems.

Weaver : The rest of the film consists of several much shorter extracts than those we have seen. These show mostly our teaching technique in the studio, and some of the television techniques. Have a look at the one called colour separation overlay, which is a fancy name for the ability to add and subtract features from a diagram as you go along.

FILM

I think we should now go back and have another look at the grids on the deformed cube, to see again this transition from plane strain at the centre to plane stress at the outside. In the centre we have no deformation in the Z direction (this is the plane strain region), but as we come out through the specimen we start to find a deformation taking place in the Z direction, until right out at the surface, where the specimen is free to contract, we have changed over to the plane stress region. So far we have been looking at deformations, but we are more interested in stress distributions. We have a contour map here, a stress contour map, which shows the shear stress distribution which runs along the crack front. This is actually an envelope of maximum shear stress. If we look along the direction of the crack front, the shear stress distribution takes the shape of two lobes which radiate from the crack tip. You will become quite familiar with this shape when you carry out the home experiments in photoelasticity.

Weaver : Now I would like to show you how a television camera can get really close into an experiment. These next two examples are taken from a programme which closely supports the students' initial learning of linear elastic fracture mechanics.

FILM

As with any load displacement curve, the area under the line represents the amount of energy stored in the specimen. If this were a simple material test we could transpose the load and displacement into stress and strain and take the slope of the line as the elastic modulus of the material. But remember, this specimen has got a crack in it, and its behaviour is a consequence of both the material it is made from and its geometry. And it is more convenient to take the displacement per unit load and this term is referred to as the "compliance" of the specimen. If I repeat the test but this time use a specimen with a longer crack in it, I again get a straight line but it is somewhat shallower. There is less energy stored in this specimen for the same displacement. And it has got a higher compliance. If I look at the shaded area, this represents the difference in energy between the specimen with the short crack in it and the one with a long crack. If you consider what has happened, if the crack has grown from a length of a_1 to a_2 with no increase in displacement, you can see that the energy released by the crack growing is represented by this area that is shaded. By carrying out a series of tests on specimens with varying crack lengths we get a family of straight lines and from this we can generate a plot of compliance against crack length. Here we have compliance plotted on the vertical axis and crack length on the horizontal and you can see from this line that the compliance increases with increasing crack length.

Weaver : I like the way that the dynamic picture of the load displacement curve suddenly becomes a teaching graphic. The next one is similar. It shows the load displacement curve again, but in a different way. It is a split screen with the sample and the load displacement curve on view simultaneously. You will see a white marker too, to indicate the position of the crack tip. You will see the toughness specimen actually fracture in this one.

FILM

Once more a plot of load against displacement and as the load increases the stress intensity at the end of the crack is increasing.

Weaver : The next excerpt is another example of going to an expert and using his professional equipment. When we wanted to explain fatigue testing we contacted Dr. Clive Richards of the C.E.G.B. to do the talking for us.

FILM

Now let's start a test and see the load cycling. As the load or stress cycles, the K value varies from a minimum to a maximum value. It is this difference, ΔK, that we are interested in in fatigue crack growth. The quantity, ΔK, is really the counterpart of stress amplitude in S-N endurance testing. Remember that K_{max} cannot exceed K_{1C}, otherwise the specimen will fracture suddenly. Now, how do we measure fatigue crack

235

growth rate in the laboratory? Well, we can use any of the types of tests normally used in the evaluation of fracture toughness. In this case we have this particular type of specimen on the machine. Usually we do tests in which the stress amplitude remains fairly constant throughout. But as the crack length increases the alternating stress intensity increases and so, too, does the fatigue crack propagation rate. We need to count the number of cycles as we reach successive crack lengths. In this particular case, we are doing it visually using scribed markings on the surface of the specimen, placed 1mm apart. The fatigue crack is moving fairly slowly but if we waited five minutes, the crack would have moved like this ...

Weaver : Finally a short example of animation.

FILM

Well, when we have a fatigue crack we are in a situation in which we can control both the rate of breaking of the oxide film and the rate of supply of the environment to the crack tip. Each time we apply a tensile cycle we are going to break the oxide film, so we are going to give the opportunity for environmental crack growth to occur. But, in addition, every time we open the crack it will suck in fresh environment and every time we close it we squeeze it out. So, in every opening cycle, we have a fresh amount of aggressive environment which is ready to really attack the exposed metal at the surface. And then, when it is used up, we close the crack and flush it out. So let's just have a look at an animation that will show this in detail. We have another crack, this time in a passivated metal : in other words the metal surface is covered with a thin film of metal oxide. As we apply tensile load, dislocations move along their slip planes and produce a slip step near the crack tip. The oxide film ruptures. The dissolution and passivation processes compete with each other at the crack tip producing a characteristic crater. A dissolution takes place in this confined region, it gradually uses up all of the corrosive strength in the solution, which will gradually get weaker. The liquid will possibly become sufficiently weak to prevent further dissolution. Reversing the loading and applying a compressive load, closes up the crack, expels the stale solution and amalgamates the crater into the crack itself. Reversing the load again opens up the crack, sucks in fresh solution and the process repeats itself. At each stress application we get a dissolution-aided increment in crack length.

Weaver : In conclusion, I think the only sadness that we have about television is that in spite of the fact that it is broadcast on the general services, and anybody can see it, the one thing we want to do is to *stop* broadcasting it on general service, because the limiting factor is the amount of transmission time available to us. The day we can have a videotaped casette in every student's home as easily as we can now have a sound recording tape in every student's home, is the day when I think television teaching will really come to the Open University.

Armstrong : For the second part of our panel discussion we would like to have short presentations, of the order of three minutes or so, by those individual panellists who wish to do so. We will begin with Professor Embury of McMaster University.

Embury : I really want to say some things which may appear rather the statements of a heretic. One of the things which strikes me about education in regard to fracture is that we should look very carefully at those who are going to actually consume this education. One has to look very carefully at the idea of whether education is complete in a three-year process or a four-year process or whatever we normally think about. It seems to me that one of the vital educational functions, in fact, for the engineering community, may be to require a very different approach than the one that we generally take for standard undergraduate or graduate students. Now with that type of audience in mind, I would like to make some comments on the information which we have received so far in the Conference as it relates to education. Let us consider concepts such as Professor Ashby's mapping approach. I think (and this is not meant to be a critical comment) that there is very little which is new in that approach in the sense of new physics. There is not obviously any new information there. The thing which is really startling is the way in which you can condense information. You can condense a great deal of engineering experience and you can distil this with the added knowledge of good constitutive equations and models. That seems to me one direction in which educators could go.

This raises, of course, the question of academic disciplines. One could perhaps say as a heretic that there might be really two branches of engineering in the future - a kind of systems engineering and materials engineering. The subject of fracture is very much at the interface of these two disciplines. Another point I would like to raise is the idea of the consumption of information specific to the fracture field. Very few of the presentations that I have heard so far have dealt in any sense with the statistical nature of fracture. There was a very important point raised this morning which concerns the kind of bases on which you approach the whole question of yield criteria and fracture criteria. The other area which strikes me as being one of great interest, and this is a point of Canadian nationalism, is the area of rock mechanics. It seems to me that there is a geological interface which is certainly of long term value of which we have heard virtually nothing.

The final point I would like to raise is concerned with the whole question of case studies. Most of the information that we think about in case studies is really a simple transmittal of information, a change of environment if you like. There is something else in the case studies system which is worth exploring : this is the use of case studies to encourage group study. It is a very valuable way of putting together civil, mechanical and metallurgical engineers and asking them jointly to solve a problem. This is often the case in design and it is a thing which we really use to a very limited extent. There is a need to force final year students, graduate students and others into the process of synthesizing their knowledge. I think that is one of the things which comes through very forcibly in a meeting such as this. There is a real need for educators to force students into this process of synthesis.

My last statement, as a heretic, is that really I think the function of universities is not to tell students what professors know, but what they do not know. Professor McClintock raised a very important point this morning concerning the almost biblical nature of the educative process. You really have to force the student, both to this interface of synthesis and, to the point, particularly at the graduate student level, where he will ask questions about the real basis of his education. Are the criteria which you give him, the premises on which arguments are based, really complete? It is in many cases really the premises of the argument which you have to attack rather than the conclusions.

Armstrong : That will certainly help to promote strong audience participation, I think. Let us move on to Dr. Knott. Let me say, too, as far as the panellists are concerned, that this 3-minute stretch is only for your opening important comments. You certainly will have another chance to enter the discussion as we go along.

Knott : As a point of information, I think that any comments that I make ought to be seen in the context of the typical British educational system, in which the University period consists of three years of undergraduate training, often rather specialized by American standards, followed in some cases by three years of research training for a Ph.D. degree. There is also a limited number of Master's courses which are taught on a one-year basis, or perhaps done by a mixture of teaching and research in two years. Speaking very generally, in undergraduate engineering courses, the vast majority of engineers leaves after three years' undergraduate training to go out into industry, and does not stay on to do research. A number of metallurgy courses and materials courses are also taught on a science basis : for example, as Professor Ashby noted in his Monday morning lecture, he is teaching in the Cambridge University Engineering Department; I am in the Faculty of Physics and Chemistry. Again, of the people we have in the final year, about 30% go into research training and the rest go into industry. I think we must separate perhaps our thoughts, in the British context, about the undergraduate training for general industrial graduates as compared with the much more limited field of research training.

As one of the interviewers of Sir Alan Cottrell, I would like to draw your attention to three points made in the interview with him which pertain to education. The first one of these is made with regard to undergraduate courses and, particularly, the teaching of materials to engineering students and, in the reciprocal sense, the teaching of smatterings of engineering to the materials students. Here, I think, there is a point which is perhaps just a British one, but I suspect is not; that is, that very often such reciprocal teaching is done as service courses to other departments and very often it is regarded as a chore to be taken on by one of the least senior of the lecturers. This often does not create a very good impression in the other department. There are exceptions, where senior people make it a point of interest and honour : for example, I know that Professor Smith at Manchester is very keen on the teaching of materials to engineers. The second point Sir Alan Cottrell makes is on case study approaches, which he generally endorses, but I think that Professor Knapper has greatly overemphasized Cottrell's dissatisfaction with traditional methods of teaching. If you look at the words that he used, he was by no means advocating that we should throw overboard traditional methods in teaching and go on to case study approaches, although he recognized their value, particularly in the process of developing the interest of people. The third point that Sir Alan makes is a very general one on the nature of research training, particularly of university research, and I believe that he has very, very clear distinctions in his own mind between what is pure science and what is engineering. He does feel that, in the fracture field, as Dr. Scully mentioned this morning, there are some real problems still to be solved as a pure science, such as stress corrosion and some of the fatigue problems. That, in the view of Sir Alan I am sure, can be done only through the traditional research training methods.

As a personal view, I think I must overwhelmingly support Professor Embury on what it is necessary to teach engineers. In the British context, they are going out into industry at the end of their undergraduate training. They may go into design or they may go into general management. Thinking only of mechanical and civil engineers, to make consideration a bit easier, we do want obviously to teach them something about materials, so that they

do not make really silly mistakes right from the outset. They are con-
ventionally taught all their general considerations of design, plastic
collapse, buckling stability and things of that nature, and we do want
to bring home to them the importance of failures. The failures should
include not just fracture mechanics but the whole concept of why things
tend to fail, for material, environmental, or mechanical reasons. This,
of course, requires a very complicated combination of knowing something
of the mechanics of stress analysis, knowing something about fracture
mechanism, and knowing quite a lot about non-destructive testing as well.
I think one of the points made on the film was that the X-ray or the isotope technique being used to do the non-destructive testing was quite
good at picking up volume defects such as slag entrapment; not very good
for cracks, where ultrasonics should be used. Engineers, even in management, should have some awareness of the problems of defects and of the
ability of techniques to detect or not to detect them.

I think, in summary, the most interesting paper with an educational aspect,
apart from the one on "teaching methods" that is in this Conference, is
the one by Professor Tetelman. He presents an overall philosophy of trying, through a combination of statistics and engineering modelling, to
quantify past experience on structures and trying to predict failures or
the chance of failures in real engineering situations. I think, for two
reasons, we want to look at this effort; we want to look at it as a framework on which to hang all these other things when we are teaching undergraduates, and, on the other hand, I think we want to use that sort of
analysis to examine our own educative methods and the substance of what
we teach. We want to examine, in fact, the percentage return on the investment we put into our undergraduates.

Hertzberg : There are a few things that I would like to comment on with
regard to what I have been doing at Lehigh University. Before I do that
I would like to call everyone's attention to the fact that The Metallurgical Society (TMS) of AIME is going to be conducting a seminar at their
meeting in Chicago this October which relates to this subject. There will
be a Mechanical Metallurgy Committee-sponsored day-length seminar on
teaching graduate mechanical metallurgy courses involving a series of
7 or 8 talks dealing with the kinds of material that should be introduced
into such courses in universities. I am not sure whether that is going
to be published in any form or whether it will simply appear as an abstract
in the Journal of Metals (JOM).

I think that the idea of a fracture course at University has to be considered in the context of other relative courses being taught and, perhaps
more specifically, with regard to the department in which one is going to
try to teach such material. With regard to the metallurgy department that
I am a member of, one tries to seek a balance between the mechanical
properties of solids, the physical properties of solids, and the chemical
properties of solids. Thus you are always fighting to create a balance
for the student so as to give him or her a perspective as to the relative
importance of these various disciplines within the general field of
materials. On that basis I have serious doubts as to whether a course
solely devoted to the subject of fracture is appropriate within an undergraduate curriculum. On the other hand, I feel that a course that deals
with mechanical behaviour of solids including fracture *is* appropriate.
About ten years ago, our department decided to completely overhaul our
undergraduate curriculum and in so doing, a course was introduced for
the first time dealing with the subject of "mechanical behaviour of solids".

I would say about 40-45% of the course was devoted to the subject of "deformation of solids" and here we were dealing with tensile testing, some elements of dislocation theory, the deformation of crystalline solids, slip and twinning behaviour in crystalline solids, a discussion of high temperature behaviour, creep behaviour, creep deformation, and, then, a section on the deformation of polymeric solids. In this last section, we introduced some elements of visco-elasticity. The other half of that course deals with the subject of "fracture" and here the primary emphasis is with regard to fracture mechanics and the role of fracture mechanics in the study of static and dynamic fracture processes.

In addition, the mechanical behaviour aspect of our curriculum includes a separate course in metal forming, which, together with the course on mechanical behaviour, is a required course in our department. A separate course in dislocation theory is offered as an optional course. There are graduate courses, as well, that deal with the subject of mechanical behaviour. The course that I am involved with deals with fracture mechanics concepts at a more advanced level, but here again, the treatment is more from the metallurgical standpoint than the continuum standpoint. Thus, my emphasis is not that of deriving stress intensity factors and talking about some of the continuum aspects, though I do present this material in a more physical sense. Rather, I focus on the material aspects, and the more continuum aspects are handled in the mechanics department with other courses in fracture mechanics that are taught there, with appropriate personnel. There is a new course that I am going to be attemping this fall which brings to mind the question of case histories. I will be bringing in a considerable body of knowledge on failure analysis and, in addition, introducing a new concept of product liability. This will involve some of the legal aspects and legal implications of failure in structures : we hope to bring in some lawyers to assist in the teaching of this part of the course.

Armstrong: Our next speaker can also give us information from the very important perspective of a university Dean, so I introduce to you Dean Liebowitz.

Liebowitz : I have been asked by the chairman to comment on the offerings of fracture courses at George Washington University and to mention some of the problems in implementing fracture mechanics courses in a School (Faculty). At first, I will interpret fracture in its broader sense and not just fracture mechanics. At the undergraduate level we have five courses being offered, including materials science, introduction to mechanics of solids, materials engineering, mechanics or materials laboratory, finite element methods and structural mechanics. All students in the engineering school are required to take the first four courses and the fifth could possibly be taken as an elective. None of the five is totally oriented to the field of fracture. At the graduate level, however, fourteen courses are offered which are directly involved with fracture and, at least, an additional 20 courses are related to the field. Examples of the ones specifically in fracture are : failure and reliability, analysis of engineering structures, fracture mechanics, fatigue and failure of materials, finite element methods, structure of materials, introduction to continuum mechanics. Examples of some of the 20 related courses include environmental effects in materials, theory of elasticity, physical ceramics, design of metal structures, automated design of complex structures, nuclear reactor engineering, transformations in materials, deformation of material, composites, and others. If we were to interpret fracture to embrace only fracture mechanics, then we would reduce the number of full courses at the undergraduate level from five to zero and, at the graduate

level, from 14 courses to six. At both the undergraduate and graduate
levels there are courses which contain particular aspects of fracture
mechanics. Certainly there is a need for greater curricular emphasis on
this important field. We have to differentiate between the different
departments and curricula in a School or College of Engineering. By this
I mean that there are certain curricula which require fracture mechanics
more than others. For example, aeronautical, mechanical and ocean engineering require more of a knowledge of fracture mechanics than electrical,
chemical and possibly civil engineering programmes. The highest priority
problems in aeronautics are fatigue and fracture. Certainly an aeronautical engineer being trained today in such a field should have a good awareness and working knowledge of fatigue and fracture in wings, engines,
fuselages and landing gears.

At the undergraduate level it is important not only to offer fracture
mechanics programmes, whether in complete courses or piggy-back on present
ongoing courses, but also to include such information in the last year of
design projects. As you appreciate, there are a number of problems
associated with implementing new courses in fracture mechanics into a
curriculum, which I would summarize thus : 1) many schools in the United
States are trying to maintain present course offerings at a constant level,
or to reduce their number; 2) faculty will try to maintain their own pet
courses; 3) there is a need to educate and train faculty for teaching
design, elasticity, plasticity, and finite element courses so as to expand,
re-orient and integrate their present courses to include fracture considerations; 4) faculty must be convinced that fracture has been growing in
importance and requires a more substantial presence in up-to-date curricula;
5) Chairmen and Deans must be persuaded to support such an effort; 6) a
course in fracture mechanics must be composed which is not too superficial
to be meaningful, even at the undergraduate level; 7) the support of
knowledgeable people in industry and government who are faced with fracture
problems in their structures should be obtained; 8) work on engineering
accreditation groups will be necessary in order to appreciate and acquire
the correct curricular emphasis to be given to fracture mechanics (this
would particularly pertain to the United States and Canada); 9) some
agreement on what constitutes a good programme in fracture mechanics must
be reached.

Since there is no single proven methodology accepted by us in the field,
I would think, on the broad level we are considering here, that there
would be a difficulty in recommending a particular programme for a student
to pursue. I would recommend a very broad background involving physics,
mathematics, continuum mechanics, computer mathematics and, also, including
materials sciences. Problems in the establishment of a new course can be
approached with some optimism, because similar problems were experienced
not too long ago with the integration of computer courses into modern
engineering curricula. Because of the newness of this field, most of the
knowledgeable people in the computer field were young and had little
seniority on university faculties. However, we were able to overcome the
formidable obstacles and consequently arrived at a satisfactory situation
in many universities. We did this with other programmes as well, and I
have no doubt that it will also be accomplished soon for fracture mechanics.
I am concerned about a remark made previously that we have enough researchers and that we should train instead people more practically oriented.
While I feel some additional effort should be made on the latter, I
certainly do not feel we have enough proven researchers to meet our needs.

Francois : What I am going to say must be seen against the background of the traditional French education, which is very much centred on theoretical aspects. Mathematics has high prestige in our educational system and this includes secondary school, where the entire selection of young people for engineering and science is done through mathematics. Thus, when these people arrive at University, they anticipate more and more mathematics, which is indeed the case. When we want to train engineers we are faced with the problem that they see no relation between this theoretical background and the reality of things : the way materials behave and larger structures behave. It is very difficult to give the proper practical basis to our courses. Students feel that they are receiving a low grade education if we attempt that. There is also reaction from mathematics teachers, who would like theory and not practice. This is the problem that we face.

First, I will describe the content of the five year training period of engineers. Two periods of six months are spent in industry. In these periods, the students are faced with really practical situations and this is very successful. Some students have carried out excellent projects in industry. They have learned a lot and have also been of considerable help. We are also trying, of course, to introduce as much practical teaching as possible into the curriculum. The difficulty here lies in the amount of money and equipment available, which is never enough for the needs of the teachers. I have tried to overcome this problem in the teaching of first year undergraduate students by employing very simple experiments to give them a feeling for practical failure problems. The approach has some similarity to the things we have seen in the film. The students are asked to build, using only a sheet of cardboard and some glue, a small bridge, which is then loaded until it breaks. They do this as a group project, and once they have seen the way it breaks they do it again a few days later. Then all the results are analyzed with the help of their teachers. We also arrange to have engineers come in to demonstrate what is really done in industry. The way that bridges are built is described so the students see the connections between the practical observations they have made in the labs and the industrial reality.

As far as fracture is concerned, there is no specific course. It is introduced in mechanical engineering, where there is a course on mechanical properties of materials. It is only later, in graduate studies, that there is a specific course on fracture mechanics. We think the teaching of fracture and failure more generally is very important, because of the inter-disciplinary approach it requires. You have to appreciate the relation between the internal structure, i.e., the microscopic structure and events, and the macroscopic behaviour of a structure. I think it is very important that students realize that a few dislocations moving somewhere can cause a crack to start and that crack can make a bridge or some other big structure fail.

Yokobori : I will talk about the fracture course in our Department at Tohoku University. It may be felt that it is better to avoid teaching students subjects which are too specialized in undergraduate courses. This may be true in some circumstances. On the other hand, in the light of the importance of fracture problems in engineering applications, we start to teach undergraduate students about fracture in a course called Fractology, by emphasizing three special considerations. The first is the understanding of the fundamentals and methodologies of fracture by an inter-disciplinary approach. The features of this approach are outlined as follows :

1) Atomistic approach, say, in terms of dislocation theory.
2) Microstructural approach, say, in terms of larger scale features, such as grain size, non-metallic inclusion parameters, etc. A fractographical approach is also included based on morphology.
3) Continuum mechanical approach, say, so-called fracture mechanics.
4) Mathematical theory approach.
5) Stochastic theory approach.
6) Approach based on thermodynamics and statistical mechanics, such as reaction rate process theory, nucleation, etc.
7) Environmental approach including chemical effects.
8) Material testing approach.
9) Design aspects approach.

Based on the many approaches mentioned above, and on the other hand, by using systems analysis, new methodologies are being explored such as :

10) Combined micro- and macro-fracture mechanics, including the interaction of cracks and dislocations.
11) Kinetic theory approach, combined micro- and macro-aspects.
12) Stochastic theory approach, combined micro- and macro-variables.

... and so on.

The second special consideration is based upon developing the science of comparative fracture. This is a science analogous to comparative anatomy, comparative literature, comparative philology or comparative psychology. Brittle fracture, fatigue fracture, creep fracture, other fracture modes and yielding behaviour are compared in a range of materials. Differences in behaviour of cracked and uncracked specimens are also considered. It should be noted that fracture mechanics concerns only cracked specimens, whereas dislocation mechanics concerns mainly uncracked specimens.

The third special consideration in the "Fractology" course is the treatment of fracture and related problems in liquids and gases as well as in solids.

That summarizes the subject of "Fractology", for which the standard teaching text is my book, "Methodologies and Fundamentals of Fracture of Matter and Solids" (Iwanami Shoten).

Westwood : I would like to make a few remarks from rather a different point of view from most of the other members of the panel. I speak as an industrial scientist and manager : the person who is concerned with hiring the products of Universities. We are concerned in industry not only with the consequences of fracture, but also with its important positive applications. I would certainly want any mechanical or materials engineer that I might hire to have been exposed to some courses on fracture for three reasons. The first, fracture is a phenomenon which can work either for you or against you; second, fracture is an issue which must be considered in relation to cost and also to safety; and third, fracture must be considered in connection with industrial productivity. It seems to me vital that any practical engineering course on fracture should deal with these three aspects.

I feel that the first aspect relating to the mechanisms and phenomenology, is probably well taught, but the other two are either less well-covered or hardly dealt with at all. In connection with the teaching of phenomenology, I certainly strongly support the Open University's view that their study of real examples of failed components, fractographic analysis and such is an extremely useful educational technique. I feel this is true both in terms of adding substance to the theories and in preparation for one of the most likely uses of the knowledge that the student will acquire in the University. As Sir Alan Cottrell has said, "It is amazing how far you can get with a little practice in diagnosis." This fact of life, of course, has long been the secret of success for most medical doctors and perhaps also for practising engineers, as indicated by Dr. Tetelman. Nowadays, we have quite sophisticated techniques for fracture studies. I urge that training in the use of scanning electronmicroscopy and Auger analysis be considered an automatic part of courses because these are certainly techniques which you find widely applied in industry. It is also important for students to realise that it is on extremely rare occasions that we find a situation in industry where we are concerned with a single crystal being deformed in uniaxial tension. More often we have, of course, a polycrystalline, multiphase, multi-component system subjected to a variety of stress modes. It is generally in contact with some other metal and, therefore, there is a potentially active electro-chemical situation arising. Its protective paint film is invariably scratched and it is probably exposed to some complex environment which, it seems to me, invariably contains some totally unexpected aggressive species such as chloride ions, sulphide ions, copper ions.

In our laboratory, which serves a quite diversified company, we are concerned with everything extending from low grade technology, such as blasting technology, intermediate technology, such as the extraction of aluminium and its production, and high grade technology, such as putting a Viking Lander on Mars or building a satellite. I think eight out of any ten times that our materials engineers are called out to investigate a problem concerned with component performance or a failure, it turns out that the problem is related to corrosion, stress corrosion, or corrosion fatigue. Very rarely is the problem concerned with strength or fracture toughness. It seems to me, then, that any course on fracture for engineers must contain an adequate consideration of environmental factors.

This brings me to the second aspect of what I think the course on fracture should contain. This has to do with the prevention and control of fracture. The keys to prevention and control are proper design of the component and appropriate material selection with respect to the environment to which it will be exposed. The third aspect is the use of protection of some sort by inhibitors, surface films, or control of the chemistry. This is another area which, I think, is not yet being adequately handled in most university courses on fracture. Design is usually well-covered and a design problem rarely is encountered in industry. Material selection is a much weaker area, and we frequently find a failure that has come about because of improper materials selection. I think it is crucial that our engineers think like engineers about fracture and analyze it in terms of a systems approach. The components of the system are the structure, the state of stress, and the environment. I think if you neglect any one of these three components it will probably be to your detriment.

Finally, I would like to turn to the application of fracture processes. This is a subject close to my own heart and, I think, largely neglected by scientists and teachers at present. Fracture is sometimes a beneficial process, although the usual concept of fracture is that expressed by Sir Alan Cottrell. He said that "fracture is a very attractive subject because of big engineering failures and that sort of thing". I think this is the common concept which sometimes leads us astray in this field. In my view it is the little fractures that are more often of account, and by little fractures I mean the small and usually not well-controlled fractures which are used in such industrial processes as blasting, comminution, grinding, drilling, machining and polishing. Here fracture is a critically useful industrial phenomenon. However, we spend most of our time attempting to prevent it. "How important is it?" you might ask,when we want it to occur. In the United States, for example, we spend about 5% of the Gross National Product each year, which in 1975 amounted to 75-billion dollars, in fracturing things. This is done by drilling holes in mountains for rapid transit systems, or by machining, grinding, polishing and other industrial processes. I wonder how many scientists here have thought about fracture in the context of machining, grinding, or blasting processes. We also spend a lot of time being concerned with the fact that active or corrosive environments enhance the risk of fracture. In this case, we give our attention to the possibility of facilitating fracture when that is the desired end by the use of those very active environments that engineers otherwise spend most of their time designing against. I hope that the teachers of fracture who are here will begin to think about this different view of fracture in what I have called applied fracture and perhaps tailor their courses differently. To cite a few examples of applied fracture involving Soviet tests as well, people in this field are showing that you can drill stainless steel, for example, 8 to 10 times faster by using certain liquid-metal environments, or you can drill titanium and certain steels four or five times faster by using an oleic acid environment. Now we do not know the reason for these effects but we do know that they occur. You can increase life by factors of four to five with the use of appropriate liquid environments. You could go on with numerous examples but very few of these are understood at all and in very rare cases have any of the sytems been optimized. My point in this regard is that there is need for consideration of the constructive use of fracture instead of thinking that it is always a detrimental phenomenon. This would be a useful concept to get across to students. If it should happen I would think of this as a means of improving productivity and so maintaining or even improving upon our standard of life.

McEvily : It seems to me that somebody has done a fairly good job of acquainting society with the subject of fracture, during the past 25 years. Recently, there was an unfortunate accident at the top of the Pan Am Building in New York City, involving a helicopter, and people were killed. To my surprise, I noted that the headline of the Daily News the next day said in bold letters, "Metal Fatigue Causes Crash", which seems indicative of a certain awareness in the world at large. As a consequence of this development, as well as of developments in product liability, and its attendant publicity, students are becoming more aware of fracture. They realize that, in fact, it can be a career unto itself, as evidenced by Alan Tetelman's group, Failure Analysis Associates, and they appreciate that prevention of fracture is a societal need in which they can become involved. I have noted this development in the 10 years I have been at the University of Connecticut.

In this ten year period, the University has offered a course entitled, "The Metallurgy and Mechanics of Fracture". It has been a course which has always drawn students from a variety of departmental backgrounds : civil engineering and mechanical engineering, as well as metallurgy. In offering the course even at the graduate level, the subject becomes somewhat simplified. The students have available to them a number of texts. I might mention the text co-authored by Alan Tetelman and myself. More recently, we have available texts by David Broek, by John Knott and by Richard Hertzberg. At the undergraduate level we are still relying very much on a broad treatment of the subject of mechanical metallurgy. I use the text by George Dieter, which is now in its second edition. We are fortunate in the sense that our department is a relatively new department at the University, and we did not have to dislodge some other course to make room for teaching fracture, and we could put fracture in right at the beginning. We have also done some other things, particularly at the under-graduate level which have been designed to inform non-engineering students about this subject. We have a course taught by Professor Gallagan entitled, "Technology in Modern Society". The objective of this is to bring up some of the concerns about nuclear reactors, to explain what the problems might be, to acquaint students with the engineering approach to solution of such problems and to consider fracture of such structures as a matter of topical concern. The subject of material selection was also mentioned. We have recently offered such a course. It was a very popular course, taken by a variety of engineers in their fourth year where they could have the satisfaction of seeing the application of the principles which they have been taught in other courses. The course, which relates very closely, I think, to the Open University approach, seemed to be a very popular and worthwhile venture.

Another new area, which is being explored by Professor Greene, whose interest is in corrosion, is the video-taping of standardized experimental techniques. This is a very valuable teaching aid, particularly at the graduate level. There are certain routine but exacting techniques involved in electro-chemistry, which can be carried out properly by a graduate student and taped and so made available to his successors. I do not know whether ASTM has been involved in this sort of activity or not. It would seem to be a very worthwhile thing to have the procedures involved in, say, the specification of fracture mechanics testing, i.e. E399, presented on video-tape so that people could understand better what these procedures are.

Another interesting feature of fracture to me is that it is a very dynamic field in contrast to some of our other graduate programmes. I think of the classical subjects, such as thermodynamics, as being rather static : one might even use class notes of some years ago in teaching such a course, even at the graduate level. In our curriculum the information is relatively new : this is evident from this particular meeting. The concepts are still under development, which makes it a very exciting field. It is important for the teachers of this course to be well-versed in the field in order to do a good job. The other rewarding aspect, my final point, is that the demand, at least from our department, for graduates with backgrounds in fracture is as high as in any other area. This seems to reflect a strong need for this type of training as far as industry and other universities are concerned.

Hornbogen : My remarks refer to the West German situation. We have no formal course on fracture in West Germany, and there is no textbook. Most Germans in the fracture field tend to regard themselves as experts. Three groups can be distinguished : those who are doing creative work in fracture; those who know the state of the art and design and apply it successfully; and, those who work without adequate knowledge. In order to improve this situation, in a country like ours, I think we should first concentrate our teaching efforts at the graduate level, by setting up schools and courses which will serve somewhat the requirements which Dr. Westwood mentioned. There is a need to set standards for what a fracture expert should have as his educational background. This could be done, perhaps, on the basis of results from the two or three competing graduate schools which would train these experts.

A second point of concern is the degree to which undergraduate engineers should be trained in this field. Here we are in conflict with the tendency to reduce, rather than increase, the number of courses. One possibility I can see is to include fracture and failure in a course given for the different types of engineer just before graduation in their third or fourth year. A course which also deals with material selection and dimensioning of parts is important. In this type of course, also, I think we could include failure analysis and some information about understanding fracture. Otherwise, I do not see much hope of including a special course on fracture in the field of engineering. There is a problem because fracture must compete with two other fields, wear and corrosion, in importance and in popularity. I think if we are to consider fracture as a field, we have to consider wear and corrosion equally, as they seem correspondingly important for society.

Armstrong : Now we should turn to the floor. We will first try to run through all those who might wish to contribute and then come back to ask the panellists for their responses.

R. Eisenstadt, Union College : In light of current events in flight liability, flight safety and malpractice insurance for engineers, what is the absolute minimum information we should supply for undergraduates or for feedback courses in this regard? I cannot see how fracture courses can be offered to undergraduates without some level of minimum information being given, in order to alleviate some of the major problems that we are having.

F.A. McClintock, M.I.T. : I wonder if we are not putting this matter somewhat backwards with respect to education. Perhaps we should look at the product, namely, a design engineer working at his board, a project engineer, or a failure analyst, and ask what he will need to know in order to solve a particular problem on fracture. This is likely to be a good deal more than fracture alone, and it is also likely to be more than he learned when he was an undergraduate or graduate student. How do we make a living body of information available so that people can draw from it the specific answers that they need, and put this information to work? If we could design such a structure of information and learn how to modify it other than by the use of short courses, by conferences, and by all the things we have now, including books, then perhaps we could turn this process around and start teaching our graduate students and our undergraduates to use that body of information. This would give them a little practice in digging into it here and there. Now it seems to me that perhaps this EMMSE project for Educational Modules for Materials Science and Engineering at Penn State falls into this category. I would like to hear what people's comments are about the success of that project, and whether it fills this sort of need. My own feeling is that it is a little too polished. It is too hard to prepare things in the beautiful detail that we have seen here from the Open University and from EMMSE. Nevertheless, perhaps these indicate the right ways to go.

V. Weiss, Syracuse University : I would like to reinforce points made by Professor McClintock and Dr. Westwood. It is clear to many educators that an undergraduate Fracture course is not very likely in the near future. It is also clear that no engineer involved in design, mechanical engineering, civil engineering or aerospace engineering should be able to graduate without knowing something about fracture. It is equally clear that the phenomenon of fracture or failure is a part of metallurgy, a part of materials if you are in ceramics, a part of mechanics in terms of stress analysis, and a part of design. Thus it is necessary that this topic be taught as a part of these courses. I feel that the problem that we address in meeting the industrial or market requirements, whatever you will, is that we are in the position of having to capture the interest of people who will

247

teach these courses in fracture. Now the reverse would clearly be true if I were to teach a course in mechanics of materials or a related area. I cannot see how I would teach it without involving fracture in an important way. However, if somebody who has not been as close to fracture as all of us have been, were teaching this course, it might be hard to include fracture. There is a textbook on mechanical behaviour of materials that is used in some undergraduate schools which does not even include a chapter on fatigue. I think correction of deficiencies of this sort is of high priority in our effort to improve the education of undergraduates. It is important, too, that they get a multi-disciplinary approach.

Armstrong : Do you want to respond directly to that, Dr. Westwood?

Westwood : I have a brief comment on what might be done to prepare undergraduates for coping with practical problems in industry. This is based on our experience in industry of being called out to look at numerous failures in the field. I have observed that people with doctorates, plus perhaps fifteen years' experience and numerous published papers, invariably use Metals Handbook at times like this. It is remarkable how the wealth of information in that book has contributed to solving so many practical problems. I feel that it would be very useful if engineers would learn to use books like this. It may be that parts of their courses should involve questions for which they would simply have to go to the book to find a solution, just as most people in industry have, in fact, been doing. In this way they would learn the trade.

D. Felbeck, University of Michigan : There is really so much to say on all that has been brought up that I hardly know where to start. A word to Deans, Associate Deans and Chairmen : if you want fracture taught, you have to start by hiring people who know something about fracture. Older staff tend to be cutters, polishers and etchers, whose only concept of failure analysis is looking for inclusions and porosity. It is unrealistic to try to teach them anything about fractography and fracture mechanics; much less expect them to teach students. I think that would be a disaster.

We have drawn several departments into the operation of teaching fracture at the University of Michigan. We now have a graduate course in fracture mechanics that is taught in the aeronautical engineering department. We have a senior level graduate course in failure analysis which involves case studies and involves practical experience on real-life, real-time, failure analysis in industries in the Detroit-Ann Arbor area. These students are able to use an SEM as a tool in their failure analysis. Lastly, we teach fracture mechanics for about two weeks in our sophomore-junior level course, the latter stages of this course being on mechanical behaviour of materials. The tragedy is that there are people in some other departments, for example, engineering mechanics, who teach a thing called "strength of materials", but never talk about cracks and, also, that in my own department there are people who teach an engineering design course but never mention fracture mechanics. They teach design according to the old standard based only on yield stress. They talk about fatigue only in terms of Goodman and related diagrams. They do not go any further than that. We are now graduating engineers whose only contact with fracture mechanics has come in a materials course rather than in a design course. I think that at this time of rapid change the people who are teaching these courses are perhaps too old to do it, because they do not intend to learn any fracture mechanics themselves. At our university, and possibly at many of the universities represented here, one of the saddest features is the low hiring rate of the last five years and probably the future ten years. I see no way of improving and enlarging the teaching of fracture when the

staff is in general approaching retirement and knows less and less about
the subject, of course, with a few notable exceptions. It is quite significant that there is virtually nobody coming into the Universities. To
reiterate, I think that the word must be carried to the Deans and Admini-
strators of the Universities. They are in need of some of the young men
in this room, perhaps, to come in and teach more about fracture. They are
the ones who know most of what is going on in this subject area and could
do the most effective job.

J.R. Rice, Brown University : I have some philosophical differences with
some of the recommendation we have heard. In fact, I am a little concerned
that one can detect an almost "trade school" mentality in some of what is
being proposed as education for engineers. By that I mean an emphasis on
rather highly specialized techniques, which are almost sure to be outdated,
probably not very long after our graduates leave us. I feel that the best
way to teach Fracture is indeed to continue to demand, as I hope we are all
doing now, that our students have the most fundamental, rigorous and demanding exposure that we can give them to things like the science of
materials and the mechanics of solids. Of course, in presenting courses
like that we must mention the fundamentals of the subject of fracture, but
mainly we want to train people who can think, who can look at ideas and
evaluate exactly where their boundaries of knowledge are. We hope they
can figure out what they do not know and can plan a route to get the additional information and answers they need. I think it might also be worthwhile to reflect a moment on those who have made really seminal contributions to this field, both at a fundamental scientific level and by pointing
the way to really innovative engineering applications, and ask whether the
production of seminal thinkers in the future would be best served by a
technique oriented education or one which continues to put the emphasis on
the fundamental engineering and physical sciences that underly fracture
mechanics.

J.A. Alic, Wichita State University : I would like to offer a partial reply
to that by repeating some of the earlier remarks. I have always felt that
the seminal thinkers in any field get that way without much help from those
of us in teaching. I think what we need to do in teaching fracture is to
reach the practitioners. There is no doubt that, in Wichita, where I teach,
fracture mechanics in the aircraft industry is at the more or less routine
day-to-day level. The students who are graduating from Wichita State
University and from many other Universities in the country will be doing
fracture work in future. I agree that there is not a place for a fracture
course in the undergraduate curriculum. I think our curricula are far too
fragmented already and I support very strongly the incorporation of fracture
topics into design courses, mechanics courses, materials courses and so on.
I have been heartened to see that most of the more recent materials science
and engineering textbooks do include material on fracture mechanics. It
is unfortunate that the design texts do not, but I think that will inevitably happen. I think what we need to do is to get the word to our
colleagues. The people we have to convince are not in this room,
unfortunately. We are all specialists to some extent and it is the rest
of the engineering community and teaching community that needs to hear our
message.

I think also that there is a place in the materials testing laboratory for
experiments that deal with fracture mechanics, particularly with the use
of feedback controlled testing machines. There has been a phenomenal growth
in the use of that sort of equipment in industry. I have brought a slide
that illustrates this. The slide shows the exponential increase in the
use of servo-controlled testing equipment over the last ten to twelve years.

I do not think it makes very much difference what kinds of experiments we do in the laboratory at the Universities, but it does matter that we use modern equipment. I think we need to begin to do that for the very good reason that there is no doubt that our graduates will be using such equipment for many kinds of applications in the future.

M.W.T. Spencer, Alberta Gas Trunk Line : I am one of the objects of this discussion : I am an ex-student. The Universities tend to feel that the teaching process ends at graduation. After that, communications between industry and University are relatively poor. They may occur on a consulting basis, but that is not very educational. The education of an engineer or of other persons who are working in industry is taken over then by other institutions such as the ASM. I believe there should be a continuing input from the University in updating industrial people in the current developments of their particular fields. It seems that most people who are here agree that there is not room in the undergraduate curriculum to include a fracture course independent of other courses. That is probably true, but when a person goes into industry there should be some attempt made in one way or another to keep the lines of communication open so that that person can get hold of the necessary expertise or the information that would help him to develop this particular expertise. I agree that if we attempt to give engineering courses on specific topics at the undergraduate level we are going to lose some of the essential fundamentals. I think that is the rule of the University now and it should be maintained. There may not be room during this time to give the person that specialist training. That has to be done afterwards, and thus it is very important to maintain effective channels of communication.

Armstrong : I propose now to take some responses from our panellists.

Knott : First, I would point out that, in Britain, fracture is being taught in materials departments and engineering departments at the undergraduate level. I think it is important that fracture be an accepted topic in an engineering department in the sense that it is thought of as part of the proper engineering teaching. I also think it is proper that materials be treated in the same way. I see these as rather similar products. It may be difficult if Fracture gets into a course, but is thought of as not being "real" Engineering. I think it is much better that it should go into the design and structural engineering parts of the course. This can be done as part of a general failures appreciation course through case studies or through formal teaching. It is very much a question of balance and selection and here I would take some issue, I think, with Professor Rice. Of course we want rigour and challenge and excitement in our undergraduate courses, but we can get carried away at the undergraduate level with oversophistication in what we teach them. I think that what we really ought to work towards on this particular topic is what Sir Alan Cottrell referred to as the "science of materials in service".

Hertzberg : I have some comments related to the question that Professor Eisenstadt raised and to some other points that have been made. I definitely feel that the idea of product liability should be, at least, introduced in the undergraduate curriculum. This could be done either in a course that deals with fracture topics or in a course that deals with engineering professionalism. This is where the student is exposed to what it means to be an engineer and is made aware of the career opportunities which are available. Some schools have this, including Lehigh, where our department is concerned with this. I think that perhaps one of the most profound things that a student can become acquainted with is the concept of strict liability, where no one has to be negligent. All you have t' do in this regard is to establish the fact that a structure contained a defect and that someone was hurt as a result of the defect. I think the student really learns something from such considerations.

With regard to a comment that Professor McEvily made, one could also use video tapes for *in situ* experiments involving electron microscopy, particularly, with regard to fractography. I have been working along these lines and I think it is very useful to bring the instruments into the classroom.

Regarding Professor Rice's remarks about fundamental versus applied concepts, I think one has to recognize the fact that the Universities provide a service, and one must therefore consider the market that seeks such service. In the future, there is going to be a demographic redistribution of the available student body. Many projections that the Deans are well aware of show that there will be fewer students going to college and, thus, the student body that one will have to attract will be increasingly a student body derived from industry. We will, therefore, have to design our courses to be more responsive to the needs of industry and that will involve some emphasis towards the more practical things.

Liebowitz : Regarding the comment concerning curriculum content and the fundamental aspects/practical aspects balance, I would be very concerned about any master plan to be used by all Colleges and Universities. Each College and University has a certain resource. It has its own specific objectives, whether it is a state or a private institution. I think it would lead to a problem if we trained all engineers to be of the same fabric. I think mixture and diversity in engineering education is very very important. I would hesitate to see any one curriculum being followed throughout. Now, concerning the question raised by Mr. Spencer on lines of communication between Universities and students upon graduation : there are some countries which have been very active in the field of continuing engineering education. I think that is one of the many ways that the University can be in touch with former students. I think there was a very interesting point raised by Professor Francois on our panel concerning the importance of work experience. Certainly those institutions that have co-operative education arrangements can look to ways in which fracture experience could be gained by their students.

Reid : I want to say that I certainly agree strongly with Mr. Spencer about the fuller involvement of Universities in what we in Britain call "post-experience education". I think, certainly, speaking of British Universities, we could do much more than we do now in this area. It is really the exception rather than the rule that Universities are active locally in this area. There is no limit, I think, to what can be achieved here and, certainly, from my own point of view, I think the ultimate future of the Open University, perhaps, lies very largely in this direction. We have no guarantee that the current very gratifying flux of students will keep coming through our gates at the same rate. Indeed, there are some plausible reasons why we might eventually come to teach our way out of business, as people's ambitions become satisfied. That is an imponderable point, but it certainly is a possibility that the undergraduate education, which at present takes up the majority of our energies, may decline in relative importance, and yet, in the post-experience area, as I say, there is no limit. I think we will probably develop in that area over the years, although certainly our brief from the government at the moment is to give absolute priority to the undergraduates. This will apply only as long as the undergraduates are there.

Things are happening already on a modest scale. For instance, some of my

colleagues in the Energy Research Group at the Open University are at
present planning a short correspondence course that will be launched
nationally quite soon, called "Energy in the Home". This will have quite
modest educational objectives and it will be concerned really with helping
people to understand the physics of heating and lighting homes, so that
they can do something about saving money and saving energy in the process.
There is no reason why another such example in the future could not be the
presentation of fracture mechanics to those who graduated and underwent
their training before that was a recognized kind of analysis.

Westwood : One of the things that I have found difficulty in learning to
live with as I have made the transition from an active scientist to a full-
time manager is my own increasing ignorance. You get used to being an
expert and suddenly you find you are ignorant in just about every area you
are supposed to be involved in, but yet you have responsibility for problems
in specific areas where you are totally ignorant. What I am really saying
is that you cannot know everything in the same sense that a teacher cannot
teach everything either. How can this be compensated for? The answer is
to have some awareness of phenomena, and some awareness of issues, but not
to know exactly all there is to know about any of them, simply to know that
they exist. In other words, I think that it is important that an engineer
knows about such things as fatigue, creep and wear but he does not have to
know all the details involving the mechanisms of those things.

Secondly, it is important to be able to follow on from knowing that these
things exist to being able to do something about them when a problem arises.
The principal way to do this is to learn to seek help. You have to learn
to ask questions and you have to learn to draw on the resources available.
I find in industry that engineers simply do not do this. They do not read
the engineering literature. One of our staff, many years ago, did an
analysis of how many times the engineers ever went to the library or ever
asked for textbooks, and it was remarkably low considering the incredible
problems which arise every day. Perhaps this indicates that more must be
done to teach our students to go to the appropriate literature and hand-
books and to make use, too, of the research laboratories and the University
professors. The students must not be afraid to ask questions. It is
rather important that we know what we do know well, but it is also important that we recognize what we do not know, and perhaps do not need to know,
because there are very probably people who do know, and, therefore, we must
learn to question them.

Knapper : I certainly endorse the last comment. I do, however, think that
the skills that have to be taught are much broader than the discussion
this afternoon has suggested to me and will no doubt be brought out in the
second Panel on the overall political ramifications of fracture including
the social, moral and ethical considerations. The engineer of the future
or the materials scientist of the future will be expected by society to
consider all these, and that is a real challenge for the educator.

Armstrong : I am sorry that we have run out of time. I surely feel that
I have learned a great deal from the discussion, from the panellists, and
from the active participation of the audience. I certainly want to thank
the audience for their attention and constructive comments and, on behalf
of the audience, I want to thank the panellists for all their effort. I
suggest that we close with a round of thanks for the panellists.

Taplin : Thank you, Dr. Armstrong, for your very constructive Chairmanship.
The second Panel on Friday afternoon will address the wider public issues
in Fracture Problems and is entitled *Fracture, Politics and Society*.

FRACTURE AND SOCIETY - PART 2

Edited Transcript of the Second Panel Discussion on

FRACTURE, POLITICS AND SOCIETY

Nichols : This afternoon's panel discussion is on *Fracture, Politics and Society*. It is perhaps appropriate that this comes at the end of the Conference, if only because we would otherwise have spent the whole week talking on this single topic. We have a varied and enthusiastic panel set up here, as you can see, but I hope that we will have a fair amount of audience participation. To encourage this participation, I intend to open the floor to general discussion after each panel speaker, and only when the steam has gone out of that part, will we move onto another panellist, who will, I hope, move us onto a different topic. Unfortunately, because there is only one plane a day to Manchester I have to leave before four o'clock. If you see me get up and sneak off it is not because I have dis-agreed completely with the panel, and I apologize in advance for having to leave you. So, to make the most of my time I will spend no more of it on introductions. You will notice that I have not said anything relevant to the topic, and the reason is that, having read Max Saltsman's paper in Volume 4 , I was left with precious little that I wanted to say, since he had said it all for me. So, since Max is one of this panel's bigger guns, I will fire that gun off first, and make a start.

Saltsman : I was a little worried when you started saying "Fire Max Saltsman", there is an election in the offing and we get sensitive about these things. The main theme of this meeting, of course, is purely technological, but it is a sign of our times that the proceedings should include consideration of theme in political, social and educational terms. Politics and public opinion (I want to say something about public opinion shortly) now play a major role in deciding the level of support that can be given to technology, and while this may be somewhat unpalatable to research engineers and scientists, it is not wholly unreasonable that he or she who pays the piper should call the tune. At any rate, for better or worse, that is the reality of politics. I think another important reality of present day politics is that public opinion is everything. There are no big levers in government that can be pressed, that you can pull down and say, "I have reached the right people and they are going to make the decision". I have never seen politics and the political system so sensitive to what the public thinks.

To the extent that I have any advice at all to give to scientists it is simply this, that you must present your case to the public rather than to the politicians, with a reasonable assurance that, if you make your case to the public, and that is not always easy to do, the politician will res-pond to that, because there is a kind of running scared of public opinion that exists in Canadian politics. I should say that I am not an elitist about public opinion. I think it is quite intelligent, or at least that part of public opinion that is capable of being shifted away from traditional loyalties. It is not venal, it is I think just confused. It is confused by the conflicting claims that are made by the experts of our times. Thus, I think it entirely proper that I should address this question in my capacity as a politician, and it is not my purpose to be political in an absolute sense, but, rather, to try to give you a politician's view on technology in general and fracture in particular.

Professor Saltsman then presented his paper on Political and Social Decision Making in Relation to Fracture, Failure, Risk Analysis and Safe Design, which is included in this volume.

Nichols : This gives us a good start to the discussions. I would like to just pick on a couple of points which could be run together. Right at the beginning you made reference to the fact that, whilst you had admiration for the public, the public can be confused, or think itself confused, either through lack of information or from too much information. It wonders whether it can trust the politicians or the media; when it comes to the scientists, the public often gets contradictory stories. One of the points that I would like to hear a little discussion on is this question of how the public can judge whether the scientist that is speaking is indeed qualified to speak on that topic. I am afraid that often scientists offer profound statements outside their own particular field of expertise : how is the public to know that? This leads to the second point, your very interesting suggestion that we might consider having a technical court. I am rather wondering whether the same sort of thing would happen there. It would probably be impossible to have a single technical court which could deal separately and competently with all forms of expertise. Someone will have to answer the same question : How relevant are the people in the technical court to the particular problems that they are tackling?

Questioner : The usual procedure, once a court has been established, is that a member of the legal profession, typically a judge, is appointed to run that court. His selection undergoes some sort of scrutiny and his competence is widely accepted. The usual procedure is then to ask the judge to select from a panel who should sit in the Court. That at least solves the selection side of the problem. The selection of competent people does not then have to go through a long *ad hoc* procedure every time, as discussed in the interview with Sir Alan Cottrell.

Saltsman : That then raises the problem of whether judges trained in the law are appropriate people to make technical judgements. It is interesting to see what has happened in society. I think in all societies it is the case that new kinds of courts that are not called courts have arisen : for example Labour Relations Boards and National Energy Boards.

Nichols : They might even be called professional institutions. Very many of these now operate as courts in some cases.

Saltsman : Often they operate together, in situations where the normal legal training is not required to supervise the court, but rather the technical training in order to make judgements. For instance, recently, in Quebec, a judge refused to hear a labour case because he said the court was not competent to hear a labour case, even though it was standing before the court. Standing is important, and you can always get standing in another court if you disagree with technical decisions.

Averbach : A few courts have been established on quite technical grounds. There is, for example, a water court in Sweden which deals, on a very technical basis, with the distribution of rights to bodies of water, and use of water. Some of the considerations are political, of course, but the courts are quite separate from the normal legal system. We have regulatory bodies with specialized judges, but I think that the Swedish experience is quite unique in this respect.

Saltsman : So they have taken riparian rights out of the legal system and turned them over to special courts.

Averbach : Yes, and they have existed for quite some time.

Hahn : In the United States the National Academy of Sciences and the National Academy of Engineering mount committees which make technical evaluations : for example, in the case of Freon used in aerosols and its effect on the environment.

However, I think the problem in these situations is that there is a sudden public controversy, and an informed and trusted opinion is needed right then and there, whereas these committees take two years to come up with a report and a collective opinion. Meanwhile, people pass judgement who perhaps are not qualified to speak, and I see this question of time scales as a real problem. Public opinion would like to be served very quickly, but the scientific process takes a long time.

Questioner : I have a question that is really very central to what the Chairman had to say about how to establish credibility, in other words, whom does the public believe? This is terribly important if we talk about nuclear energy. On the one hand you have a group of scientists who say, "Do not move in that direction at all"; you have another group of scientists who say, "There is no harm, we have everything under control". Whom does the ordinary layman, and a politician is to some extent that kind of a layman, believe? We have a very good organization in Canada : the Science Council of Canada, although they tend to be rather circumspect, perhaps a little less today than formerly. Even with such a body in existence, how does a scientific community agree amongst itself on what it should put before the public and whom the public should accept as authoritative. This may be a tentative authoritativeness, able only to say : "Up until this moment this is the state of scientific knowledge."

Questioner : I see a problem in the tendency for the public to ask for absolute truth. As Max Saltsman commented, the general public has a different time scale from people in science. We cannot produce answers immediately, as the public requires. Thus opinions are often taken as being absolute truths, whereas, in fact, many are highly arguable : argument being one of the bases of our profession.

Nichols : Certainly people may read more into statements than is meant by the speaker. I think the classic case of this is the man that goes along to his physician and then comes away and analyses virtually every sentence, and reads into it more than was intended, and, in all probability, his physician was speaking off the top of his head, giving an instant opinion. I think we should now turn to another topic, and I am going to call on Dr. John Knott to talk about his interview with Sir Alan Cottrell.

Knott : Before I do so, I might point out, apropos of the earlier discussion on the membership of technical courts, that it is technically quite possible at the University of Cambridge to read two years' engineering followed by one year of law and become acceptable both to lawyers and to the professional engineering institutions.

The aim was initially to try to get Sir Alan Cottrell to come and address the meeting, as being a person who has spent a lot of time in high level scientific research, and more recently has been very much involved in government decisions on scientific matters. Unfortunately for the Conference, he has become Vice-Chancellor Elect of the University and is unable to come at this time, and presents his apologies. I am going to pick out some of the points which I think need to be mentioned, although some may not be particularly relevant to the broader theme. The interview itself was very loosely structured. We started with a list of questions which had been provided by various people and these were in no particular order, so that the logical connections between various parts of the interview are perhaps not as good as they would be had we had three or four recording sessions, but we touched on a large number of points.

The first thing that we were concerned with was his views on the science of fracture. Here, with the exception of some remaining interest in the surface energy of iron in fracture processes, Sir Alan tends to regard the *basic science* of low temperature fracture as more or less complete, whereas he still feels that there is a lot of work to be done in understanding fatigue, stress corrosion and interactions of various sorts. Sir Alan, of course, is a man whose science is of a rather broad nature. He likes to treat materials as fairly simple continua and it may be that there are details in low temperature fracture, particularly where embrittlement is involved, where his overall judgement may not be correct. But the point behind this, I think, is one of the support of university research, particularly in Britain. He says that, for the more complicated problems of fatigue and stress corrosion, "I think the only way that you can make progress with that sort of problem is to have a healthy university research environment and let people get around pretty freely to exchange ideas." So that what he is saying there is that there is a need still for some fundamental research in universities to try to understand some of the more complicated problems. That is one point on the science.

The second one is on the application of scientific knowledge to engineering design. His view here is that, with one or two notable exceptions, our application of scientific knowledge to design is not all that good. There are some major exceptions : the plastic design theory (initiated by Sir John Baker) and fracture mechanics (due to George Irwin) - a concept that he thinks is of considerable value is "leak before break". He is, I think, not as enthusiastic as I expected on the philosophy of probabilistic design and things of that nature. I was also rather disappointed that he had nothing to add in the way of examples of areas into which effort could be usefully put to improve the translation of scientific concepts into design. I think there may be lots of fields which one might regard as being semi-empirical still - where we are using data from e.g. SN-curves in fatigue - where perhaps one may say there are still things where the science can aid in engineering design. Cottrell is less strong on these points than I envisaged.

The third point is the teaching of engineers about fracture and materials. This was covered in Wednesday's discussion and I do not intend to bring it up any further except to mention the point that he makes about the materials men being brought in at the engineering design stage. The factor that he emphasizes is that the materials man has not found it

really attractive to make a career in helping the engineer to design
things because, "I think the materials man has known that he would always
be only an assistant in that kind of work. He would never become the Chief
Designer and he would never become the head of the firm. It does not
prove such an attractive avenue for materials people as some of the other
careers." That is perhaps a point to bear in mind. Towards the end of
his interview there is specific discussion on the areas into which research effort should be put : whether one in fact should be dealing with
fracture toughness and fracture mechanics or whether the stress analysis
or non-destructive testing sides are the areas where we would get the
most benefit from directing our research.

I think that the point that will be of most interest to the present
audience is his view generally on the avoidance of failures. This is in
two parts, one of them being on duty : whose duty is it to ensure that
things are done properly and that safety standards are maintained? Does
one use inspectorates, the institutions or professional bodies? What
should be done about design codes and what is the responsibility of large
companies in maintaining their own safety standards? Some points were
made on the use of materials and on the rather large number of fairly
similar materials that we use and whether it would not be better to try
to rationalize this by concentrating on specific materials and learning
a lot about them. I think there will be most interest in Sir Alan
Cottrell's views on nuclear pressure vessels and on alternative forms of
energy supply and the safety of these in current use. This follows his
comments on the *leak before break* concept and the importance of this and
the way in which he felt that this was a protecting feature which the
light water reactor did not have. He had commented in letters to *The
Times* and elsewhere that the situation was such that it was necessary to
be ultra-critically careful in terms of inspection and quality. Sir Alan
says, "I think that the specifications that the Americans have set for
their water reactor pressure vessels are extremely rigorous, there is no
doubt about that. If human frailty is able to achieve that degree of
rigour in practice then they will be all right. But you must always have
a question mark against human frailty and this is the thing that worries
me, whereas with the pressure tube kind of reactor, again you have to be
just as good as you can be against human frailty. Nevertheless, if you
are let down by human frailty then you have got a natural back-up, the
leak before break. That is where the difference is, and I still feel
strongly about that point."

We followed with a somewhat more searching set of questions as follows :
"Do you think, because of the emotive word *nuclear*, that more attention
is given to your commentary on the nuclear reactor case, than is given in
the equally worrying ecological case of having large pipelines running
hundreds of miles across the bottom of the North Sea with large amounts
of oil running through them, where a split could again be equally disastrous?" Sir Alan replied : "I think so, yes. My own position on that
specific reactor problem does not reflect any sort of general position
that I have about nuclear power. In general I feel that politicians and
the general public are being taken for something of a ride by the environmentalist lobby which has been going very hard against nuclear energy. I
feel that this is an extremely unfortunate development because the only
assured new major source of energy for the world in thirty years'
time or so is nuclear energy. And to turn one's back on that without
very very good reasons, could, I think, be a disastrous step for mankind.
I think that the fossil fuel position, certainly in the Western World, is
really alarming. It is much worse than it is said to be in the newspapers.

We in Britain are locally in a good position for oil since the North Sea will give us what we need for the next twenty or possibly forty years. But if you go outside Britain then the position is really alarming and we may already have left it too late. The only way out of this situation is the nuclear one. I think that the environmentalists have served the Western World badly with their overdone campaign against nuclear energy."

The next point that we put was whether in fact a double standard is applied in the assessment of risk, in that we are asking the nuclear people to fulfil criteria of safety which are much more stringent than for equally worrying problems such as pipelines. Sir Alan makes a very good point : "This is true, and it is true of other things. A highly dangerous source of energy is hydroelectricity. You have the big dams and if a big dam bursts it could not only take out enormous acreages of ground but could drown large numbers of people. On the whole a big dam bursts about once a year and these as incidents are large scale, even by the standards of the worst imagined nuclear reactor incident." I have quoted these parts because I think they are most pertinent and of specific interest to the present session. I am sorry that you have had to receive these views by proxy, but I hope I have been able to convey Cottrell's points fairly.

Nichols : I happen to know that Jim Justice from Trans Canada Pipelines is here, and I wonder if he has any views on the treatment he receives relative to the nuclear people in this respect.

J.T. Justice, Trans Canada Pipelines : I know very little about the nuclear regulations. I feel that we get a considerable amount of regulation, and in most cases I think it is very fair and just regulation. I understand that nuclear regulations are much tougher, but I think that Dr. Mills is probably far more qualified to answer the question than I.

Nichols : You do not feel as though the public ignores any risks in your pipelines?

Justice : No, I do not think so.

Nichols : Dr. Mills, would you like to say anything about either big dams or nuclear?

Mills : I have a few remarks on nuclear issues prepared, and this seems an appropriate moment to make them. When I looked at the title of this discussion, I made a note of my own as to what it might be. My title was "Fracture, Ontario Hydro and Society", and I seem to have that right with two fracture experts, then Ontario Hydro, and Max Saltsman representing society on my right. There are certainly many in Ontario who would find Ontario Hydro synonymous with politics. Ontario Hydro's interest in fracture is shown readily in its concerns with major plant. From the very first conception of a plant design Ontario Hydro seeks to ensure four things:
1) Safety of the public.
2) Safety of operating personnel.
3) Avoidance of economic loss due to plant shutdown and repair.
4) Enhanced reliability due to fracture resistant design.
I would like to comment briefly on each of these topics with examples (illustrated by slides).

My first topic is public safety, and we, in particular, as proponents of nuclear power, have been very sensitive to the need to assure the public

that our plants are safe. In the Canadian CANDU reactor system we employ
5 boundaries between the public and the nuclear heat source. Firstly, we
clad our nuclear fuel in an alloy designed specifically for that purpose.
Secondly, we contain the primary coolant of that fuel in a pressure
boundary which is subject to rigorous quality control during construction
and which receives regular periodic in service inspection. Thirdly, we
contain that primary pressure boundary inside a thick concrete reactor
building. Fourthly, should this reactor building ever become slightly
above atmospheric pressure, it is automatically connected to a vacuum
building maintained at low pressure and containing a dousing system to
condense steam. And fifthly, we build an exclusion fence around the plant
such that no one at the fence receives radiation doses higher than he or
she would receive from natural sources. Thus our philosophy is to put as
many boundaries as is economically possible between the public and the
nuclear heat source.

Regarding the safety of our operating personnel, we cannot expect them to
do their job in anything but a safe working environment. And there are
now very specific regulations in Ontario which lay the onus for plant
safety on the employer. This slide shows 2.44m sections of Schedule 20,
340L stainless steel pipes which were hydraulically pressure tested to
failure for our heavy water plant operations to determine their burst
pressures and fracture mode. A stress raiser was present along the longitudinal seam in these pipes. In one case an internal surface flaw was
introduced using a milling cutter. These experiments showed that the
pipes can tolerate large plastic deformation and that final failure is
controlled by plastic instability. The burst pressures were also predicted using a plastic instability analysis to be well above anything
that they might see during abnormal operation.

Our third concern is economic loss, and the next slide shows a fractured
forged 'T' from a small boiler pressure equalizing line at our Pickering
station. The pipes running into the 'T' are about 10mm diameter. The
fracture of that 'T' resulted in a forced outage of 36 hours, resulting
in a replacement energy cost of $75,000. The downgrading of the heavy
water caused a further direct loss of $80,000. That fracture was thought
to be caused by a combination of stress corrosion cracking and fatigue
from tube vibration.

Our final concern is enhanced reliability, and we are engaged in an extensive programme of testing the fracture properties of pressure boundary
materials and both nuclear and conventional thermal heat transport
systems. The ASME Boiler & Pressure Vessel In-service Inspection Code,
Section XI, states that : components with flaws exceeding the normal
allowable standard may be considered acceptable for continued service if
a fracture mechanics analysis performed in accordance with the recommended
procedures of that section shows that the structural integrity of the
component is not impaired. An extensive fracture mechanics data base is
required to perform these analyses. We therefore have an experimental
programme in progress to determine fracture properties of primary and
secondary pressure boundary materials.

As with any large electrical utility, in operation there are always
failures of components, and we do have many interesting case histories,
although this panel is certainly not the forum for discussing these in
detail. Let me conclude by saying that Ontario Hydro, as, I hope, a responsive crown corporation, with significant influences on the economy
and technology of this province, has as one of its objectives the desire

to ensure that in the operation of its major plant there is safety for the public, safety for the operating personnel, avoidance of economic loss and enhanced reliability.

Nichols : I would like to ask you a question which, I think, relates both to your comments and to a point in Sir Alan Cottrell's interview; that is the comment about the difficulty of communications, of getting the information across. This is an area in which I have learned something over the past 6-9 months. Criticism of the nuclear power industry has at times reached such a pitch that it appears to be something that the public just does not want anymore. We are reaching a stage when virtually every person in my laboratory is encouraged to go out and talk about nuclear power. We all are presented with facts outside our own area of special expertise in a way that might interest the general public, and to enable us to answer questions that might come up in formal and informal discussions. Everyone is encouraged to talk over the garden fence, in the pubs, in the clubs, and as I told you last night, to actually go out to the Women's Institute, the schools and so forth. We have come to the conclusion that we must try to get our message across; we believe now that not only is it our duty to make nuclear reactors safe, but also it is our duty to convince the public that we are so doing. I do not know whether many of you have done this sort of thing. I believe that scientists as a whole tend not to regard it as part of their duty to talk to the public, and perhaps this is a problem to be faced, we will talk amongst ourselves but not to the public. Therefore, I wonder if you have any organization in Ontario Hydro for getting this message, on your attention to safety, across to the public. In other words safety must not only be done. It must be believed to be done.

Mills : There are two things that Ontario Hydro does. First, it provides a technical information service at its main head office where any member of the public can go to examine research reports and safety reports, which are submitted to our regulative authority. In fact, any document in the company, unless it is to do with a commercial contract, is open to the public. However, in the last two years there has been a Royal Commission sitting in Ontario and its duty has been to examine what Ontario Hydro should be doing in terms of electric power planning over the next ten to fifteen years. This Commission has received submissions from the general public, from environmentalists, from the many concern groups who have the general label 'anti-nuclear', and from Ontario Hydro itself, which is trying to make a case for an expanded or continued nuclear power programme. In order to understand Ontario Hydro's case on nuclear power we should consider what it costs Ontario Hydro to make 1 kilowatt hour of power. It costs us about five thousandths of a dollar to make a kilowatt hour with hydraulic power, and there is very little left in Ontario now. It costs us at Pickering approximately 9 thousandths of a dollar to generate a kilowatt hour. On our best coal-fired station it costs us about eighteen thousandths of a dollar to generate a kilowatt hour. At an oil fired station it costs us 28 thousandths of a dollar to generate a kilowatt hour. Thus, any movement that we are directed to make towards greater use of fossil fuel, as a large utility, is going to have severe impact on electric power costs in the future. This is a major concern as Ontario's economic base is its relatively cheap electric power, and, if that base is destroyed, there are going to be severe economic consequences, not only for Ontario, but also perhaps for the whole of Canada.

Nichols : We should not allow this to turn into the nuclear debate itself, but thank you for giving us those examples and indicating how your public relations is handled.

E. Von Bezold, University of Waterloo : I have a question on the matter of
public education. Mr. Saltsman pointed out that in this country the
decisions made by politicians are rather sensitive to public opinion, and
in some respects I would agree with him. My concern is with the facts
which are available to the public in formulating its opinions : how can
the intelligent lay person obtain the information to enable him to assess
properly, for example, the options in allocating resources for research
or the choice of nuclear reactors? Can scientific workers be relied upon
to act in the public interest and bring controversial questions to public
attention? Will the public receive all the facts?

Averbach : The point which you raise is central to our discussion here.
That is, how do we go about assuring the public that it is hearing something which is impartial? For example, we have just heard somebody from
Ontario Hydro and somebody from the U.K. Atomic Energy Commission. Each
may be impartial, but their affiliations are such that there will be some
suspicion of self interest. On the other hand, if we hear someone from
the Ralph Nader Office, he is also labelled with a certain tag, which all
may not accept as impartial. This categorization is an extremely difficult thing to avoid. Until we establish some kind of arms length approach
to these technical problems, which we can make understandable to the
public, we will not really ever come to grips with this situation. We
will simply argue ourselves to the point where the decision is made,
willy-nilly, by some political body which has made an assessment on some
quite different grounds.

Saltsman : I have a serious comment on establishing credibility : how we
should go out to the public and explain out position. I think the first
thing that is involved is to develop a better mental attitude towards
the public. This applies to politicians, the media, and I think to
scientists as well. There is a tendency to think of the public as children. Nobody is going to say : "We think the public is a bunch of 12-
year olds." but very often in private conversations or little asides this
is the kind of impression you get. There is a feeling that they are not
really going to understand what we have got to say and really we know
better than they know. I think you have to purge your mind. I, as a
politician, have tried to do this, and I think, in fact, that I have been
reasonably successful as a politician, because I have done that. It is
certainly true that there are 12-year olds out there, adult 12-year olds,
but there are many intelligent people as well. Those people are important and can be addressed in a very straightforward way. You do not
have to tell them : "There is nothing to worry about on this issue."
You can say to them : "There are certain benefits and there are certain
problems. We do not know everything about these problems, and there is
some risk involved". The people will understand that. They do not
understand it when you simply say that there is nothing to worry about
regarding a problem that is highly technical. That immediately loses
you credibility. People are prepared to accept certain risks. They
accept a risk when they drive a motor car, they accept a risk the minute
they start to live in society. That is not the only answer of course,
but to start with I think you have to develop a certain mental attitude,
or at least come to the conclusion that there is a sophisticated, intelligent public that can be addressed.

A.N. Sherbourne, University of Waterloo : I would like to make one point.
That is to say that not all the 12-year olds are to be found in the public

domain. Many of them are in universities and other organizations which should know better.

Nichols : We are all 12-year olds with regard to some subjects, are we not?

N.I. Adams, Nuclear Installations Inspectorate, London : First I should state that what I am going to say reflects my own opinion and should be in no way interpreted as a part of the policy of the Nuclear Licensing Authority of the United Kingdom. I have, in the past two years, whilst working in London, been intimately concerned with the U.K. assessment of the light water reactor system. I am also intimately concerned with assessment of the Steam Generating Heavy Water Reactor, and I would like to make a few remarks which stem from John Knott's interview with Sir Alan Cottrell and perhaps go on to mention public opinion. To sketch the relevant background, Sir Alan Cottrell was the Government Chief Scientific Advisor, and in 1974 a decision was made in the form of a government white paper that the next U.K. power system would be the home grown SGHWR. Sir Alan stated that he did not believe that it could be shown that a pressurized water reactor pressure vessel could be convincingly shown as safe for its entire lifetime. I have since concluded that that was his instinctive decision as a metallurgist with a great deal of experience. I believe he was right, at the time, in making that judgement, and I think that judgement was supported by the outcome of the Marshall report, which Dr. Nichols, in fact, helped to write. This report took something of the order of 2½ years to produce and concluded that, with quite a considerable number of improvements to U.S. Technology, it should be possible to have a pressure vessel that would be safe at the start of life. It further concluded that, given a great deal of in-service inspection, and provided that this can be shown to detect defects, the vessel should be safe during its working life, and that, I believe, vindicated Sir Alan's views expressed in 1974.

Sir Alan has, however, written to *The Times* more recently and expressed further views which show that on some other matters, whatever his feeling, he has got part of the technical story incorrect, and is making judgements on the basis of an incorrect interpretation of the real situation. This leads me to wonder whether we can really communicate effectively with the public at large. If eminent scientists cannot really get to grips with technical issues that are outside their own particular scientific sphere, can we really expect the public, without any scientific understanding at all, really to come to grips with the technical issues? I would agree with Mr. Saltsman that we can put across the major issues - the benefits and the problems - but I do not think that we can expect them to come to terms with the real technical issues. I think it is difficult enough for those of us who have to try to do it as part of our livelihood.

Mills : I would like to come back on this business of public acceptability. Perhaps we in Ontario Hydro have really suffered very little with regard to the nuclear issue as compared to many of the large private utilities in the U.S.A. We have made some attempt to increase public awareness of nuclear power. When the Pickering construction was half complete, and we had two reactor units operating as a nuclear island, the entire station was thrown open to any member of Ontario Hydro and his family. They toured the whole station including the turbine hole and the reactors under construction, the only exception being the nuclear island, to avoid any danger of contamination. I feel that this has probably done more to show that we have a reasonable system, which is put

together with care, than anything Ontario Hydro could ever do in print or
by publicity. I think this philosophy of showing the public should be
carried out more. If it is shown inside the system, while it is still
possible, I think it is much more acceptable to them.

Nichols : I would like to take the matter of communications one stage
further. I believe, Dr. Hahn, that you have a few points on this in
relation to the difficulty of one man speaking to another if they have
different backgrounds. I will take a point from the floor first though.

Dr. C.F. Old, AERE, Harwell, U.K. : I would like to offer a piece of information which surprised me when I first read it and which is particularly germane to the points which Prof. Averbach and Dr. Adams
raised. A survey was done, the results of which were published in " New
Society" and recently also in the House Journal of the UKAEA.   Although
it addressed the topic of nuclear power, in fact it dealt on a much more
fundamental basis with the credibility of engineers which Prof. Averbach
mentioned. A question was asked of the total sample, which was over a
thousand people spread throughout the population : "Whom would you believe if you were reassured as to the safety of a nuclear installation?"
I remember the figures roughly and something like 3 to 5% percent of the
sample said that they would actually believe the news media. Around 5
to 7% said that they would believe politicians. Approximately 17 - 20%
would believe the manufacturers of the installation and something in
excess of 60% would believe the qualified engineers and scientists in the
field concerned. This surprised me, and suggested that our standing may
not be as low as we might suppose.

Nichols : It might also suggest that the public is, as we have said, a
very wise public.

Would you like to go ahead Dr. Hahn?

(At this point Dr. Nichols left the meeting and Dr. Knott took the chair.)

Hahn : I wonder if it may pay to examine for a moment why there are
differences of opinion. I think that in some cases these are honest
differences of interpretation, but there are other cases, I think, where
the differences of opinion derive from differences in the interests of
the parties involved. I think that, in this general problem of science
and politics, we are dealing with many sub-cultures which have really
quite different interests, quite different values, quite different jar-
gons, we have already mentioned different time scales, and quite different
scientific IQ's, by which I mean the ability to discriminate between two
different scientific arguments.

I would take the public to be one sub-culture, and industry, technology,
management and invested capital together form another. It seems to me
that the basic interests of the public and of industry, say, are not
quite the same. I think industry is looking for a guaranteed return on
investment whereas the public is looking for the highest guaranteed
quality of life, and there are differences. Industry must protect its
competitive position. The public does not have this particular interest.
I see science and research and development as being another sub-culture,
and I think that within specialized fields people can have the scientific
IQ to enable them to talk with one another. Outside these fields dialogue
is very difficult, and I do not know that a great deal can be done about
that. Government, the regulatory boards and agencies, and the standards

associations constitute a further culture with somewhat different interests. I think that the Government interest in the last few decades has
been primarily in terms of supporting the economy and preventing unemployment, and while they talk about safety and environment, when employment and the economy are threatened, these other things lose importance,
to some extent with public approval.

Thus, it seems to me that there are some basic differences and conflicts
of interest : if, for example, you present the different cultures with
the proposition that you would like to make, say, a pressure vessel more
fracture resistant, you will receive a range of responses. The science
and research and development culture would say : "Yes, that is a great
idea", and they would present a bill for millions of dollars. (The world
wide investment in fracture mechanics, for example, must be at least 100-
million dollars, so that there is a tremendous amount of money involved
in technology, in science and in changing things.) The public would say :
"Yes, but please do not pass on the cost to us or raise taxes". Industry
might say : "No, it would hurt our sales and our competitive position",
and I think that is a perfectly valid viewpoint. Government may take the
position : "We see both sides, so let us move slowly, or let us do nothing
at all." This exemplifies the fundamental conflicts of interest, which,
I think, stand in the way of more rapid solutions to some of our problems.

It seems to me that it is first of all a task for the political system to
reconcile these cultures and to try to bring them together, and I see
several necessary elements in this process. A possible solution must be
found that is not totally destructive of existing investments, be they
emotional or capital. Leadership will be needed, and it is not clear
whence the best leadership can come - science, industry or government -
and frequently a lot of money will be required which may be a stumbling
block. Regarding possible innovations, at our private research institute
we have found that one mechanism for bringing together money to solve
research problems is to try to develop industrial associations, either
formal or informal, to bring many companies with a common problem together, so that each may provide some funding and make it possible to
tackle certain problems. In the past, some of these associations have
not been legal, for antitrust reasons. This is a situation where other
interests and political considerations stand in the way of pooling research money. Certainly a phenomenon of the last three decades has been
the tremendous growth of government involvement in and support of
research.

Another solution that I would like to raise in connection with failures
and safety is the possibility that industry and government might consider
purchasing hardware, aeroplanes or nuclear plants, not for delivery on a
certain date and then to be taken over, but for the whole lifetime. You
would purchase the ability to generate electricity, say, for 20 years,
and then if there were an outage, if the plant were to fail, the vendor
would have to come in and bear the expenses; consequently he would have
the incentive to worry about these things more, and perhaps spend more
on research and development. I think there is a difficulty there, in
that Government, at least in the United States, restricts the way the
cost of power is carried over to the consumer, and this makes extra expenditure more difficult. It would cost more to buy a plant on that
basis initially, and with the cost of capital, and the way these costs
are passed on, it is difficult, for reasons which are not too clear to
me, for the utilities to operate in this way. I think the government is
moving in that direction in, for example, the purchase of aeroplanes,

where it buys a unit on the clear understanding that it last for the expected lifetime, and here again, perhaps changes of laws are needed to facilitate different ways of doing business.

Knott : I think that you have highlighted two problems there, one being that of communication and the other the weighing together of conflicts of interest. Sir Alan Cottrell, speaking from the British viewpoint, clearly passes the buck to the Ministers for balancing all these various factors. Here is where the communications problem can enter, because the Minister himself, the decision maker, may have to become familiar with the technical arguments. Sir Alan says that he does not. He has to trust his advisors and they have to put it into language that he can understand. That is in terms like : "If you build it this way, there is a real chance of the thing breaking; that way, the chance no longer exists." The British system then, appears to be the technical advisor going to the Minister and the Minister weighing a technical argument, together with all the other public arguments. Perhaps I could ask Mr. Saltsman how it is done in Canada? Then perhaps one of you gentlemen will tell us about the American method of decision making.

Saltsman : I think that it is one of the great mysteries of Canadian politics, exactly how people arrive at these decisions. I think mostly by accident. I think one of the severe restrictions on public policy in this country is that nobody has sat down and tried to formulate long term plans in a conscious way, and we tend to live on crisis politics, simply responding to crises as they arise. I think, in that sense, North American experience has been considerably different from European experience, and mostly because we could allow ourselves the luxury not to plan. The resources seemed to be so vast, that we could squander, we could be wasteful, we could move from crisis to crisis, without really being in any serious trouble. When the decision was taken it was usually by government. We do have a strong form of government. Decision is usually taken by the executive, again in response to public opinion, and usually after the establishment of a Royal Commission to tell the government what they wanted to hear in the first place, but did not have the courage to say themselves.

I want, if I may be permitted this indulgence here, to talk about leasing, or rather the letting arrangement that Dr. Hahn mentioned. We do not have that problem, and, at the risk of preaching to our American cousins, I would point out that that is because the utilities are under public ownership in Canada. We are the boss and the user and everything else at the same time, and the kind of a conflict you describe does not arise. You also raised something else that was extremely interesting to me, in asking where the leadership comes from. It can come from any source, but I think one of the most effective examples of leadership in our time has been not an *either or*, in other words the politician or the scientist, but rather what we saw in the early years of the Kennedy regime in the United States. Here was a very articulate politician, whose credibility was, I think, bolstered very very considerably by the fact that he had surrounded himself with well-known and highly regarded academics. I think, getting back to what you were saying about public opinion polls, you will find the same thing about academics generally, that, while people may make a target of academics, whenever a political scientist polls the public to find out how people feel about professors and how credible professors are, they usually come out with a high degree of credibility. The cartoonists may have a field day depicting the eggheads advising the politicians, but, in fact, experience has shown that that combination of a sensitive politician with expertise from the intellectual and the university communities is really the kind of leadership that the public will respect.

Knott : This is done then, by setting up a Royal Commission that has these people on it and acts as a single scientific advisor?

Saltsman : Yes, this is true in Canada because of the different political system, and I think it is true in Britain too, as ours is very much like the British system. In America, of course, the executive can directly take upon itself advisors. They do not have to stay within the legislative process in order to pick advisors, they can select expertise from outside. In our system, that is a little more difficult to do, and whenever it has been attempted, all kinds of terrible things have happened, as with Walter Gordon's budget a few years ago. Therefore, we tend to use Royal Commissions, Select Committees, and we tend to pick generally very good people. Two of the references I had today were of such groups.

Averbach : We do not have Royal Commissions but we do have Congressional Committees which serve quite the same functions. It is interesting to note that, until very recently, Congressional Committees had no scientific staff at all to advise them. Scientific and technical questions were frequently not answered or turned over to the National Academy of Science for advice. I have testified before some of these Congressional Committees and I am mystified as to how a decision is ever reached. A wide range of opinions is usually presented, with some very technical, and others not technical at all. Someone in the backroom eventually sorts it all out, and a report is eventually published. This is a situation where we have failed to help the public to understand the problem, or to help our Congress and our Executive to arrive at procedures to assist them in scientific matters.

N.A. Sinclair, IBM, U.S.A. : First a point on leasing, the possibility of which you mentioned. IBM, as you may know, do lease their equipment, and, largely because of that, they have set up a whole division to handle product assurance, one aspect of its activities being the use of physical models to predict lifetime.

Before going to IBM I was a nuclear specialist, and I engaged in many debates about nuclear power with people that worked in the development of it as scientists. I saw the fallibility of inspection techniques, where the system may easily break down because of the human element. It has been known that, when the lunch whistle blew, an inspector has wiped off the magnetic particle inspection indications and gone to lunch. In other words we get back to human frailty. In teaching these inspectors I dismissed three students for cheating because I upheld a puritan standard that said cheats are not allowed in the inspection team. However, that had to be reconciled with the general liberalism of the organisation. The plant psychologist of this nuclear agency for which I worked said that we could not do that; everyone cheats because he is motivated to succeed. Thus we have contradicting values in that the drive for success may induce the neglect of an obvious crack in a weld, regardless of the rigour of the inspection techniques or the qualifications of the inspectors. Therefore, it becomes a matter of probability as to whether failure of a plant or a nuclear incident occurs.

Turning to political issues, I write to Jimmy Carter and ask what explanation he has ready for the inevitable nuclear accident. I think this thought must be in our minds throughout the nuclear safety debate. The former chief of the Atomic Energy Agency made a statement to the effect that the public should be prepared for the eventual nuclear accident. There is another point of view that says that no politician is willing to accept a probability of failure; I find these somewhat irreconcilable. The question of the public, which thinks in absolute terms, being able to comprehend the concept of the probability of failure must also be ad- dressed. Is the general level of education such that the public is able to think as a scientist does, in terms of probability of failure?

I would like to make another comment on the nuclear debate. I saw a television programme in the States on the nuclear question in which the anti-nuclears put forward their fifteen best points, and the pro-nuclears theirs. There was no point of contact between the groups, and I found that format extremely destructive, and I feel that we should pay attention to this point when we consider such questions. A more positive approach may be found in Futures magazine, or in social modelling techniques, where statistical concepts such as decision theory can be used in an attempt to weight and evaluate all the factors involved and their cross interactions. It seems to me that the credibility of engineers is a very tenuous quality. I have knowledge of more than one occasion when an engineer has been threatened with being released or fired if he should speak up against the best interests of corporate policy. Public aware- ness of this situation will inevitably undermine the credibility of the scientist and engineer.

Knott : I will just hold this point for a moment, before we look at whether it is possible to educate the general public to accepting a failure probability. Dr. Reid, do you want to say something first?

Reid : It may be exactly the same point, but I feel that one of the by- products of failures is that they provide a tremendous stimulus to either re-design or re-invest, and it is important to capitalize on this. Yet, as Mr. Saltsman pointed out, there is an opposing tendency, and that is the quite spontaneous tendency to extreme confidentiality, a shyness about discussing these things, and it is here that very big, often fatal failures, are important, because they force the disclosure of information and there often has to be a public court of inquiry. This, in turn, publishes a full report, which is normally available to the public, al- although even that has its problems, as we saw in Britain within the last couple of years over the big chemical factory explosion at Flixborough. There was a court of inquiry, but there was certainly a hint in the media after this court of inquiry reported that full justice perhaps had not been done. Really it was a somewhat inconclusive matter, and certainly some people had gone away from the inquiry feeling that their advice had not been taken into account. The question I would like to pose is : What is the optimum way to run a court of inquiry? There is a lot of international experience here and there may be people present who have some experience of serving on these courts of inquiry. What is the best way to let all the experts have their say, and then, when the experts tend to neutralize one another, what do you do about it? Do you have a jury system, and if so what kind of a jury? These are the questions to which I would like to draw some response.

Knott : That gives us another question to answer now.

Old : I do not want to offer an answer to Dr. Reid's question, but I would like to raise a point related directly to what he said. It often seems to me that there is a scale factor in the interaction of fracture or failure politics and society. He mentioned the accident at Flixborough, which provoked an enormous inquiry. It must have used up an enormous amount of time and resources. There is a point which I am not clear about, and which I would like to put to Mr. Saltsman as a politician, because he can perhaps give an explanation. *Which?*, the consumer magazine in the U.K., quoted that there were something like 26 people killed last year by the failure of electric blankets. I cannot understand why failures which cost lives at a steady slow rate attract no attention whatsoever, whereas the single isolated incident, for example Flixborough or the aeroplane accident in Tenerife, will attract an enormous amount of attention. Can you perhaps help me to understand why the one is politically acceptable and the other is not.

Knott : I think, therefore, that we have three points. One is whether you think that the general public can be educated into realizing that engineers, when they design things, may have a failure probability; perhaps a one in ten to the tenth power chance of failing. The second point is the technical one of general information on how a court of inquiry should be run, and the third is why it is that the small scale accidents, that occur all the time, receive far less publicity than the very big catastrophic events. Perhaps you would like to take the public reaction one first of all Mr. Saltsman.

Saltsman : Whenever you are faced with this kind of a problem, ask yourself how you would put it in a headline, if you were an editor. For instance, I see no difficulty in producing an eyecatching headline for the electric blanket story. Therefore, I feel that, if you get electrocuted in bed, that is news, and not only that, you can possibly sue somebody for a lot of money, and I come to the conclusion that the consumer magazine does not know what it is talking about. If what the magazine said could be supported by fact, it would have made a great story and would indeed have received publicity.

Old : That is actually a statistic.

Saltsman : Statistics are curious things. It is a question of how they are obtained and what kind of information is used. It may be their statistic, but it does not mean that it is a valid statistic or a provable statistic. As a scientist, you immediately assume that because it is a statistic and somebody said it, it is true. I am a politician. I do not make that kind of decision. It is a verifiable statistic, but I would like to see how it was determined.

Old : They count the number of bodies!

Saltsman : You know a person could have died for all kinds of other reasons.

Averbach : I think the point is that there is an individual remedy, and it does not mean that the victim has no recourse. The accidents like this do not create an outcry in that they occur over a long period. When a disaster occurs which involves a lot of people, it gets a lot of attention and is subject to official inquiry. I think both types of event do receive attention, but in different ways.

Knott : To come back to the point on snappy headlines, I suppose a headline which said : "Engineers Expect There to be a Chance of Failure in such and such a Reactor", would attract the public interest.

Saltsman : "Engineers Expect.." does not show up as a headline. "Engineer Charges Failure of Nuclear Reactor", "Engineer Warns of Dire Conse-quences", is the required form of words : "may be", "perhaps", "on the other hand", "later", do not get into headlines.

Knott : On the concept of deliberate design for a chance of failure, do you think the public will accept that or not?

Saltsman : Yes, I think so. At the risk of being considered naive by harping on this, I think that if people are basically honest in what they say, or at least are perceived to be honest, because it is pretty hard to test honesty, then they will be listened to. Let me give you an example of what is happening here in Canada. At the moment we have what is supposed to be rather a scandal with our police. The minister is accused of almost abetting a break in. He rose in the House and straightforwardly stated the facts, although some of the facts were somewhat damaging to him and the opposition tried to make a great issue. This ended the affair because he appeared to be honest. I think Nixon would have been forgiven if he had publicly admitted his involvement in Watergate. The classic example is the Profumo case in Great Britain. John Profumo had to leave the House, not because of his liason with a beautiful girl, or because her reputation was somewhat questionable, but because he lied to the House of Commons when he was asked about it. There are certainly risks attached to being honest : you know that when you are honest, everybody is going to attack you as much as possible, but you can be sure that this will only last so long. But, if you are not honest, you will be hounded indefinitely until finally the truth is dragged out. This is a terrible situation. Thus, I think that, even with science as with politics, if you lay all the facts on the table, the good and the bad, and it is perceived that you have, that is probably the best policy.

Mills : I would like to add a comment on public acceptability. I ask you to picture a steel box with four wheels, containing up to 20 gallons of very very inflammable liquid and a battery and a means to produce sparks. I pump the inflammable fluid into a cylinder, compress it with air, and spark it. Now, if anyone were to ask you to sit inside that steel box, your first reaction would be : "No, I would be crazy." Yet General Motors sell hundreds of thousands of these steel boxes every year, and because they have been around for a long time, they are acceptable to the public. I think that, once nuclear reactors have been around for about 75 years, they will achieve the same public acceptability. Further, if you consider acceptable design lives, the public buys some-thing from General Motors which probably has a design life of about 1,500 hours. For any sort of nuclear installation or major generating facility the minimum design life is 100,000 hours. I think that public accepta-bility is really a conditioning process dependent upon how long these things have been around. There are too many people who still remember nuclear meltdown of an American test reactor.

Questioner : This is a general background comment. During the course of the Conference, I have noticed that, quite often, the one thing that does not come out, and it has been commented on, is the understanding of probability and risk. We should understand this better than the majority of the public and we do not. It shows up in our work. We draw a straight line through a series of points and half the audience will say : "Yes, a straight line". This, I think, carries over to public acceptability

also. I like the analogy of the steel box and the flammable substance.
The probability of being injured in it is tremendously high, but the
public is fairly aware of the risk that they are taking. However, when
we put a number on the risk of a nuclear reactor or an oil pipeline or a
gas pipeline, failing, the number really does not mean anything to them,
or to us, because we really have no reference with which to compare it.
Thus, one aspect of public acceptability, when it comes to fracture,
say, is simply an inability on the part of most of us to understand what
risk and probability really is.

Knott : I think that there is a slight problem, particularly in Britain,
that the only really large number attached to chance is something like
the chance of winning the football pools, where a large number of people
bet and one wins. Unfortunately, there is always the feeling that "yes,
it is a very large number, but someone is going to win", and if that sort
of thinking carries across to failures, then there is always the thought
that the number is very large but one is going to break. I think it
might be a real task to get across large numbers in a really compre-
hensible way.

Saltsman : An interesting thought occurred to me as you were speaking,
that you might be able to write a political formula for acceptability
and it would read something like this : "Acceptability = familiarity +
alternatives". Familiarity is an important thing, and we live with all
kinds of dangers. I was thinking of, say, a loaded jumbo jet flying over
a major city and the destruction that would be caused if that jet crashed,
and I was thinking of alternatives. We have fossil fuels, and we have
gas and oil and, while their prices do not reach the point which is un-
acceptable, as long as you have those alternatives, people will not want
to take the risks associated with nuclear energy. As those alternatives
start to run out, or become increasingly expensive, then the whole formula changes. The figures in the formula get changed and you are in a
different position. I want to say something else about the question
raised about the risk of the people who operate the system. There is an
advantage, I think, to some of the publicly owned facilities in terms of
risk, for this reason. I think that it brings together a different breed
of people, the bureaucracy. There are many things wrong with a bureaucrat. You can be very critical about his lack of imagination and all the
rest of it, but as Weber once wrote about the bureaucrat, he is honest,
he is generally reliable and the lack of imagination turns out to be an
asset. He is not climbing all the time, as he might be doing in private
industry work, he is getting ahead differently. I think that this is
probably one of the arguments for public ownership of some of these
chancy and dicey things, because in fact you need that habit of mind,
that bureaucratic attention to little tiny details, and to making sure
that everything gets checked off and all appears on your report.

Knott : I think at this stage I would like to ask Professor Yokobori to
give a short description of the implications of the failure of a Japanese
oil tank.

Yokobori : The fracture of a very large oil tank occurred in Japan in
1974. The tank had an inner diameter of about 52m, and height of 23.7m.
Some of the audience will know of this accident. It involved spillage
from the fractured part of the tank of 7,500-9,500 kiloliters of heavy
oil into an inland sea. This caused great damage to the fishing industry
of the two prefectures on the coast of the inland sea. I was a member of
the government inspection committee. At the conclusion of its inspection,

several possible causes of the failure were suggested by the Committee. One was the digging out of part of the foundations when constructing a staircase along the side of the tank, which led to local subsidence. Another was that compacting was not properly carried out in view of the state of the ground on which the tank was built. In addition, some of the Committee considered some defects in welding the side plate and the bottom plate. The crack initiation path was traced back with some difficulty, as the surface of the fractured part was so heavily covered by a layer of oxidized scale or rust. When we removed these oxidized layers, we found a characteristic intergranular fracture surface. Naturally, we could not observe the exact fracture surface, but only the surface exposed after removal of the oxide film.

As can be seen from this example, a large scale fracture accident throws up problems requiring an inter-engineering, interdisciplinary approach and must be considered in terms of the interaction between engineering, economy and politics. I feel that not only design engineers but all other engineers and even the public should be educated in, at least, the fundamentals of fracture. On the other hand, a standing investigation system for such a large scale fracture accident may be necessary in order to ensure the correct interdisciplinary basis.

Knott : Before the session comes to a close, I would like to ask Professor Averbach to make general comments on the field, and particularly perhaps if he could say something about American Courts of Inquiry that might help Dr. Reid on his question earlier.

Averbach : The American Court of Inquiry is a traditional system, in that the matter is usually settled in a court of law. The operation of these courts is interesting in that no witness is allowed to give an opinion except an expert witness. As a result there are expert witnesses for both sides with each stating impartial opinions. A decision is eventually reached by a judge or a jury on the basis of a lay interpretation of what has been presented by the experts. Perhaps we can develop a system whereby we have special engineering courts presided over by masters who might have some technical competence and be able to call in impartial experts to advise the court.

Knott : I think that is taking responsibility, which is a good thing. I can understand the people's feelings over the particular incidents of the Flixborough report, because there there was an awful lot of work done, and the conclusion was basically that it was a patched-up job by a non-qualified engineer which led to the failure. Does anybody have burning points from the audience?

G.L. Dunlop, Chalmers University, Sweden : Dr. Mills disappointed me somewhat, because it seems that he is trying to form an acceptability in the eyes of the public; taking people on a tour of a power plant and showing them that the walls of a pressure vessel are very thick is very impressive, because who can imagine that several inches of steel can be broken by a reasonable sized force. I think we really have to do as Mr. Saltsman suggests : we have to lay all the cards on the table and be completely honest. This is exemplified by my own experience in Sweden, where there is a very large political debate concerning nuclear power generation. I work in an institute where there is a very large group of nuclear physicists, many of whom are anti-nuclear power. That is rather surprising, but it occurs because they know very little about engineering, and it is thus very important for engineers to put all the cards on the table.

Mills : We do, in fact, before any construction starts, put all the cards
on the table regarding our nuclear plant by presenting a safety report to
a government agency, the Atomic Energy Control Board. This is usually a
report of considerable magnitude, which details all the design calculations, all the risk factors, which are put into that plant. That presentation to the Board is not necessarily a public affair, I am not too
sure if and when the public is involved, but that government agency is the
regulatory body for Canada. We cannot proceed with a plant constructional
design before we get their approval. The purpose of the plant tours was
to avoid the sort of confrontation which the U.S. Utilities have experienced with very virulent groups who want to stop nuclear power development at any cost. By trying to inform some of the public, we can
perhaps turn away some of these fears.

Dunlop : I do not doubt that the plants as designed are reasonably safe,
but I think that it is not just the government or a decision making body
which has to be convinced of this. The public must also be convinced,
and we must, therefore, in a straightforward way, make then understand
the design principles and the engineering principles behind the construction. It is not sufficient to detail all the information in very
thick volumes, but, we have a duty to make it much clearer, and more
easily accessible to the public.

Knott : We are back to the communications problem again.

D.F. Watt, University of Windsor : I hate to see us go away congratu-
lating ourselves on our credibility on the basis of the poll to which
Dr. Old has referred. I noticed that in the list of people whom you
would believe there was no category that said "none of the above". I
think that if we had rephrased the questions to read : "What is the
probability that this person would mislead you if his personal interests
were involved?" a rather different result, and perhaps a better reflection of public opinion would have been obtained.

Questioner : I would like to reinforce Dr. Dunlop's comment that laying
the books open for public scrutiny is not good enough. We have to take
the information to the public, as otherwise only interest groups, frequently of preconceived opinions, will go and get it. We have to help
the public to form an opinion because it has to make the decision.

Mills : The activities of a group in Ontario called the "Electronic &
Electrical Manufacturers' Association" may be relevant here. Over recent
months they have been putting very small ads into the business section of
The Globe and Mail which say "When oil and gas run out, what about
electricity?" and other ads to the effect that electricity is vital to
the economy and is vital to various parts of the public. Recently, in
the electric power hearings, that group has been accused of putting forward its point of view to the detriment of the credibility of the anti-
nuclear or the anti-electrical society people, and they have objected to
these tiny ads. I think that again we find ourselves in the middle. If
we go out and advertise, and try to sell our product on a wide basis we
are criticised, and similarly if we say nothing. It seems to be the
fate of the utilities to be whipping boys for both the public and sometimes politicians.

Knott : I must now bring discussion to a close. I cannot really try to sum up such a wide ranging conversation in any brief statement. Points have, I think, been well made, on various topics. Some possible roles for ICF are emerging from some of these discussions. That is presumably a matter for the Executive to consider. Before we finish I will ask Mr. Saltsman if he wants to say anything else.

Saltsman : Not really, except to say that I have enjoyed the discussion, I have appreciated the invitation, and I found a lot of what you had to say very helpful.

Knott : I suggest that we close with a round of thanks to the panellists, after which Professor Taplin will take the Chair for the final closing of the Conference.

Taplin : Let me record my own appreciation to the co-chairmen of these two panels on *Fracture and Society* - Ron Armstrong, Roy Nichols and John Knott - to the members of both panels and to the other participants in this venture. I have spent my time during these discussions in the control box, with the Conference Secretary, Dr. Richard Smith. As you know, the entire discussion of the two panels has been taped - using two separate systems to allow for any failures or erasures - and we shall transcribe and edit the entire discussion for publication in the Pergamon Edition of the Proceedings. I can say now that the taping was successful and I believe we can look forward to an interesting written document. Let me also say how very pleased and honoured we have all been here at Waterloo to host this Conference. It has been hard work - and I would like to mention just two of the many people who have been particularly unstinting in their efforts - Dr. Richard Smith, Conference Secretary, and Professor Roy Pick, Registration Chairman. It has also been a totally fulfilling and realizing experience and now that the Conference is over I wish to record my appreciation to all the various participants of ICF4 for permitting this to occur. Many friendships have been made and renewed here in Waterloo and we look forward to their continuance and further renewal in the next four years.

Before turning over to Professor Ben Averbach, President of ICF 1973-77, Dr. van Elst and Professor Francois would like to say something.

van Elst : I would like to make an announcement, principally to my ICFEA (ICF - European Association) Colleagues. I take great pleasure in announcing that a European Group on Fracture was founded here on the Waterloo campus last Thursday. As Chairman of this European Group on Fracture, Professor Kerkhof of the Institute of Mechanics of Solids at Freiburg, Germany was elected, as Secretary Dr. Brughofen of Delft University in the Netherlands. The objectives of this Group are very similar to those of ICF, but on a more modest continental, rather than global, scale. The Group will apply to ICF for membership, demonstrating its affiliation to ICF, with whom it seeks further co-operation and will consult in relevant matters. The European Council members will all receive a letter of invitation to their country to join the European Group on Fracture.

The Group's initial activites will involve the organization of advanced courses on fracture mechanics given by invited lecturers and the organization of colloquia or symposia on fracture, papers for which will be solicited among European research workers. It is envisaged that these will take place at least annually, and care will be taken to avoid overlap with other ICF activities or any other fracture meetings. I might

remind you that the seminar organized in Italy, October 1975, and the
1st European Colloquium on Fracture in France in November 1976 were both
great successes, that of the latter demonstrating the talents of Professor
Francois for organizing such meetings. A second seminar with the theme
"Elastoplastic Fracture Mechanics" is planned for the early Spring or
the late Autumn of 1978, and a second European Colloquium on Fracture
will take place at Imperial College in London in September 1978. The
Congress is happy to see this integration of European efforts in the
study of fracture at ICF4. It feels that it will certainly promote
progress and dissemination of information and will stimulate research
on fracture. I am sure you all will share the European feelings of
content with this development.

Taplin : Now may I ask Professor Dominique Francois, Chairman of ICF5,
France, 1981 to say a few words.

Francois : I feel it a great honour that the Executive of ICF has decided that the next ICF Conference should be held in France. I want to
tell you that you will all be welcome in our country, and that we expect
you all to come and to bring your friends to ICF5.

Averbach : At 9:00 o'clock last Monday morning we started - 5:00 o'clock
on Friday afternoon we have finished. During the period we have talked
about almost everything including politics and we have even spawned at
least one new society. We have had a marvellous time here at Waterloo
and I would like to close by thanking the Canadian Organizing Committee
which has done a tremendous job in arranging and running this Conference.
I would like to suggest that we give them all a standing ovation. Good-
bye and good luck!

Taplin : ICF4 stands adjourned.

*   *   *

CITATION INDEX

This index lists each author to whom reference is made, and gives the page of the reference list in which his name appears.

Corti, C.W., 1-280
Coster, M., 2-232
Cottrell, A.H., 1-81, 1-228,
    1-302, 1-373, 1-398,
    1-581, 1-739, 1-740,
    1-764, 2-4, 2-52, 2-145,
    2-162, 2-168, 2-178,
    2-186, 2-547, 2-862,
    2-1095, 3-10, 3-306,
    3-316, 3-422, 3-449,
    3-690, 3-823, 3-850,
    4-14, 4-15, 4-18, 4-22,
    4-53, 4-69, 4-92, 4-219
Coughlan, J., 4-16
Courts, A., 1-620
Coutts, L.H., 3-1139
Cowan, A., 3-50, 3-709, 3-717
Cowling, M.J., 4-16
Cowper, G.R., 2-418
Cox, B., 4-79
Cox, P.A., 1-184
Cox, T., 3-348
Cox, T.B., 2-152, 2-245, 2-346
Cracknell, A., 2-245
Craggs, J.W., 3-37, 3-975
Craig, J.V., 1-247, 2-721
Crandall, G.M., 2-1011
Crane, J., 2-529, 2-547
Crane, R.L., 3-43
Crawford, R.J., 3-1130
Creager, M., 2-946, 3-1002
Crews, J.H., 2-1121, 2-1149, 3-144
Crimmins, P.P., 3-333
Crist, B., 1-482
Crook, A.W., 4-33
Crooker, T.W., 2-914, 4-42
Cropley, A.J., 4-219
Crose, J.G., 1-546, 3-917
Crosley, P.B., 1-199, 3-534
Crosley, R.P., 3-753
Crossman, F.W., 1-9
Crouthamel, C.E., 3-727
Cruden, A.K., 2-446
Cruse, T.A., 1-584, 3-365
Crussard, C., 3-823
Cullen, W.H., 4-70
Culver, L.E., 1-417, 1-460,
    3-1116, 3-1130
Curry, D.A., 1-81, 1-740, 2-4,
    2-285, 4-70
Curry, J.D., 1-620, 1-621
Cuthbert, W.L., 1-184
Cutler, C.P., 2-547

DABELL, B.J., 2-970
Dahl, W., 2-19, 2-168
Dahlberg, E.P., 2-317
Dahlberg, L., 3-519

Dahmen, U., 2-152
Dalgleish, B.J., 3-911
Dally, J.W., 2-1335, 3-82
Damali, A., 2-1204
Dance, S.H., 3-520
Daniels, H.E., 1-583
Dannenberg, H., 1-654
Darken, L.S., 2-252, 2-900, 3-761
Darken, L.W., 1-373
Darlaston, B.J.L., 2-656, 2-834
Darlington, H., 2-446
Das, S.C., 3-182
Date, E.H.F., 2-675
Datsko, J., 2-581
Davenport, A.T., 2-68
Davidenkov, N.I., 3-856
Davidenkov, N.N., 2-186
Davidge, R.W., 1-546, 3-887, 3-896,
    3-911, 3-1019
Davidson, D.L., 2-760, 2-900
Davies, G.J., 1-582, 1-583
Davies, P.W., 1-346
Davies, R., 3-619
Davis, E.A., 1-345
Davison, J.K., 2-656
Dawes, M.G., 1-764, 3-333, 3-348,
    3-642, 3-744
Dawson, J.K., 3-717
Dean, G.O., 1-418
de Andrade, S.L., 2-471
de Arvicar, R.J., 3-1158
DeBruyne, N.A., 1-654
Decker, R.F., 2-564
Decroix, J., 2-572
Dedrick, J.H., 1-228
DeFerran, E.M., 1-581
De Fouquet, J., 2-721, 2-869
Dehoff, R.T., 2-399
Delange, B.G.M., 3-1121
Delamore, E., 3-1069
de Luca, B., 2-471, 2-506
De Meester, B., 2-589
Dennison, J.P., 1-346
Denton, K., 2-1252
Denys, R., 1-784
De Pierre, V., 2-446
Derby, R.W., 3-422
Deruyettere, A., 1-398
Desalvo, G.J., 3-955
Deschanvres, A., 2-224
DeSilva, A.R.T., 1-582
Desisto, T.S., 2-581
de Souza, M.C.B., 2-471
DeVekey, R.C., 1-582
Devereux, O.F., 2-862
Devison, D.C., 2-1310
DeVries, K.L., 1-482, 1-495, 1-654
DeWexler, S.B., 1-418

Masubichi, K., 2-186
Matera, R., 1-302
Mathews, J.R., 3-727
Mathur, V.D., 1-345
Matlock, D.K., 2-617
Matsuda, Y., 2-984
Matsuiski, M., 2-970
Matsumoto, T., 2-1329
Matsumoto, Y., 2-976
Matsuoka, S., 2-1277
Matsushige, K., 3-1083
Matzer, F., 2-751
Maurer, K., 2-751
Maxey, W.A., 1-199, 2-656, 3-50, 3-422
Maxwell, B., 1-459
Maxwell, D.H., 2-998
Maxwell, L.H., 3-937
Maxwell, P.C., 2-138
May, B.J., 2-1086
May, M.F., 3-324
May, M.J., 2-285, 3-552, 3-576
Mayer, G., 2-145
Mayson, H.J., 1-546
Mazanec, K., 2-258
Mazars, J., 3-1208
Mazzio, V.F., 1-583
McBridge, F.H., 1-764
McBurney, G.W., 2-1020
McCabe, D.E., 1-98, 3-597, 3-1034
McCammond, D., 3-1027, 3-1101
McCaughey, J.M., 1-398, 2-335
McClaren, S.W., 2-506
McClean, A.F., 1-545
McClintock, F.A., 1-8, 1-82, 1-98, 1-317, 1-461, 1-546, 1-691, 2-245, 2-373, 2-381, 2-387, 2-418, 2-464, 2-480, 2-547, 2-998, 3-62, 3-297, 3-315, 3-316, 3-441, 3-744, 3-745, 3-834, 4-14, 4-17, 4-33
McCombs, J.B., 2-145
McConnelee, J.E., 1-279
McCoy, H.E., 2-564, 2-617
McCullough, R.L., 1-482
McDonough, W.J., 1-545
McElhaney, J.H., 1-621
McElroy, 3-43
McEvily, A.J., 1-82, 1-248, 2-697, 2-794, 2-864, 2-998, 2-1317, 3-365, 3-745, 3-833, 3-842, 4-42
McGowan, J.J., 3-744
McGarry, F.J., 1-582, 1-583
McGrath, J.T., 1-751, 1-763, 2-697, 3-776
McGregor, J., 3-297
McIntyre, P., 2-272

McIvor, I.D., 2-68, 2-145, 2-346, 2-729
McKinney, K.R., 3-937
McLachlan, D.F.A., 2-645
McLaren, J.R., 3-896
McLean, D., 1-302, 1-345, 1-346, 1-763, 2-118, 2-206, 2-292, 2-547, 2-675, 4-18
McMahon, C.J., Jr., 1-81, 1-82, 1-280, 1-345, 1-372, 1-373, 2-45, 2-68, 2-126, 2-152, 2-258, 2-285, 2-292, 2-300, 2-852, 3-760, 4-18
McMahon, J.A., 2-285
McMeeking, R.M., 2-367, 2-1077, 3-144, 3-510
McMillan, J.C., 1-247, 2-272, 2-285, 2-869
McNicol, R.C., 2-963
Mears, D.R., 3-1083
Mears, R.B., 1-417
Mecholsky, J.J., 3-937, 3-962
Mehl, R.F., 2-37
Meinel, G., 1-482
Meisel, J.A., 3-896
Melbourne, S.H., 2-515
Melton, K.N., 2-547
Melvin, J.W., 1-621
Menczel, J., 1-621
Mengelberg, H.D., 1-417
Menges, G., 3-1096
Merchant, R.H., 3-528
Merinov, G.N., 3-770
Merker, L., 3-989
Merkle, J.G., 1-512, 1-691, 3-51, 3-269, 3-275, 3-282, 3-297, 3-422, 3-534, 3-586, 4-14, 4-16
Merrick, H.F., 2-998
Meshii, M., 2-353
Metcalf, A.G., 1-583
Meuris, M., 2-152
Meyer, R.A., 3-932
Meyers, G.A., 3-776
Meyn, D.A., 1-247, 2-224
Miannay, D., 2-265
Michel, D.J., 2-564
Mihashi, T., 2-1182
Mileiko, S.T., 1-583
Miles, J.P., 1-113, 2-429
Miller, D.R., 1-302, 2-346
Miller, G.A., 1-672, 2-317, 2-998, 2-1220, 3-833
Miller, K.J., 1-54, 1-740, 2-816, 2-932, 3-243, 3-744, 4-14

Miller, M., 3-650
Miller, W.A., 2-101
Miller, W.R., 2-1303
Mills, W.J., 2-1310
Milne, I., 3-422, 3-690, 3-745, 4-22
Mimura, H., 3-850
Minani, K., 1-764
Mincer, P.N., 3-1138
Mindlin, H., 2-825
Miner, M.A., 2-970
Minty, D.C., 3-1069
Mirabile, M., 1-81
Mitchell, A.B., 2-697
Mitsche, R., 2-751, 3-823
Miyake, K., 2-218, 2-232
Miyamoto, H., 2-1039, 3-449, 3-496
Miyamoto, T., 2-976
Miyata, T., 3-324, 4-42
Miyoshi, T., 3-273
Moberly, J.W., 2-335
Mochizuki, T., 2-984
Mogford, I.L., 2-406, 3-823
Moghe, S.R., 3-989
Mohamed, F.A., 2-529, 2-547
Moller, H., 2-126
Monden, Y., 3-760
Monks, H.A., 2-707
Monteiro, S.N., 2-138
Montgrain, L., 1-418
Montulli, L.T., 2-1039, 2-1297, 2-1358, 4-17
Monthulet, A., 3-1208
Mooder, L.E.J., 3-708
Moon, D.M., 3-348
Moore, C.T., 3-823
Moore, D.M., 2-92
Moore, G., 2-480
Moore, G.A., 2-245
More, C.C., 1-208
Morgan, J.D., 3-88
Morgenthaler, K.D., 3-1027
Mori, K., 4-42
Mori, T., 1-582, 2-45, 3-760
Morillon, Y., 3-823
Morimoto, S., 2-1182
Morimitsu, T., 2-1103
Morita, M., 3-677
Morlet, J.G., 3-760
Morley, J.G., 1-581, 1-584
Morozumi, F., 1-316
Morris, J.W. Jr., 3-850
Morris, P.F., 1-302
Morris, S., 1-584
Morris, W.L., 2-794, 2-900, 2-1297
Morrissey, R.J., 1-317

Morrison, J.A., 3-382
Morrison, W.B., 2-68, 2-537
Morrow, J., 1-279, 2-684, 2-729, 2-932, 2-1149, 2-1165, 2-1182, 2-1204, 2-1209, 2-1303
Morse, P.M., 2-464
Mortimer, D.A., 2-118
Morton, J., 1-581, 1-582
Mosca, S.R., 2-471
Moses, R.L., 1-546, 3-1002
Moskowitz, A., 2-138
Mossakovski, I., 1-714
Mossakovskii, V.I., 4-121
Mostovoy, S., 1-654, 2-100, 2-317
Moteff, J., 1-279, 2-564
Mott, N.F., 2-1317, 3-937
Mowbray, D.F., 1-279, 2-807
Moyar, G.J., 2-771
Mubeen, A., 3-744, 3-745
Mukherjee, A.K., 3-528, 3-619
Mukerjee, B., 3-504, 3-612, 3-1130
Mulford, R.A., 1-372, 2-292
Muller, S.A., 1-621
Muller, T., 3-297, 3-348
Muller, T.L.F., 2-572
Muller, W., 1-621
Mullin, J.V., 1-583
Muncher, L., 3-625, 3-770
Munro, H.G., 3-597, 4-14
Munro, M., 1-621
Munz, D., 1-691, 2-76, 2-735, 2-1297, 4-42
Mura, T., 3-195, 3-449, 4-69
Murakami, Y., 2-984, 2-1182
Murphy, B.M., 1-460, 2-834
Murphy, M.C., 1-581, 1-621, 2-666, 3-568, 3-569
Murray, J., 1-460, 3-1108
Murray, J.D., 1-763
Murray, M.J., 2-224
Murrell, S.A.F., 3-1195
Murzewski, J., 2-1174
Muscati, A., 3-152
Musiol, C., 2-68, 2-317
Muskhelishvili, N.I., 1-739, 2-557, 3-137, 3-189, 3-205, 3-650, 3-1970, 4-121
Myers, F.A., 1-460
Myers, J., 1-763, 2-645
Mylonas, C., 2-126

NABARRO, F.R.N., 1-373, 1-740, 2-232, 4-53
Nadai, A., 2-507, 3-360

Seireg, A., 1-621, 3-1158
Sejnoha, R., 3-753
Sekiguchi, S., 2-976
Seikino, S., 3-850
Sellars, C.M., 1-316, 3-317
Sendeckyj, G.P., 3-1177
Sengupta, M., 2-1243
Seo, M., 1-419
Serensen, H.V., 2-788, 2-800
Serensen, S.V., 3-804
Sergeant, R.M., 2-843
Sertour, G., 2-1310, 3-365
Server, W.L., 3-552
Sessler, J.G., 3-289
Seth, B.B., 3-558
Sethi, P.S., 2-852
Severud, L.K., 1-280
Shah, D.C., 1-316
Shah, R.C., 2-1226, 2-1365,
    3-144
Shah, R.T., 4-149
Shah, S.P., 3-1201
Shahinian, P., 1-248, 2-1243
Shand, E.B., 3-937
Shannon, R.W.E., 1-199
Shapiro, E., 1-316, 2-529,
    2-547
Shaw, F., 1-184
Shaw, G.G., 2-735, 2-794
Shaw, K.G., 1-125, 4-128
Shaw, M.C., 3-955, 4-16
Shchukin, E.D., 1-398
Sheets, E.C., 2-1021
Sheffler, K.D., 2-760
Sheffler, K.S., 1-280
Shei, S.A., 2-529
Shen, H., 2-869
Shepard, O.C., 2-581
Shephard, L.A., 1-582
Sheppard, M.F., 3-717
Sherbakov, E.H., 3-1201
Sherby, O.D., 2-581
Shesterikov, A., 1-113
Shewmaker, A.P., 3-43
Shibano, Y., 2-1039
Shibuya, T., 4-121
Shiels, S.A., 2-862
Shih, C.F., 3-727
Shih, T.T., 2-272, 2-1039,
    2-1095, 2-1112, 2-1297
Shimizu, M., 2-26, 2-715
Shimizu, S., 2-984
Shimizu, T., 2-381
Shinkal, N., 3-937
Shinozuka, M., 2-1174, 3-917
Shirasaki, Y., 3-1177
Shiratori, M., 2-1182, 3-273
Shmuely, M., 1-199, 3-227, 3-753

Shoemaker, A.K., 1-98, 3-348,
    3-534, 3-677
Shor, J.B., 1-169
Shorb, A.M., 3-1002
Shore, B.M., 4-219
Shortall, J.B., 3-1121
Shroder, K., 2-1243
Shuhinian, P., 2-1243
Shunk, F.A., 2-335
Sidey, D., 2-816, 2-843
Sidey, M.P., 2-666
Siebel, E., 2-506
Siegfried, W., 1-345
Sih, G.C., 1-714, 2-178, 2-788,
    2-932, 3-18, 3-37, 3-75,
    3-95, 3-116, 3-137,
    3-189, 3-195, 3-205,
    3-365, 3-399, 3-496,
    3-559, 3-624, 3-639,
    3-744, 3-997, 3-1027,
    3-1052, 3-1076, 3-1116,
    3-1149, 4-128, 4-149
Sikka, V.K., 2-564
Silano, A.A., 3-1083
Sillwood, J.M., 1-582
Silverman, A., 1-459
Simkin, A., 1-621, 3-1169
Simon, R., 2-1220
Simons, D.A., 1-54
Simpson, L.A., 3-414, 3-709,
    3-937, 3-887, 3-911,
    4-70
Sims, C.E., 2-245
Sinclair, G.M., 2-771, 2-984,
    2-1121, 2-1182, 2-1226,
    3-534, 3-677
Sinclair, J.E., 2-557, 4-53
Sines, G., 2-715
Singh, B., 1-582, 2-1243
Siverns, M.J., 1-764, 2-631,
    2-645, 2-656, 2-666,
    4-91
Skat, A.C. Jr., 2-1011
Skelton, R.P., 1-280, 2-843,
    2-852
Skibo, M.D., 3-1130
Skinner, D.W., 3-1130
Skylstad, K., 2-693
Slate, F.O., 3-1201
Slater, R.A.C., 2-433
Slot, T., 1-280, 2-825
Slutsker, A.I., 3-237, 4-160
Smallman, R.E., 1-418, 2-127
Smeaton, D.A., 1-208
Smith, C.I., 2-529, 2-537
Smith, C.L., 2-292
Smith, C.N., 3-850
Smith, C.S., 2-37

Tate, A.E.L., 2-697
Tateishi, T., 3-1177
Tattersall, H.E., 1-545, 1-621
Tattersall, H.G., 3-414, 3-887
Tauscher, H., 2-1262
Tavernelli, J.F., 2-843, 2-932, 3-348
Taylor, A.J., 3-870
Taylor, G.I., 2-486, 4-53
Taylor, L.C., 4-219
Taylor, L.G., 1-763
Taylor, L.H., 1-83
Tchoupnovsky, A.I., 3-466
Teer, D.G., 2-697
Tefft, W.E., 3-937
Tegart, W.J. McG., 1-317
Teller, E., 2-118
Tenge, P., 1-184
Tenhaagen, C.W., 3-569
Terada, H., 3-116
Terai, K., 2-126
Terasaki, F., 2-26, 2-37, 3-761
    3-823
Terlinde, G., 2-10, 2-589
Terry, E.L., 2-507
Tetelman, A.S., 1-81, 1-148, 1-248,
    1-417, 1-584, 1-749, 1-813,
    2-26, 2-60, 2-285, 3-306,
    3-347, 3-365, 3-534, 3-551,
    3-745, 3-760, 3-842, 4-18
Theissen, A., 1-740
Theocaris, P.S., 3-199, 3-624, 4-17
Thomas, A.T., 2-92
Thomas, C., 1-581
Thomas, G., 1-82, 2-26, 2-285,
    2-1243, 3-850
Thomas, W., duB., 1-184
Thomason, P.F., 1-82, 1-316, 3-583,
    2-547, 2-843, 3-441
Thommek, H., 2-81
Thompson, A.W., 1-247, 1-418, 2-10,
    2-239, 2-252, 2-275, 2-381
Thompson, D.S., 2-108, 2-152, 4-70
Thompson, K.R.L., 1-247, 2-721
Thompson, N., 1-248, 2-684, 2-794,
    2-1317
Thompson, R.B., 1-582
Thomson, R., 1-81, 1-545, 1-672,
    2-162, 3-392, 4-18, 4-53
Thornton, D.V., 1-764, 2-631
Thresher, R.W., 2-1365
Throop, J.F., 2-998
Tien, J.K., 2-252, 2-900
Tiffany, C.F., 3-43, 3-559
Tikhomirov, P.V., 2-237
Timbres, D.H., 1-81
Timo, D.P., 2-781
Timoshenko, S.P., 1-512, 1-654,
    2-1138, 2-1262, 3-88, 3-552,

Timoshenko (continued)
    3-989
Tiner, N.A., 1-419
Tipler, H.R., 1-83, 1-280, 1-346,
    2-666, 2-834
Tipnis, V.A., 2-507, 4-33
Tipper, C.F., 3-823
To, K.C., 2-645
Tobler, R.L., 3-842
Tobolsky, A., 1-482
Tomashevskii, E.E., 1-482, 3-50,
    3-237, 4-160
Tomita, Y., 2-464
Tomkins, B., 1-228, 1-248, 1-280,
    2-729, 2-744, 2-998,
    3-833, 4-79, 4-84, 4-85
Tomoda, Y., 2-26
Tong, P., 3-660
Tonti, E., 2-464
Topper, T.H., 1-279, 1-280,
    2-825, 2-932, 2-1121,
    2-1149, 2-1209, 3-289
Torronen, K., 1-83, 2-60, 2-126
Townley, C.H.A., 3-51, 3-152,
    3-422, 4-15
Toyosada, M., 2-1365
Tracey, D.M., 1-8, 1-25, 1-81,
    1-125, 1-740, 2-45,
    2-265, 2-418, 2-557,
    2-572, 2-900, 3-25,
    3-365, 3-510, 3-1057,
    4-17
Tralda, P.M., 1-248
Trantina, G.G., 3-727, 3-924
Trapesnikov, L.P., 3-1215
Trauble, H., 3-449
Trebules, V.W., 2-1112
Tressler, R.E., 3-937, 4-70
Trevana, P., 2-335
Troiano, A.R., 1-373, 2-240,
    2-285, 3-760
Tromans, D., 2-327
Troshchenko, V.T., 3-685
Truesdell, C., 1-714, 3-70
Truszkowski, W., 2-486
Truss, K.J., 2-852
Tsai, S.W., 3-1063, 3-1178
Tsukuda, H., 2-976
Tsuya, K., 3-850
Tu, L.K.L., 3-558
Tuba, I.S., 3-251, 3-557
Tucker, P., 1-495
Tuliani, S.S., 1-763
Turkalo, A.M., 2-37, 2-292,
    2-300, 3-823
Turner, A.P.L., 2-794

EDITORIAL NOTE

Citation indexes provide an indicator of the usefulness and value of an author's work. One outstanding feature of the current Citation Index is the number of times the work of Professor J.R. Rice is cited (85 citations). Surely Rice must be regarded as the first name in fracture today - especially as Rice must be one of the youngest of the renowned fracture researchers. Another feature of the present index is the fact that the only well-cited author who was not a participant at ICF4 was Professor G.C. Sih. Dr. Sih was invited to attend but unfortunately had to send his apologies.

These indexes were prepared with assistance from A.L.W. Collins, A. Miyase and C. Gandhi.

Author Index

The following provides an index of all the authors of plenary and workshop papers in the full proceedings. The first number indicates in which volume of the proceedings the paper is published and the second number the page in that volume.

A

Abdel-Latif, A.I.A., 3-933
Abdel-Raouf, H., 2-1207
Aberson, J.A., 3-85
Achenbach, J.D., 3-97
Adams, N.J., 3-593
Aifantis, E.C., 3-257
Aksogan, O., 3-177
Albrecht, P., 2-959
Alic, J.A., 3-1031
Amstutz, H., 2-943
Amzallag, C., 2-873
Andersen, O., 2-569
Anderson, A.F., 2-919
Anderson, G.P., 1-643
Anderson, J.M., 3-85
Antolovich, S.D., 2-919, 2-995
Aoki, M., 2-173, 3-687
Argon, A.S., 1-445, 2-595
Arita, M., 2-1375
Armstrong, R.W., 2-1, 4-61
Arnott, J.A., 2-513
Arone, R., 3-549
Asada, Y., 2-767, 2-1195
Ashby, M.F., 1-1, 2-603
Astiz, M.A., 3-395
Atluri, S.N., 3-457
Aurich, D., 2-183
Averbach, B.L., 1-201
Awatani, J., 2-695, 2-1153

B

Backfisch, W., 2-73
Baer, E., 3-1079
Balandin, Y.F., 2-797, 3-633
Ball, A., 3-971
Bandyopadhyah, S., 3-741
Banerjee, S., 3-293, 3-343
Banerji, S.K., 1-363
Barker, L.M., 2-305
Barrachin, B., 3-361
Bartlett, R.A., 2-831
Bartolucci Luyckx, S., 2-223
Bartos, J., 2-995
Bathias, C., 2-1283, 2-1307
Baudin, G., 2-1353
Bazant, Z.P., 3-371
Beardmore, P., 3-1105

Beaumont, P.W.R., 3-1015
Beevers, C.J., 1-239
Beinert, J., 3-751
Benson, J.P., 2-65
Berger, C., 3-687
Bernath, A., 3-541
Bernstein, I.M., 2-33, 2-249
Berry, J.T., 2-565
Berryman, R.G., 2-195
Beste, A., 2-943
Besuner, P.M., 1-137
Bhandari, S.K., 3-361
Bilby, B.A., 1-821, 3-197, 4-1
Bilek, Z., 3-531
Bily, M., 2-1143
Birch, M.W., 1-501
Blauel, J., 3-751
Bluhm, J.I., 3-409
Bouchet, B., 2-867
Boutle, N.F., 2-1065, 2-1233
Bowne, A.W., 2-1217
Boyd, J.D., 2-377
Bradt, R.C., 3-933
Bratina, W.J., 3-773
Brenneman, W.L. 2-123
Brett, S.J., 2-719
Briant, C.L., 1-363
Brinkman, C.R., 2-561
Brook, R., 2-313
Brown, D.K., 3-507
Brown, T., 1-173
Buch, A., 2-1057
Bui, H.D., 3-91
Bunk, W., 2-105
Buresch, F.E., 3-939
Burns, D.J., 1-173
Byrne, J.G., 2-1287

C

Calil, S.F., 2-1267
Cantor, B., 2-719
Carlsson, A.J., 1-683
Cardew, G.E., 3-197
Cartwright, D.J., 3-647
Chawla, K.K., 3-1039

335

Author Index